Lecture Notes in Mathematics

Edited by A. Dold, F. Takens and B. Teissier

Editorial Policy
for the publication of monographs

1. Lecture Notes aim to report new developments in all areas of mathematics – quickly, informally and at a high level. Monograph manuscripts should be reasonably self-contained and rounded off. Thus they may, and often will, present not only results of the author but also related work by other people. They may be based on specialized lecture courses. Furthermore, the manuscripts should provide sufficient motivation, examples and applications. This clearly distinguishes Lecture Notes from journal articles or technical reports which normally are very concise. Articles intended for a journal but too long to be accepted by most journals, usually do not have this "lecture notes" character. For similar reasons it is unusual for doctoral theses to be accepted for the Lecture Notes series.

2. Manuscripts should be submitted (preferably in duplicate) either to one of the series editors or to Springer-Verlag, Heidelberg. In general, manuscripts will be sent out to 2 external referees for evaluation. If a decision cannot yet be reached on the basis of the first 2 reports, further referees may be contacted: the author will be informed of this. A final decision to publish can be made only on the basis of the complete manuscript, however a refereeing process leading to a preliminary decision can be based on a pre-final or incomplete manuscript. The strict minimum amount of material that will be considered should include a detailed outline describing the planned contents of each chapter, a bibliography and several sample chapters.
Authors should be aware that incomplete or insufficiently close to final manuscripts almost always result in longer refereeing times and nevertheless unclear referees' recommendations, making further refereeing of a final draft necessary.
Authors should also be aware that parallel submission of their manuscript to another publisher while under consideration for LNM will in general lead to immediate rejection.

3. Manuscripts should in general be submitted in English.
Final manuscripts should contain at least 100 pages of mathematical text and should include
- a table of contents;
- an informative introduction, with adequate motivation and perhaps some historical remarks: it should be accessible to a reader not intimately familiar with the topic treated;
- a subject index: as a rule this is genuinely helpful for the reader.

Lecture Notes in Mathematics 1733

Springer
Berlin
Heidelberg
New York
Barcelona
Hong Kong
London
Milan
Paris
Singapore
Tokyo

Klaus Ritter

Average-Case Analysis of Numerical Problems

 Springer

Author

Klaus Ritter
Fachbereich Mathematik
Technische Universität Darmstadt
Schlossgartenstrasse 7
64289 Darmstadt, Germany

E-mail: ritter@mathematik.tu-darmstadt.de

Cataloging-in-Publication Data applied for

Die Deutsche Bibliothek - CIP-Einheitsaufnahme

Ritter, Klaus:
Average case analysis of numerical problems / Klaus Ritter. - Berlin ;
Heidelberg ; New York ; Barcelona ; Hong Kong ; London ; Milan ; Paris
; Singapore ; Tokyo : Springer, 2000
   (Lecture notes in mathematics ; 1733)
   ISBN 3-540-67449-7

Mathematics Subject Classification (2000): 60-02, 60Gxx, 62-02, 62Mxx, 65-02, 65Dxx, 65H05, 65K10

ISSN 0075-8434
ISBN 3-540-67449-7 Springer-Verlag Berlin Heidelberg New York

Springer-Verlag is a company in the BertelsmannSpringer publishing group.
© Springer-Verlag Berlin Heidelberg 2000
Printed in Germany

Typesetting: Camera-ready T_EX output by the author
Printed on acid-free paper   SPIN: 10725000   41/3143/du   543210

Acknowledgments

I thank Jim Calvin, Fang Gensun, Stefan Heinrich, Norbert Hofmann, Dietrich Kölzow, Thomas Müller-Gronbach, Erich Novak, Leszek Plaskota, Oleg Seleznjev, Joe Traub, Greg Wasilkowski, Art Werschulz, and Henryk Woźniakowski for valuable discussions and comments on my manuscript in its various stages. My special thanks are to my coauthors for the opportunity of joint research.

This work was completed during my visit at the Computer Science Department, Columbia University, New York. I am grateful to Joe Traub for his invitation, and to the Alexander von Humboldt Foundation and the Lamont Doherty Earth Observatory for their support.

Contents

CHAPTER I

Introduction

Our subject matter is the analysis of numerical problems in an average case setting. We study the average error and cost, and we are looking for optimal methods with respect to these two quantities.

The average case analysis is the counterpart of the worst case approach, which deals with the maximal error and cost and which is more commonly used in the literature. The average case approach leads to new insight on numerical problems as well as to new algorithms. While the worst case approach is most pessimistic, the average case approach also takes into account those instances where an algorithm performs well, and the results should match practical experience more closely.

In the average case setting we use a probability measure to specify the smoothness and geometrical properties of the underlying functions. Equivalently, we consider a random function or stochastic process. The measure is the prior distribution of the unknown functions, and thus the average case approach is sometimes called Bayesian numerical analysis. The choice of the measure in the average case setting corresponds to the choice of the function class in the worst case setting. We stress that deterministic methods, in contrast to randomized or Monte Carlo methods, are analyzed in these notes. Probabilistic concepts only appear in the analysis.

We study continuous problems on infinite dimensional function spaces, where only partial information such as a finite number of function values is available. The average case and worst case setting are among the several settings that are used in information-based complexity to analyze such problems.

The average case analysis is closely related to spatial statistics, since a probability measure on a function space defines a stochastic model for spatially correlated data. Part of the average case results therefore deals with estimation, prediction, and design problems in spatial statistics, which has applications, for instance, in geostatistics.

We focus on the following problems:

(1) numerical integration,
(2) approximation (recovery) of functions,
(3) numerical differentiation,
(4) zero finding,
(5) global optimization.

Moreover, we present some general results for linear problems in the average case setting, which apply in particular to (1)–(3).

In the following we describe the problems (1)–(5) in the average case setting. Let $D = [0,1]^d$ and $X = C^k(D)$ for some $k \in \mathbb{N}_0$. In problem (1) we wish to compute weighted integrals

$$S(f) = \int_D f(t) \cdot \varrho(t) \, dt$$

of the functions $f \in X$, and in problems (2) and (3) we wish to approximate the functions

$$S(f) = f$$

or derivatives

$$S(f) = f^{(\alpha)}$$

in a weighted L_p-space over D. Here $1 \leq p \leq \infty$. These problems are linear in the sense that $S(f)$ depends linearly on f. In problems (4) or (5) we wish to approximate a zero or global maximum $x^* = x^*(f)$ of $f \in X$. There is no linear dependence of $x^*(f)$ on f, and several zeros or global maxima of f may exist.

The data that may be used to solve one of these problems consist of values

$$\lambda_1(f), \ldots, \lambda_n(f)$$

of finitely many linear functionals λ_i applied to f. In particular, we are interested in methods that only use function values

$$\lambda_i(f) = f(x_i),$$

but derivatives or weighted integrals of f are studied as well. The data may be computed either adaptively (sequentially) or nonadaptively (in parallel), and the total number $n = n(f)$ of functionals may be the same for all $f \in X$ or may depend on f via some adaptive termination criterion. Based on the data $\lambda_i(f)$, an approximate solution $\widetilde{S}(f)$ is constructed. Here $\widetilde{S}(f)$ is a real number for integration, a function for problems (2) and (3), and a point from the domain D for problems (4) and (5).

Let P be a probability measure on X. For the linear problems (1)–(3) the average error of $\widetilde{S}$ is defined by

$$\left(\int_X \| S(f) - \widetilde{S}(f) \|^q \, dP(f) \right)^{1/q}$$

for some $1 \leq q < \infty$. Here $\| \cdot \|$ denotes a weighted L_p-norm or the modulus of real numbers. For zero finding the average error is either defined in the root sense by

$$\int_X \inf\{ |x^* - \widetilde{S}(f)| : f(x^*) = 0 \} \, dP(f),$$

assuming that zeros exist with probability one, or in the residual sense by

$$\int_X |f(\widetilde{S}(f))| \, dP(f).$$

For global optimization we study the average error

$$\int_X \left(\max_{t \in D} f(t) - f(\widetilde{S}(f)) \right) dP(f).$$

As a lower bound for the computational cost we take the total number $n = n(\widetilde{S}, f)$ of functionals λ_i that are applied to f during the construction of $\widetilde{S}(f)$. However, for most of the problems in these notes, good methods $\widetilde{S}$ exist with computational cost proportional to $n(\widetilde{S}, f)$. The average cardinality of $\widetilde{S}$ is defined by

$$\int_X n(\widetilde{S}, f) \, dP(f).$$

The main goal in these notes is to find methods $\widetilde{S}$ with average cost not exceeding some given bound and with average error as small as possible. Optimal methods, i.e., methods with minimal average error, are only known for some specific problems. Hence we mainly study order optimality and asymptotic optimality, where the latter also involves the asymptotic constant.

In the sequel we outline the contents of the following chapters, and then we give some historical and bibliographical remarks.

1. Overview

In Chapters II–VI we study (affine) linear methods for linear problems. We formally introduce the average case setting and derive some general results in Chapters II and III. Chapters IV–VI are devoted to the analysis of the problems (1)–(3).

A linear problem is given by a linear operator

$$S : X \to G,$$

where G is a normed linear space, and a class Λ of linear functionals on X that may be used to construct approximations to S. We have $G = \mathbb{R}$ for integration, and G is a weighted L_p-space over D for approximation or differentiation. A linear method applies the same set of linear functionals $\lambda_1, \ldots, \lambda_n \in \Lambda$ to every function $f \in X$ and then combines the data $\lambda_i(f)$ linearly to obtain an approximation $\widetilde{S}(f)$ to $S(f)$. Obviously the (average) cardinality of $\widetilde{S}$ is n.

We mainly study the case where Λ consists of all Dirac functionals, so that $\widetilde{S}$ uses function values $\lambda_i(f) = f(x_i)$. Then the key quantity in the average case analysis is the nth minimal average error

$$\inf_{x_i \in D} \inf_{a_i \in G} \left(\int_X \left\| S(f) - \sum_{i=1}^n f(x_i) \cdot a_i \right\|_G^q dP(f) \right)^{1/q}.$$

This minimal error states how well S can be approximated on the average by linear methods using n function values. Clearly one is interested in optimal methods, i.e., in the optimal choice of the knots x_i and coefficients a_i.

The worst case counterpart is the minimal maximal error

$$\inf_{x_i \in D} \inf_{a_i \in G} \sup_{f \in F} \left\| S(f) - \sum_{i=1}^{n} f(x_i) \cdot a_i \right\|_G$$

on a function class $F \subset X$. Thus $f \in F$ is the a priori assumption in the worst case setting that corresponds to f being distributed according to P in the average case setting.

Chapter II provides the basic concepts for the analysis of linear problems in the average case setting. Furthermore, it contains some definitions and facts about measures on function spaces. It concludes with a detailed presentation of a classical example, integration and L_2-approximation in the average case setting with respect to the Wiener measure.

In Chapter III we mainly consider linear problems with a Hilbert space G. Integration and L_2-approximation are examples of such problems. As long as we study (affine) linear methods, average case results depend on the measure P only through its mean

$$m(t) = \int_X f(t) \, dP(f)$$

and its covariance kernel

$$K(s, t) = \int_X (f(s) - m(s)) \cdot (f(t) - m(t)) \, dP(f).$$

The Hilbert space with reproducing kernel K is associated with P. We study basic properties of reproducing kernel Hilbert spaces and their role in the average case analysis. Moreover, we derive optimality properties of splines in reproducing kernel Hilbert spaces, if G is a Hilbert space or P is a Gaussian measure. The splines are minimal norm interpolants of the data $\lambda_i(f)$. Finally we discuss kriging, a basic technique in geostatistics, and its relation to the average case analysis.

Chapters IV–VI are concerned with minimal average errors for integration, approximation, and differentiation. Optimality of linear methods refers to the selection of coefficients and knots, and the choice of the knots is in general the more difficult task. We are interested in results that do not depend on the particularly chosen measure P, but hold for a class of measures. Such classes are often defined by smoothness properties of the covariance kernel K. Equivalently, we derive results from smoothness in mean square sense of the underlying random function. For L_p-approximation with $p \neq 2$ we additionally assume that P is Gaussian.

Chapter IV deals with integration and approximation in the univariate case $D = [0, 1]$, where a variety of smoothness conditions are used for the analysis of integration or approximation. We focus on Sacks-Ylvisaker regularity conditions, which yield sharp bounds for minimal errors. We also discuss Hölder conditions and local stationarity, but these properties are too weak to determine the asymptotic behavior of minimal errors. Regular sequences of knots, i.e., quantiles with respect to a suitably chosen density, often lead to good or even asymptotically optimal methods.

In Chapter V we address linear problems with noisy data

$$\lambda_i(f) + \varepsilon_i.$$

Stochastic assumptions concerning the noise $\varepsilon_1, \ldots, \varepsilon_n$ are used. We derive optimality properties of smoothing splines in reproducing kernel Hilbert spaces, and then we turn to the problems (1)–(3) for univariate functions, with emphasis on the differentiation problem.

In Chapter VI we study integration and L_2-approximation in the multivariate case $D = [0, 1]^d$. We distinguish between isotropic smoothness conditions and tensor product problems, and we study the order of the nth minimal errors for fixed dimension d and n tending to infinity. In the isotropic case, Hölder conditions lead to upper bounds for the minimal errors. Moreover, if P corresponds to a stationary random field, then decay conditions for the spectral density yield sharp bounds for the minimal errors. We also derive sharp bounds for specific isotropic measures corresponding to fractional Brownian fields.

In a tensor product problem we have

$$K\left((s_1, \ldots, s_d), (t_1, \ldots, t_d)\right) = \prod_{\nu=1}^{d} K_\nu(s_\nu, t_\nu)$$

for the covariance kernel of P, and the smoothness is defined by the smoothness properties of the kernels K_ν. We present results under several different smoothness assumptions for the kernels K_ν. A classical example is the Wiener sheet measure, and in this case the average error of a cubature formula coincides with the L_2-discrepancy of its knots (up to a reflection). The Smolyak construction often yields order optimal methods for tensor product problems, while tensor product methods are far from being optimal. The order of minimal errors only partially reveals the dependence of the minimal errors on d, in particular, when d is large and only a moderate accuracy is required. This motivates the study of tractability, which we briefly discuss at the end of the chapter.

Chapter VII deals with nonlinear methods for linear problems. In order to analyze these methods, which may adaptively select functionals λ_i and use an adaptive termination criterion, it is not enough to know the covariance kernel K and the mean m of P. For linear problems with Gaussian measures it turns out that adaption does not help and linear methods are optimal. Furthermore, adaptive termination criteria do not help much in many cases. In particular, the optimality results from Chapters IV and VI hold in fact for the class of adaptive nonlinear methods with average cardinality bounded by n, if the measure P

is Gaussian. Geometrical properties of Gaussian measures lead to these conclusions. This is similar to the worst case setting where, for instance, symmetry and convexity of the function class F implies that adaption may only help by a multiplicative factor of at most two. Furthermore, it is crucial that the measure P or the function class F is completely known.

Adaption may be very powerful for linear problems if one of the previous assumptions does not hold. We discuss integration and approximation of univariate functions with unknown local smoothness. Adaptive methods that explore the local smoothness and select further knots accordingly are asymptotically optimal and superior to all nonadaptive methods. We also discuss integration of monotone functions. The Dubins-Freedman or Ulam measure, which is not Gaussian, is used in the average case setting, and adaption helps significantly for integration of monotone functions. There are no worst case analogues to these positive results on the power of adaption. A worst case analysis cannot justify the use of adaptive termination criteria, and adaption does not help for integration of monotone functions in the worst case setting. Moreover, one cannot successfully deal with unknown local smoothness in the worst case.

The number $n(\widetilde{S}, f)$ is only a lower bound for the computational cost of obtaining the approximate solution $\widetilde{S}(f)$. We conclude the chapter with the definitions of the computational cost of a method and of the complexity of a numerical problem. These definitions are based on the real number model of computation with an oracle, and they are fundamental concepts of information-based complexity. In these notes, upper bounds of the minimal errors are achieved by methods $\widetilde{S}$ with computational cost proportional to $n(\widetilde{S}, f)$ for $f \in X$.

In Chapter VIII we briefly present average case results for the nonlinear problems of zero finding and global optimization. For zero finding, we consider in particular smooth univariate functions having only simple zeros. In this case, iterative methods are known that are globally and superlinearly convergent, but the worst case error of any method is still at least $2^{-(n+1)}$ after n steps, so that bisection is worst case optimal. Because of this significant difference, an average case analysis is of interest. We present a hybrid secant-bisection method with guaranteed error ε that uses about $\ln\ln(1/\varepsilon)$ steps on the average. Furthermore, this method is almost optimal in the class of all zero finding methods. The change from worst case cost to average case cost is crucial here, while the analogous change for the error is of minor importance. For global optimization we consider the average case setting with the Wiener measure. Here adaptive methods are far superior to all nonadaptive methods. In the worst case setting there is no superiority of adaptive methods for global optimization on convex and symmetric classes F. We feel that the average case results for zero finding and global optimization match practical experience far better than the worst case results.

2. Historical and Bibliographical Remarks

These notes are based on the work of many researchers in several areas of mathematics, such as approximation theory, complexity theory, numerical analysis, statistics, and stochastic processes. Most of the original work that we cover here was published during the last ten years. The notes slowly evolved out of the author's Habilitationsschrift, see Ritter (1996c). Original work is quoted in 'Notes and References', which typically ends a section. Here we only present a small subset of the literature.

Information-based Complexity. The complexity of continuous problems with partial information is studied in several settings. The monograph Traub and Woźniakowski (1980) primarily deals with the worst case setting. Traub, Wasilkowski, and Woźniakowski (1988) present the general abstract theory in their monograph on 'Information-Based Complexity'. In particular, the average case and probabilistic settings are developed, and many applications are included as well.

Novak (1988) studies, in particular, integration, approximation, and global optimization in his monograph on 'Deterministic and Stochastic Error Bounds in Numerical Analysis'. He presents numerous results on minimal errors, mainly in the worst case and randomized settings.

Further monographs include Werschulz (1991) on the complexity of differential and integral equations and Plaskota (1996) on the complexity of problems with noisy data. An informal introduction to information-based complexity is given by Traub and Werschulz (1998). This book also contains a comprehensive list of the recent literature.

Now we point to a few steps in the development of the average case analysis.

Linear Methods for Linear Univariate Problems. Poincaré (1896, Chapter 21) makes an early contribution, by using a stochastic model to construct interpolation formulas. The model is given by $f(x) = \sum_{k=0}^{\infty} c_k \, x^k$ with independent and normally distributed random variables c_k.

Suldin (1959, 1960) studies the integration and L_2-approximation problems in the average case setting with the Wiener measure, and he determines optimal linear methods. In the worst case setting the integration problem was already addressed by Nikolskij (1950) for Sobolev classes of univariate functions. It seems that Nikolskij and Suldin initiated the search for optimal numerical methods using partial information.

Sacks and Ylvisaker (1970b) open a new line of research. They provide an analysis of the integration problem not only for specific measures P, but for classes of measures that are defined by smoothness properties of their covariance kernels. They also exploit the role of reproducing kernel Hilbert spaces in the average case analysis. Kimeldorf and Wahba (1970a, 1970b) discover optimality properties of splines in reproducing kernel Hilbert spaces in the average case setting with exact or noisy data.

Integration and L_p-approximation of univariate functions in the average case setting is still an active area of research. Meanwhile a wide range of smoothness properties is covered in the analysis of linear methods. Most of the results aim at asymptotic optimality, but often only regular sequences of knots can be analyzed so far. Ritter (1996b) shows the asymptotic optimality of suitable regular sequences for integration under Sacks-Ylvisaker regularity conditions.

Average Case Complexity of Linear Problems. Based on their prior research, Traub *et al.* (1988) present the general theory, which deals with nonlinear and adaptive methods. Mainly, Gaussian measures are used in the average case setting, and then linear nonadaptive methods are proven to be optimal. Werschulz (1991) presents the general theory for ill-posed problems. Plaskota (1996) develops the general theory for linear problems with noisy data. Many applications are given in these monographs.

Adaptive Methods for Linear Problems. For some important linear problems it is not sufficient to consider linear or, more generally, nonadaptive methods. Superiority of adaptive methods in the average case setting is shown by Novak (1992) for integration of monotone functions, by Wasilkowski and Gao (1992) for integration of functions with singularities, and by Müller-Gronbach and Ritter (1998) for integration and approximation of functions with unknown local smoothness. In the latter case the estimation of the local smoothness arises as a subproblem. This problem is analyzed, for instance, by Istas (1996) and Benassi, Cohen, Istas, and Jaffard (1998).

Linear Multivariate Problems. The search for optimal knots and the respective minimal errors in the linear multivariate case has been open since Ylvisaker (1975). The first optimality results are obtained by Woźniakowski (1991) and Wasilkowski (1993). Woźniakowski determines the order of the minimal errors for integration using the Wiener sheet measure. In the proof, he establishes the relation between this integration problem and L_2-discrepancy. Wasilkowski uses the isotropic Wiener measure and determines the order of minimal errors for integration and L_2-approximation. For integration, the upper bounds are proven non-constructively; constructive proofs are given by Paskov (1993) for the Wiener sheet and Ritter (1996c) for the isotropic Wiener measure.

Smolyak (1963) introduces a general construction of efficient methods for tensor product problems, and he gives worst case estimates. Wasilkowski and Woźniakowski (1995, 1999) give a detailed analysis of the Smolyak construction, and they develop a generalization thereof. This serves as the basis of many worst case and average case results for linear tensor product problems.

Smoothness properties and their impact on minimal errors and optimal methods for integration and L_2-approximation are studied by Ritter, Wasilkowski, and Woźniakowski (1993, 1995) and Stein (1993, 1995a, 1995b). The monograph Stein (1999) provides an analysis of kriging, in particular with an unknown second order structure. For methods using function values, all known optimality results in the multivariate case are on order optimality. Optimal asymptotic constants seem to be unknown.

Woźniakowski (1992, 1994a, 1994b) opens a new line of research for linear multivariate problems. He introduces the notion of (strong) tractability to carefully study the dependence on the dimension d. Wasilkowski (1996) and Novak and Woźniakowski (1999) give surveys of results.

Nonlinear Problems. Polynomial zero finding and linear programming are nonlinear continuous problems on finite dimensional spaces. Therefore complete information is available in both cases. Borgwardt (1981, 1987) and Smale (1983) study the average number of steps of the simplex method. The average case and worst case performances of the simplex method differ extremely, and the average case results explain the practical success of the method. Polynomial zero finding is studied by Smale (1981, 1985, 1987), Renegar (1987), and in a series of papers by Shub and Smale. See Blum, Cucker, Shub, and Smale (1998) for an overview and new results.

Upper bounds for problems with partial information are obtained by Graf, Novak, and Papageorgiou (1989), and Novak (1989) for zero finding of univariate functions, and by Heinrich (1990b, 1991) for integral equations. It seems that the first lower bounds for nonlinear problems are established in Ritter (1990) for global optimization with the Wiener measure and by Novak and Ritter (1992) for zero finding of univariate functions. In these papers, however, only nonadaptive methods are covered for global optimization and only adaptive methods with fixed cardinality are covered for zero finding. For the latter problem, the use of varying cardinality is crucial. Novak, Ritter, and Woźniakowski (1995) provide the general lower bound for zero finding and construct a hybrid secant-bisection method that is almost optimal.

Calvin (1997, 1999) constructs new adaptive global optimization methods in the univariate case. His analysis reveals the power of adaption for global optimization methods. Similar results in the multivariate case are unknown.

Objective functions for global optimization are modeled as random functions by Kushner (1962) for the first time. A detailed study of this Bayesian approach to global optimization is given by Mockus (1989) and Törn and Žilinskas (1989). In these monographs the focus is on the construction of algorithms.

3. Notation and Numbering

Some notations are gathered in an index at the end of these notes. Here we only mention that $\mathbb{N} = \{1, 2, 3, \ldots\}$ and $\mathbb{N}_0 = \mathbb{N} \cup \{0\}$. Moreover, $\subset$ denotes the inclusion of sets, which is not necessarily strict.

Propositions, Remarks, etc., are numbered in a single sequence within each chapter. The number of a chapter is added in a reference iff the material is found outside of the current chapter.

CHAPTER II

Linear Problems: Definitions
and a Classical Example

In this chapter we define linear problems in an average case setting. The average case analysis involves a probability measure on a function space, or equivalently a random function or stochastic process. Some definitions concerning measures on function spaces are gathered in Section 1. Linear problems, in particular the integration and approximation problems, as well as the average case and worst case settings are defined in Section 2. Section 3 is devoted to a classical example, the average case setting with respect to the Wiener measure. For illustration, we present a complete solution of the integration problem and the L_2-approximation problem on the Wiener space.

1. Measures on Function Spaces

In an average case approach we study average errors of numerical methods with respect to a probability measure P on a class F of functions. In these notes we always have

$$F \subset C^k(D), \qquad k \in \mathbb{N}_0,$$

with a compact set $D \subset \mathbb{R}^d$, being the closure of its interior points.

Let $e(f) \geq 0$ denote the error of a numerical method for a particular function $f \in F$. For instance,

$$e(f) = \left| \int_0^1 f(t)\, dt - \sum_{i=1}^n f(x_i) \cdot a_i \right|,$$

with $f \in C([0,1])$ is the error of the quadrature formula with knots $x_i \in [0,1]$ and coefficients $a_i \in \mathbb{R}$. The average error of a numerical method can be defined as the qth moment of e with respect to the measure P. For convenience we take the qth root of this quantity, i.e., we study

$$\left(\int_F e(f)^q\, dP(f) \right)^{1/q}, \qquad 1 \leq q < \infty.$$

Of course, the choice of the measure is a crucial part of our approach. In numerical linear algebra or linear programming the problem elements, e.g., the matrices and vectors defining systems of linear equations, belong to finite dimensional spaces. However, problems like integration or approximation are usually studied on infinite dimensional classes F of functions.

This difference is important if we want to study the average case. While the Lebesgue measure and corresponding probability densities can be used in spaces of finite dimension, there is no analogue to the Lebesgue measure on infinite dimensional spaces. In the following we present some basic facts on measures on function spaces of infinite dimension.

1.1. Borel Measures.
Let us introduce the following conventions. The class F is equipped with the metric that is induced by the norm $\max_\alpha \|f^{(\alpha)}\|_\infty$ on $C^k(D)$. Here the maximum is over all partial derivatives of order $|\alpha| \leq k$.

A metric space, such as F or D, is equipped with the corresponding σ-algebra $\mathfrak{B}$ of Borel sets. By definition, $\mathfrak{B}$ is the smallest σ-algebra that contains all open subsets of the metric space. Hereby we ensure that $(f, t) \mapsto f^{(\alpha)}(t)$ is measurable on the product σ-algebra in $F \times D$, and this property is sufficient in further questions regarding measurability.

A *measure* on a metric space is understood to be a probability measure on the σ-algebra $\mathfrak{B}$ of Borel sets.

1.2. Stochastic Processes, Random Functions.
Associated with every measure P on F there is the canonical stochastic process $(\delta_t)_{t \in D}$ of evaluation functionals

$$\delta_t(f) = f(t).$$

Indeed, these functionals form a collection of random variables that are defined on the probability space $(F, \mathfrak{B}, P)$. The parameter set of the process is D, the domain of the functions from F.

Conversely, every stochastic process $(X_t)_{t \in D}$ with underlying probability space $(\Omega, \mathfrak{A}, Q)$ induces measures on certain function spaces. At first, consider the space $\mathbb{R}^D$ of all real-valued functions on D, equipped with the product σ-algebra. For fixed $\omega \in \Omega$ the function $t \mapsto X_t(\omega)$ belongs to $\mathbb{R}^D$, and it is called a sample function or sample path of the process. The image of Q under the mapping $\omega \mapsto X_\cdot(\omega)$, which assigns to each ω the associated sample path, is a probability measure P on $\mathbb{R}^D$. Depending on the smoothness of the sample paths, it is possible to restrict the class $\mathbb{R}^D$ in this construction of the measure P. In particular, if the paths are continuously differentiable up to the order k with probability one, then the process induces a measure on a class $F \subset C^k(D)$.

A stochastic process is also called a *random function*, or, if $D \subset \mathbb{R}^d$ with $d > 1$, a *random field*. It is mainly a matter of convenience to state average case results in terms of probability measures on functions spaces or in terms of random functions.

1.3. Covariance Kernel and Mean. Average errors are defined by second moments, $q = 2$, in most of the results. Hence we assume

$$\int_F f(t)^2 \, dP(f) < \infty, \qquad t \in D,$$

throughout these notes. Equivalently, we are dealing with a second-order random function. In this case the *mean*

$$m(t) = \int_F f(t) \, dP(f), \qquad t \in D,$$

and the *covariance kernel*

$$K(s,t) = \int_F (f(s) - m(s)) \cdot (f(t) - m(t)) \, dP(f), \qquad s, t \in D,$$

of P are well defined. Note that m and K are real-valued functions on D and D^2, respectively.

Let $t_1, \ldots, t_n \in D$, put $T = (t_1, \ldots, t_n)$, and consider the joint distribution of $f(t_1), \ldots, f(t_n)$ with respect to P. The first two moments of this probability distribution on $\mathbb{R}^n$ are given by the mean vector

$$\mu = (m(t_1), \ldots, m(t_n))'$$

and the *covariance matrix*

$$\Sigma = (K(t_i, t_j))_{1 \leq i, j \leq n}.$$

As long as we only study affine linear methods, results for certain linear problems like integration and L_2-approximation depend on the measure P only through the mean m and the covariance kernel K, see Chapter III. Hence these results are part of the second-order theory of random functions.

1.4. Gaussian Measures. So far mainly *Gaussian measures* have been used in the literature for the average case analysis on infinite dimensional function spaces. A measure P on a closed subset $F \subset C^k(D)$ is called Gaussian if every finite number of function evaluations $f(t_1), \ldots, f(t_n)$ is normally distributed on $\mathbb{R}^n$ with respect to P. This property implies that every bounded linear functional on $C^k(D)$ is normally distributed with respect to P. Moreover, a Gaussian measure is uniquely determined by its mean m and its covariance kernel K.

Assume that the covariance matrix Σ is invertible, and let $\langle \cdot, \cdot \rangle$ denote the Euclidean scalar product on $\mathbb{R}^n$. The joint distribution of $f(t_1), \ldots, f(t_n)$ with respect to the Gaussian measure P is given by

$$P(\{f \in F : f(t_i) \leq c_i \text{ for } i = 1, \ldots, n\})$$
$$= (2\pi)^{-d/2} \cdot (\det \Sigma)^{-1/2} \int_{-\infty}^{c_1} \cdots \int_{-\infty}^{c_n} \exp\left(-\tfrac{1}{2} \langle y - \mu, \Sigma^{-1}(y - \mu) \rangle\right) dy_1 \ldots dy_n.$$

Gaussian measures have strong integrability properties, see Vakhania, Tarieladze, and Chobanyan (1987, p. 330). For instance,

$$\int_F \|f^{(\alpha)}\|_\infty^q \, dP(f) < \infty$$

for every Gaussian measure P on F, and every q and $|\alpha| \leq k$. The support of P is a closed affine subspace of F, and P enjoys certain symmetry and translation-invariance properties.

EXAMPLE 1. As a classical example, which also plays an important role in these notes, we consider the *Wiener measure* w on the space

$$F = \{f \in C([0,1]) : f(0) = 0\}.$$

The associated canonical stochastic process is called the *Brownian motion* or Wiener process.

The measure w is uniquely determined by the following properties. Increments $f(t_i) - f(s_i)$ with

$$0 \leq s_1 \leq t_1 \leq \cdots \leq s_n \leq t_n \leq 1$$

are independent with respect to w. Each increment is normally distributed with zero mean,

$$\int_F (f(t_i) - f(s_i)) \, dw(f) = 0,$$

and variance

$$\int_F (f(t_i) - f(s_i))^2 \, dw(f) = t_i - s_i.$$

Together with $f(0) = 0$ the latter property implies that w is a Gaussian measure.

We compute the mean and covariance kernel of the Wiener measure. For every t,

$$m(t) = \int_F f(t) \, dw(f) = \int_F (f(t) - f(0)) \, dw(f) = 0$$

and

$$K(t,t) = \int_F f(t)^2 \, dw(f) = \int_F (f(t) - f(0))^2 \, dw(f) = t.$$

If $s \leq t$ then

$$K(s,t) = \int_F f(s)^2 \, dw(f) + \int_F f(s) \cdot (f(t) - f(s)) \, dw(f)$$

$$= s + \int_F (f(s) - f(0)) \, dw(f) \cdot \int_F (f(t) - f(s)) \, dw(f)$$

$$= s.$$

Thus w is the zero mean Gaussian measure on F with covariance kernel

$$K(s,t) = \min(s,t), \qquad s,t \in [0,1].$$

1.5. Notes and References. For further readings we refer to Gihman and Skorohod (1974), Adler (1981), Vakhania *et al.* (1987), and Lifshits (1995).

2. Linear Problems

We start with the definitions of the integration and approximation problems. These are the principal linear problems that are studied in these notes. For linear problems we define average as well as maximal errors, and we discuss a relation between error bounds for integration and L_2-approximation. We also indicate how the integration and approximation problems occur as prediction problems in spatial statistics, and we refer to geostatistical applications.

2.1. Integration. Let

$$\varrho \in L_1(D).$$

Approximate the integral

$$\text{Int}_\varrho(f) = \int_D f(t) \cdot \varrho(t) \, dt$$

for $f \in F$ by means of a quadrature formula

$$S_n(f) = a_0 + \sum_{i=1}^{n} f(x_i) \cdot a_i.$$

Here $x_i \in D$ are called the knots and $a_i \in \mathbb{R}$ are called the coefficients of the formula S_n. If $\varrho = 1$ then we denote Int_ϱ by Int.

Besides integrability of the weight function ϱ, we also assume $\varrho \geq c > 0$ on a nonempty open subset of D throughout these notes. Only additional assumptions on ϱ are stated explicitly in the sequel.

2.2. L_p-Approximation. Let

$$\varrho \in L_1(D), \qquad \varrho \geq 0.$$

Approximate $f \in F$ with respect to the weighted L_p-norm

$$\|f\|_{p,\varrho} = \left(\int_D |f(t)|^p \cdot \varrho(t) \, dt \right)^{1/p}, \qquad 1 \leq p < \infty,$$

or

$$\|f\|_{\infty,\varrho} = \sup_{t \in D} (|f(t)| \cdot \varrho(t)), \qquad p = \infty,$$

by means of an affine linear method

$$S_n(f) = a_0 + \sum_{i=1}^{n} f(x_i) \cdot a_i.$$

As for integration, $x_i \in D$ are called the knots and a_i are called the coefficients of S_n. However, now the coefficients are real-valued functions on D having finite norm. We use $\text{App}_{p,\varrho}$ to denote this problem, and we simply write $\| \cdot \|_p$ and App_p if $\varrho = 1$.

The weight function ϱ is always assumed to be integrable, nonnegative on D, and bounded away from zero on a nonempty open subset of D. Only additional assumptions on ϱ are stated explicitly in the sequel.

Recovery of functions is another term that is used for the approximation problem. This problem should not be confused with the problem of best approximation, where an approximating subspace or only its dimension is fixed and methods are nonlinear in general, see Section VII.2.5.

2.3. Linear Problems. For integration as well as for recovery of functions we approximate a bounded linear operator

$$S : X \to G,,$$

with X and G being normed linear spaces, on a class $F \subset X$ by methods S_n. Namely, $X = C^k(D)$, $G = \mathbb{R}$, and $S = \mathrm{Int}_\varrho$ for integration, and $S = \mathrm{App}_{p,\varrho}$ is the embedding of $X = C^k(D)$ into a (weighted) L_p-space for approximation. The error of a method S_n, applied to an element $f \in F$, is

$$e(f) = \|S(f) - S_n(f)\|_G.$$

So far we have considered affine methods S_n that use function values $\lambda_i(f) = f(x_i)$. Important generalizations are possible by dropping the affine linearity of S_n and by allowing different kinds of functionals λ_i. For instance, derivatives $\lambda_i(f) = f^{(\alpha_i)}(x_i)$ may be used for integration and integrals $\lambda_i(f) = \int_D f(t) \cdot \varrho_i(t)\, dt$ may be used for approximation.

In general we fix a class Λ of *permissible linear functionals* on X. We wish to approximate S on F by methods that apply functionals $\lambda_1, \ldots, \lambda_n \in \Lambda$ to $f \in F$ and thereafter construct an approximation $S_n(f)$ to $S(f)$. An affine method S_n is of the form

$$S_n(f) = a_0 + \sum_{i=1}^{n} \lambda_i(f) \cdot a_i, \qquad \lambda_i \in \Lambda,\ a_i \in G.$$

In these notes we always have $X = C^k(D)$ and

$$\Lambda \subset \Lambda^{\mathrm{all}}$$

with

$$\Lambda^{\mathrm{all}} = \{\lambda : \lambda \text{ bounded linear functional on } C^k(D)\} = (C^k(D))^*.$$

We are mainly interested in the class

$$\Lambda^{\mathrm{std}} = \{\delta_t : t \in D\},$$

which consists of all Dirac functionals

$$\delta_t(f) = f(t).$$

The data $f(x_1), \ldots, f(x_n)$ are also called standard information.

If we combine the data $\lambda_i(f)$ nonlinearly or if we choose the functionals λ_i adaptively, then S_n is no longer affine linear in general. Such methods are discussed in Chapter VII. Chapters III, IV, and VI deal with linear methods. In this case a linear problem is specified by the class F of functions, the operator $S :$ $X \to G$ to be approximated on F, and by the class Λ of permissible functionals.

Further examples of linear problems arise from linear integral or differential equations, formally written as $Af = u$. Suppose that a finite number of values $u(x_i)$ may be used in the computation of an approximate solution $\widetilde{f}$, whose error $f - \widetilde{f}$ is measured in some norm $\|\cdot\|_G$. Then we are facing an approximation problem, the embedding of $C^k(D)$, say, into G where Λ consists of the functionals $\lambda(f) = Af(x)$. Examples of nonlinear problems are zero finding and global optimization, see Chapter VIII.

It is natural to compare methods S_n that use the same number n of functionals from Λ, and to look for good or even optimal methods in this class. Hence we deal with an optimization problem in the functionals λ_i and coefficients a_i, if a quality criterion for the approximating methods is specified.

2.4. Average Errors. In these notes we study the average performance of methods over a class F. Let S_n be an affine method to approximate the operator S. The *q-average error* of S_n with respect to the measure P on F is defined by

$$e_q(S_n, S, P) = \left(\int_F \|S(f) - S_n(f)\|_G^q \, dP(f) \right)^{1/q}$$

with $1 \leq q < \infty$. In particular,

$$e_q(S_n, \mathrm{Int}_\varrho, P) = \left(\int_F |\mathrm{Int}_\varrho(f) - S_n(f)|^q \, dP(f) \right)^{1/q}$$

for integration, and

$$e_q(S_n, \mathrm{App}_{p,\varrho}, P) = \left(\int_F \|f - S_n(f)\|_{p,\varrho}^q \, dP(f) \right)^{1/q}$$

for L_p-approximation. For the latter problem it is often convenient to study the case $p = q$ if $p < \infty$.

The *nth minimal average errors*

$$e_q(n, \Lambda, S, P) = \inf_{S_n} e_q(S_n, S, P)$$

indicate how well S can be approximated by affine methods S_n that use n functionals from Λ. Here minimization is with respect to the coefficients $a_i \in G$ as well as the functionals $\lambda_i \in \Lambda$. Clearly one is interested in methods S_n with

$$e_q(S_n, S, P) = e_q(n, \Lambda, S, P).$$

These methods are called *average case optimal*.

Optimal methods are known only for a few problems, and therefore asymptotic or order optimality is often studied in the literature. We use $\approx$ to denote the strong equivalence of sequences of real numbers,

$$u_n \approx v_n \qquad \text{if} \qquad \lim_{n \to \infty} u_n/v_n = 1.$$

By $\asymp$ we denote the weak equivalence of sequences,

$$u_n \asymp v_n \qquad \text{if} \qquad c_1 \leq u_n/v_n \leq c_2$$

for sufficiently large n with positive constants c_1, c_2. We add that $u_n = O(v_n)$ means $|u_n| \leq c \cdot |v_n|$ for sufficiently large n with a positive constant c. In the average case setting a sequence of methods S_n is called *asymptotically optimal* if

$$e_q(S_n, S, P) \approx e_q(n, \Lambda, S, P).$$

We use the term *order optimal* if

$$e_q(S_n, S, P) \asymp e_q(n, \Lambda, S, P).$$

2.5. Maximal Errors. It is interesting to compare the average performance of a method with its worst performance. Moreover, worst case results are sometimes useful for the average case analysis. In a worst case approach we study *maximal errors* of methods S_n over a class F. For a linear problem with operator S these errors are given by

$$e_{\max}(S_n, S, F) = \sup_{f \in F} \|S(f) - S_n(f)\|_G.$$

Analogously to the average case setting, the *nth minimal maximal errors* are defined as

$$e_{\max}(n, \Lambda, S, F) = \inf_{S_n} e_{\max}(S_n, S, F),$$

where S_n varies over all affine methods that use n functionals from Λ. We wish to determine the nth minimal maximal errors and methods S_n with

$$e_{\max}(S_n, S, F) = e_{\max}(n, \Lambda, S, P).$$

These methods are called *worst case optimal*. The notions of asymptotic optimality and order optimality are also defined as in the average case.

REMARK 2. In a worst case approach linear problems are often studied on unit balls F with respect to some (semi) norm. Moreover, C^∞-functions are often dense in F, whence it suffices to consider such functions with norm bounded by one. In an average case analysis we may embed F into a class of functions with low regularity, say $C(D)$, and consider the image measure of P on this class. In a way, the choice of the norm for the worst case analysis corresponds to the choice of the measure in the average case analysis. The underlying space of functions is of minor importance in both settings.

Still we state properties such as $F \subset C^k(D)$ in these notes. This is done for convenience, e.g., in order to let a functional $\lambda(f) = f^{(\alpha)}(x)$ with $|\alpha| \leq k$ be well defined for all functions $f \in F$. The parameter k does not necessarily capture the entire smoothness of the random function f.

REMARK 3. A class F is called *symmetric* (with respect to the origin) if $f \in F$ implies $-f \in F$. Symmetry holds, for instance, if F is a unit ball in some normed space. Analogously, a measure P is called *symmetric* if $P(A) = P(-A)$ for every measurable set A. Necessarily $m = 0$ for the mean of a symmetric measure. Symmetry is equivalent to $m = 0$ for Gaussian measures, since the latter are uniquely determined by their mean and covariance kernel.

Let $S_n(f) = \sum_{i=1}^{n} \lambda_i(f) \cdot a_i$ denote a linear method and put $\widetilde{S} = S - S_n$. Observe that

$$\phi_q(a_0) = \left(\int_F \|\widetilde{S}(f) - a_0\|_G^q \, dP(f) \right)^{1/q}$$

and

$$\phi_{\max}(a_0) = \sup_{f \in F} \|\widetilde{S}(f) - a_0\|_G$$

define convex functionals on G. Furthermore, $\phi_q(a_0) = \phi_q(-a_0)$ if P is symmetric, and $\phi_{\max}(a_0) = \phi_{\max}(-a_0)$ if F is symmetric.

Given symmetry we conclude that ϕ_q and $\phi_{\max}$ are minimized for $a_0 = 0$, and therefore it is sufficient to study linear methods instead of affine linear ones. The same conclusion holds for every zero mean measure P, if G is a Hilbert space and $q = 2$, see Section III.2.

2.6. A Relation between Integration and L_2-Approximation. Since Int_ϱ is a continuous functional with respect to the L_p-norm with weight $|\varrho|^p$, the minimal errors for integration and approximation satisfy

$$e_q(n, \Lambda^{\mathrm{std}}, \mathrm{Int}_\varrho, P) = O(e_q(n, \Lambda^{\mathrm{std}}, \mathrm{App}_{p,|\varrho|^p}, P)).$$

According to the following result, this estimate can be improved substantially in many cases.

PROPOSITION 4. *Suppose that $\varrho \in L_2(D)$. Let $S_n(f) = a_0 + \sum_{i=1}^{n} f(x_i) \cdot a_i$ be a method for L_2-approximation, and let c denote the Lebesgue measure of D. Then there are knots $x_{n+1}, \ldots, x_{2n} \in D$ such that*

$$e_2(S_{2n}, \mathrm{Int}_\varrho, P) \le \left(\frac{c}{n} \right)^{1/2} \cdot e_2(S_n, \mathrm{App}_{2,\varrho^2}, P)$$

holds for the quadrature formula

$$S_{2n}(f) = \mathrm{Int}_\varrho(S_n(f)) + \frac{c}{n} \cdot \sum_{i=n+1}^{2n} (f - S_n(f))(x_i) \cdot \varrho(x_i).$$

PROOF. Let $\chi = (\chi_{n+1}, \ldots, \chi_{2n})$ consist of independent random variables χ_i that are uniformly distributed in D, and define

$$S_{2n}(f, \chi) = \mathrm{Int}_\varrho(a_0) + \sum_{i=1}^{n} f(x_i) \cdot \mathrm{Int}_\varrho(a_i) + \frac{c}{n} \cdot \sum_{i=n+1}^{2n} (f - S_n(f))(\chi_i) \cdot \varrho(\chi_i).$$

Here the first part is the integral of $S_n(f)$ and the second part is the classical Monte Carlo method to approximate Int_ϱ, applied to $\widetilde{f} = f - S_n(f)$. Note that, for each realization of χ, $S_{2n}(\cdot, \chi)$ is a deterministic quadrature formula using $2n$ evaluations of f.

Let E denote the expectation with respect to χ. Then $\mathrm{Int}_\varrho(\widetilde{f}) = c \cdot E(\widetilde{f}(\chi_i) \cdot \varrho(\chi_i))$ and

$$E((\mathrm{Int}_\varrho(f) - S_{2n}(f, \chi))^2) = E\left(\left(\mathrm{Int}_\varrho(\widetilde{f}) - \frac{c}{n} \sum_{i=n+1}^{2n} \widetilde{f}(\chi_i) \cdot \varrho(\chi_i)\right)^2\right)$$

$$= \frac{1}{n} \cdot E((\mathrm{Int}_\varrho(\widetilde{f}) - c \cdot \widetilde{f}(\chi_1) \cdot \varrho(\chi_1))^2)$$

$$\leq \frac{c}{n} \cdot \int_D \widetilde{f}(t)^2 \cdot \varrho(t)^2 \, dt$$

for every $f \in F$. Fubini's Theorem yields

$$E(e_2(S_{2n}(\cdot, \chi), \mathrm{Int}_\varrho, P)^2) = \int_F E((\mathrm{Int}_\varrho(f) - S_{2n}(f, \chi))^2) \, dP(f)$$

$$\leq \frac{c}{n} \cdot \int_F \int_D (f(t) - S_n(f)(t))^2 \cdot \varrho(t)^2 \, dt \, dP(f) = \frac{c}{n} \cdot e_2(S_n, \mathrm{App}_{2,\varrho^2}, P)^2,$$

and the statement follows by a mean value argument, applied to χ. $\square$

REMARK 5. As the proof is given non-constructively, it does not tell how to find the additional knots $x_{n+1}, \ldots, x_{2n}$. The proof is based on variance reduction for Monte Carlo methods by means of separation of the main part, see, e.g., Ermakov (1975), Novak (1988), and Heinrich and Mathé (1993). Non-constructive proofs for the existence of good deterministic quadrature formulas are also known in the worst case setting, see, e.g., Niederreiter (1992) and Sloan and Joe (1994) for the existence of good lattice rules.

In the proof above, a good Monte Carlo method yields the existence of a deterministic method that is good on the average. A classical argument, which is due to Bakhvalov (1959), uses the reverse conclusion to obtain lower bounds for the maximal errors of Monte Carlo methods. An unfavorable (discrete) measure is constructed that yields a large nth minimal average error. Due to Fubini's Theorem the latter quantity is a lower bound for the nth minimal maximal errors of Monte Carlo methods. See Novak (1988) and Mathé (1993) for details.

We will use Proposition 4 in both ways, to derive upper bounds for integration from upper bounds for L_2-approximation, and to obtain lower bounds for $e_2(n, \Lambda^{\mathrm{std}}, \mathrm{App}_{2,\varrho^2}, P)$ from lower bounds for $e_2(n, \Lambda^{\mathrm{std}}, \mathrm{Int}_\varrho, P)$.

REMARK 6. Of course, Proposition 4 is of interest only if L_2-approximation is possible with finite error. The latter is true, however, under the very mild assumptions that $\varrho \in L_2(D)$ and that K is bounded on D^2. In this case the zero approximation $S_0 = 0$ already yields

$$e_2(S_0, \mathrm{App}_{2,\varrho^2}, P)^2 = \int_F \|f\|_{2,\varrho^2}^2 \, dP(f) = \int_F \int_D f(t)^2 \cdot \varrho(t)^2 \, dt \, dP(f)$$

$$= \int_D K(t,t) \cdot \varrho(t)^2 \, dt < \infty.$$

Proposition 4 shows that

$$e_2(2n, \Lambda^{\mathrm{std}}, \mathrm{Int}_\varrho, P) \le \left(\frac{c}{n}\right)^{1/2} \cdot e_2(n, \Lambda^{\mathrm{std}}, \mathrm{App}_{2,\varrho^2}, P).$$

Usually, if P is placed on functions of finite regularity,

$$e_2(n, \Lambda^{\mathrm{std}}, \mathrm{Int}_\varrho, P) \asymp e_2(2n, \Lambda^{\mathrm{std}}, \mathrm{Int}_\varrho, P),$$

so that the minimal errors for integration and approximation differ at least by a factor proportional to $n^{1/2}$. This bound is sharp in many examples.

Clearly, analogous estimates hold if the number of additional knots is different from n. In particular, a zero approximation $S_0 = 0$ with finite error yields

$$e_2(n, \Lambda^{\mathrm{std}}, \mathrm{Int}_\varrho, P) = O(n^{-1/2})$$

for the nth minimal average errors for integration.

A worst case analysis typically yields different results. For many classes F, for instance unit balls in Hölder or Sobolev spaces, the minimal errors for integration and L_2-approximation have the same asymptotic behavior, at least up to logarithmic terms. See, e.g., Novak (1988) and Temlyakov (1994). Moreover, the nth minimal maximal errors for the problem Int_ϱ often satisfy

$$e_{\max}(n, \Lambda^{\mathrm{std}}, \mathrm{Int}_\varrho, F) \asymp n^{-r/d}$$

for isotropic classes F of functions on $D \subset \mathbb{R}^d$ having a regularity r.

This difference is important in particular when the number d of variables is large relative to r. In this case $n = O(\varepsilon^{-2})$ function values are sufficient to obtain an average error ε for Int_ϱ, while a much larger number $n \asymp \varepsilon^{-d/r}$ is needed to guarantee an error ε for all $f \in F$.

2.7. Average Case Analysis and Spatial Statistics. The average case analysis is based on a probability measure on a space of real-valued functions, or equivalently a random function or stochastic process. We assume that the domain $D \subset \mathbb{R}^d$ of these functions is the closure of its interior points, see Section 1. Thus we employ a stochastic model for spatially correlated data, viz., function values, where the spatial index $t \in D$ varies over a continuous subset of $\mathbb{R}^d$. Stochastic models of this kind are used, for instance, in geostatistics and computer experiments.

In both of these fields a main concern is prediction from a partially observed realization of the random function. The observations often consist of function values $\lambda_i(f) = f(x_i)$ or local averages $\lambda_i(f) = \int_{B_i} f(t) \, dt$. These are used to predict a functional $S(f)$ of the corresponding realization. Important examples are *point prediction*, where $S(f) = f(x_0)$, or *block prediction*, where $S(f) = \int_B f(t)\,dt$. Frequently, affine linear predictors are studied. Clearly, block prediction coincides with the integration problem when $\varrho = 1_B$, where 1_B is the indicator function of B. Furthermore, point prediction at all points $t \in D$ coincides with the approximation problem.

EXAMPLE 7. For instance, D may be a three-dimensional underground region and $f(t)$ describes the unknown ore content at $t \in D$, which is considered as a realization of a stochastic process. Core samples, taken at volumes $B_i \subset D$, yield discrete data $\int_{B_i} f(t)\,dt$. The data are used to predict $f(t)$ at some points t or to predict the total ore content $\int_D f(t)\,dt$. In a two-dimensional example, D is a surface region and $f(t)$ describes the depth of the horizon of a petroleum reservoir underneath $t \in D$.

In computer experiments, $f(t)$ is the output of a computer code on input t from some input space D. The code models a complex physical system, and, because of its computational cost, cheaper predictors for $f(t)$ are of interest. The unknown function f is again considered as a realization of a stochastic process.

Predictors are often compared according to their mean-squared prediction error. The latter coincides with the squared q-average error where $q = 2$. For point prediction at every point $t \in D$ the integrated mean-squared prediction error coincides with the squared q-average error for L_p-approximation if $q = p = 2$.

A main theme in this notes is the optimal choice of points $x_1, \ldots, x_n \in D$, where observations of f are made. Any such set of points is called a *sampling design* in spatial statistics, and the search for optimal n-point designs is known as the design problem. In the previous examples a good design should consist of a small set of inputs or a small number of drilled wells that allow prediction with reasonable accuracy.

Instead of the covariance kernel K, the *variogram*

$$\Gamma(s, t) = \int_F \left((f(s) - m(s)) - (f(t) - m(t)) \right)^2 dP(f)$$
$$= K(s, s) - 2K(s, t) + K(t, t)$$

is often used in the geostatistical literature. Among the parametric models for K or Γ that are used in practice are $\Gamma(s, t) = c\|s - t\|$, which corresponds to an isotropic Wiener measure, and Bessel-MacDonald kernels K. We derive error bounds in terms of the smoothness of Γ, and in particular for the previous examples, in Section VI.1. Covariance kernels in tensor product form are studied in Section VI.2. In the univariate case, Taylor, Cumberland, and Sy (1994) consider parametric versions of the Brownian motion kernel and the kernel of an integrated Ornstein-Uhlenbeck process to model longitudinal AIDS data. We derive error bounds for these kernels in Chapter IV.

In geostatistical applications certain characteristics of the random function, such as the mean or variogram, are fitted to the observational data. Thereafter predictors are constructed on the basis of the estimated characteristics. A standard procedure in geostatistics is kriging. We will return to this issue in Sections III.4 and VII.3.

2.8. Notes and References. 1. Suldin (1959, 1960) is the first to determine average case optimal methods in numerical analysis. He considers the class $C([0,1])$, equipped with the Wiener measure, and obtains optimal quadrature formulas. Sarma (1968) and Sarma and Stroud (1969) use the Eberlein measure to study integration of smooth functions on $D = [-1,1]^d$, see also Stroud (1971). For fixed knots x_i the optimal coefficients a_i are obtained, and several known cubature formulas are compared with respect to their average error. Larkin (1972) uses finitely additive Gaussian measures on Hilbert spaces to define an average error. Sacks and Ylvisaker (1966, 1968, 1970a, 1970b) consider regression design problems that turn out to be equivalent to the integration problem, see Section III.4.3. See also Cambanis (1985) for a survey.

2. A general framework for the average case analysis is formulated in Traub, Wasilkowski, and Woźniakowski (1988), see also Micchelli and Rivlin (1985) and Novak (1988). The relation between Bayesian statistics and average case analysis is discussed in Kadane and Wasilkowski (1985), see also Diaconis (1988). Approximation of Gaussian random elements is studied in Richter (1992).

3. Numerous papers deal with the worst case analysis of the integration or approximation problem. This field of work is initiated by Sard (1949) and by Nikolskij (1950), who determines optimal quadrature formulas for certain classes F. Further pioneering papers are Kiefer (1957) and Golomb and Weinberger (1959). A partial list of monographs on the integration problem is Braß (1977), Levin and Girshovich (1979), Nikolskij (1979), Engels (1980), and Sobolev (1992). Optimal recovery of functions and related questions in approximation theory are studied, e.g., in Tikhomirov (1976, 1990), Korneichuk (1991), and Temlyakov (1994).

4. A general framework for the worst case analysis of numerical problems is formulated in Micchelli and Rivlin (1977, 1985), Traub and Woźniakowski (1980), Traub, Wasilkowski, and Woźniakowski (1983, 1988) and Novak (1988). One also may find many results and references for the integration and the approximation problem there.

5. Optimal methods for linear problems with noisy data $\lambda_i(f) + \varepsilon_i$ are studied in Plaskota (1996); see also Chapter V. Werschulz (1991) and Pereverzev (1996) study optimal methods for operator equations.

6. Proposition 4 is due to Wasilkowski (1993), see also Woźniakowski (1992).

7. For spatial statistics and geostatistical applications we refer to Christensen (1991), Cressie (1993), and Hjort and Omre (1994). For computer experiments we refer to Sacks, Welch, Mitchell, and Wynn (1989), Currin, Mitchell, Morris, and Ylvisaker (1991), and Koehler and Owen (1996).

3. Integration and L_2-Approximation on the Wiener Space

We illustrate the concepts from the previous sections for a particular example, namely, integration and L_2-approximation in the average case setting with respect to the Wiener measure. From Example 1 we know that the Wiener measure w is the Gaussian measure on

$$F = \{f \in C([0,1]) : f(0) = 0\}$$

with zero mean, $m = 0$, and covariance kernel $K(s,t) = \min(s,t)$. Remark 3 allows us to restrict our considerations to linear methods.

We take the problems Int and App_2 and the Wiener measure for illustration for several reasons. Historically, the first average case analysis was done for the integration problem and the Wiener measure, see Suldin (1959, 1960). Moreover, for both problems, Int and App_2, average errors with respect to w can be calculated explicitly and optimal methods are known. Finally, the Wiener measure serves as a particular example at several places in these notes.

3.1. Average Errors of Quadrature Formulas. We consider the integration problem with constant weight function $\varrho = 1$ and q-average error for $q = 2$. This problem was the starting point in Suldin's work.

Let us compute the error of a quadrature formula

$$S_n(f) = \sum_{i=1}^{n} f(x_i) \cdot a_i.$$

In the sequel we frequently use Fubini's Theorem. By definition,

$$e_2(S_n, \text{Int}, w)^2 = \int_F (\text{Int}(f) - S_n(f))^2 \, dw(f)$$

$$= \int_F \text{Int}(f)^2 \, dw(f) - 2 \sum_{i=1}^{n} a_i \int_F \text{Int}(f) \cdot f(x_i) \, dw(f) + \sum_{i,j=1}^{n} a_i a_j K(x_i, x_j).$$

We calculate the covariance of Int and a function evaluation δ_t,

$$\int_F \text{Int}(f) \cdot f(t) \, dw(f) = \int_0^1 \int_F f(s) \cdot f(t) \, dw(f) \, ds = \int_0^1 \min(s,t) \, ds = t - \tfrac{1}{2}t^2,$$

and the variance of Int,

$$\int_F \text{Int}(f)^2 \, dw(f) = \int_0^1 \int_F \text{Int}(f) \cdot f(t) \, dw(f) \, dt = \int_0^1 (t - \tfrac{1}{2}t^2) \, dt = \tfrac{1}{3}.$$

Summarizing we obtain

$$(1) \qquad e_2(S_n, \text{Int}, w)^2 = \tfrac{1}{3} - 2 \sum_{i=1}^{n} a_i \cdot (x_i - \tfrac{1}{2}x_i^2) + \sum_{i,j=1}^{n} a_i a_j \cdot \min(x_i, x_j).$$

This formula can be minimized explicitly with respect to the coefficients a_i as well as to the knots x_i.

3.2. Optimal Coefficients of Quadrature Formulas. In a first step we fix the knots x_i and minimize $e_2(S_n, \text{Int}, w)$ with respect to the coefficients a_i. We assume

$$0 < x_1 < \cdots < x_n \leq 1$$

since $f(0) = 0$ for $f \in F$.

Let

$$(2) \qquad \Sigma = (\min(x_i, x_j))_{1 \leq i,j \leq n} = \begin{pmatrix} x_1 & x_1 & \cdots & x_1 \\ x_1 & x_2 & \cdots & x_2 \\ \vdots & \vdots & \ddots & \vdots \\ x_1 & x_2 & \cdots & x_n \end{pmatrix}$$

denote the covariance matrix corresponding to the knots x_i. Furthermore, let

$$b = (x_i - \tfrac{1}{2}x_i^2)_{1 \leq i \leq n}$$

denote the vector of covariances between Int and δ_{x_i}. Then (1) reads

$$(3) \qquad e_2(S_n, \text{Int}, w)^2 = \tfrac{1}{3} - 2 \cdot a' \cdot b + a' \cdot \Sigma \cdot a$$

if $a = (a_1, \ldots, a_n)'$. The right-hand side is a quadratic functional with respect to a, and the unique minimizer is given by

$$(4) \qquad a = \Sigma^{-1} \cdot b.$$

Let us compute a explicitly. Put $x_0 = 0$ and

$$D = \begin{pmatrix} x_1 - x_0 & & 0 \\ & \ddots & \\ 0 & & x_n - x_{n-1} \end{pmatrix}, \qquad B = \begin{pmatrix} 1 & & 0 \\ \vdots & \ddots & \\ 1 & \cdots & 1 \end{pmatrix}.$$

We have the decomposition

$$(5) \qquad \Sigma = B \cdot D \cdot B',$$

which reflects that the Brownian motion has independent and stationary increments, see Example 1. Since

$$B^{-1} = \begin{pmatrix} 1 & & & 0 \\ -1 & \ddots & & \\ & \ddots & \ddots & \\ 0 & & -1 & 1 \end{pmatrix},$$

we obtain

$$a = (B^{-1})' \cdot D^{-1} \cdot \begin{pmatrix} (x_1 - x_0) \cdot (1 - \frac{1}{2}(x_1 + x_0)) \\ \vdots \\ (x_n - x_{n-1}) \cdot (1 - \frac{1}{2}(x_n + x_{n-1})) \end{pmatrix}$$

$$= (B^{-1})' \cdot \begin{pmatrix} 1 - \frac{1}{2}(x_1 + x_0) \\ \vdots \\ 1 - \frac{1}{2}(x_n + x_{n-1}) \end{pmatrix} = \begin{pmatrix} \frac{1}{2}(x_2 - x_0) \\ \vdots \\ \frac{1}{2}(x_n - x_{n-2}) \\ 1 - \frac{1}{2}(x_n + x_{n-1}) \end{pmatrix}.$$

Hence we have determined the best quadrature formula that uses the knots x_i.

LEMMA 8. *The quadrature formula*

$$S_n(f) = f(x_1) \cdot \frac{1}{2}x_2 + \sum_{i=2}^{n-1} f(x_i) \cdot \frac{1}{2}(x_{i+1} - x_{i-1}) + f(x_n) \cdot \left(1 - \frac{1}{2}(x_n + x_{n-1})\right)$$

has minimal average error $e_2(S_n, \text{Int}, w)$ among all formulas that use given knots $0 < x_1 < \cdots < x_n \leq 1$.

Note that S_n is a trapezoidal rule, which is based on the data $f(x_i)$, the fact $f(0) = 0$, and the assumption $f(1) = f(x_n)$.

3.3. Optimal Quadrature Formulas. In a second step we determine the optimal knots x_i for the trapezoidal rule. Due to Lemma 8, this is equivalent to determining optimal quadrature formulas with respect to the Wiener measure.

Substituting (4) in (3) we get

$$e_2(S_n, \text{Int}, w)^2 = \frac{1}{3} - a' \cdot b$$

$$= \frac{1}{3} - \sum_{i=1}^{n-1} \frac{1}{2}(x_{i+1} - x_{i-1}) \cdot \left(x_i - \frac{1}{2}x_i^2\right) - \left(1 - \frac{1}{2}(x_n + x_{n-1})\right) \cdot \left(x_n - \frac{1}{2}x_n^2\right)$$

for the error of the trapezoidal rule. We use

$$\sum_{i=1}^{n-1} \frac{1}{2}(x_{i+1} - x_{i-1}) \cdot x_i = \frac{1}{2}x_n\, x_{n-1},$$

$$\sum_{i=1}^{n-1} \frac{1}{4}(x_{i+1} - x_{i-1}) \cdot x_i^2 = \frac{1}{12}\sum_{i=1}^{n}(x_i - x_{i-1})^3 - \frac{1}{12}x_n^3 + \frac{1}{4}x_n^2\, x_{n-1},$$

and

$$\left(1 - \frac{1}{2}(x_n + x_{n-1})\right) \cdot \left(x_n - \frac{1}{2}x_n^2\right) = x_n - x_n^2 - \frac{1}{2}x_n\, x_{n-1} + \frac{1}{4}x_n^3 + \frac{1}{4}x_n^2 \cdot x_{n-1}$$

to obtain

$$e_2(S_n, \text{Int}, w)^2 = \frac{1}{3}(1 - x_n)^3 + \frac{1}{12}\sum_{i=1}^{n}(x_i - x_{i-1})^3.$$

Thus, if we fix the knot x_n, the error is minimal iff

$$x_i - x_{i-1} = \frac{x_n}{n},$$

and in this case

$$e_2(S_n, \mathrm{Int}, w)^2 = \tfrac{1}{3}(1 - x_n)^3 + \tfrac{1}{12}\frac{x_n^3}{n^2}.$$

Here the minimum

$$e_2(S_n, \mathrm{Int}, w)^2 = (3 \cdot (2n+1)^2)^{-1}$$

is attained for

$$x_n = \frac{2n}{2n+1}.$$

We use these knots x_i in the best formula according to Lemma 8 to get the following result.

PROPOSITION 9. *The trapezoidal rule*

$$S_n(f) = \frac{2}{2n+1} \cdot \sum_{i=1}^{n} f\left(\frac{2i}{2n+1}\right)$$

is average case optimal with respect to the Wiener measure and $q = 2$. The nth minimal error is

$$e_2(n, \Lambda^{\mathrm{std}}, \mathrm{Int}, w) = e_2(S_n, \mathrm{Int}, w) = \frac{1}{\sqrt{3} \cdot (2n+1)}.$$

3.4. Average Errors for L_2-Approximation. As we did for integration, we consider the constant weight function $\varrho = 1$ and the q-average error for $q = 2$. Let

$$S_n(f) = \sum_{i=1}^{n} f(x_i) \cdot a_i$$

be a linear method for L_2-approximation with knots $x_i \in [0,1]$. Recall that the coefficients of S_n are functions $a_i \in L_2([0,1])$. Fubini's Theorem yields

$$e_2(S_n, \mathrm{App}_2, w)^2 = \int_F \|f - S_n(f)\|_2^2 \, dw(f) = \int_0^1 \int_F (f(t) - S_n(f)(t))^2 \, dw(f) \, dt.$$

Consider the inner integral for fixed $t \in [0,1]$. Analogously to (1) we obtain

$$\int_F (f(t) - S_n(f)(t))^2 \, dw(f)$$

$$= t - 2\sum_{i=1}^{n} a_i(t) \cdot \min(t, x_i) + \sum_{i,j=1}^{n} a_i(t)a_j(t) \cdot \min(x_i, x_j),$$

so that

(6) $\quad e_2(S_n, \mathrm{App}_2, w)^2$

$$= \int_0^1 \Big(t - 2 \sum_{i=1}^n a_i(t) \cdot \min(t, x_i) + \sum_{i,j=1}^n a_i(t) a_j(t) \cdot \min(x_i, x_j)\Big)\, dt.$$

Explicit minimization of this formula is possible with respect to the coefficients a_i as well as to the knots x_i.

3.5. Optimal Coefficients for L_2-Approximation. We proceed as for integration and fix increasingly ordered knots $x_i > 0$ at first, while minimizing the error with respect to the coefficients $a_i \in L_2([0,1])$. These coefficients are optimal iff they minimize the integrand in (6) for almost every t. For fixed t the problem is similar to the integration problem, since we wish to approximate a bounded linear functional, namely δ_t, by linear methods that use function values only.

In terms of the covariance matrix Σ, see (2), and the vector

$$b(t) = (\min(t, x_i))_{1 \leq i \leq n}$$

of covariances between δ_t and δ_{x_i} formula (6) reads

(7) $\quad e_2(S_n, \mathrm{App}_2, w)^2 = \int_0^1 (t - 2 \cdot a(t)' \cdot b(t) + a(t)' \cdot \Sigma \cdot a(t))\, dt$

if $a(t) = (a_1(t), \dots, a_n(t))'$. The integrand, which is again a quadratic functional, becomes minimal for

(8) $\qquad\qquad\qquad a(t) = \Sigma^{-1} \cdot b(t).$

In an explicit computation of $a(t)$ we use the decomposition (5) of Σ and we distinguish three cases. Assume that $t \in [x_{i-1}, x_i]$ for some $i \geq 2$. Then $b(t) = (x_1, \dots, x_{i-1}, t, \dots, t)'$ and

(9) $\qquad\qquad a_i(t) = \dfrac{t - x_{i-1}}{x_i - x_{i-1}}, \qquad a_{i-1}(t) = 1 - a_i(t),$

while $a_j(t) = 0$ for $j \notin \{i-1, i\}$. For $t \in [0, x_1]$ we have $b(t) = (t, \dots, t)'$ and

(10) $\qquad\qquad\qquad a_1(t) = \dfrac{t}{x_1},$

while $a_j(t) = 0$ for $j > 1$. If $t \in [x_n, 1]$ then $b(t) = (x_1, \dots, x_n)'$ and

(11) $\qquad\qquad\qquad a_n(t) = 1,$

while $a_j(t) = 0$ for $j < n$.

Summarizing, the best coefficients a_i are continuous and piecewise linear with breakpoints $x_1, \dots, x_n$ and $a_i(x_i) = 1$. Put $x_0 = 0$. If $i < n$ then a_i vanishes outside of $]x_{i-1}, x_{i+1}[$, i.e., a_i is a hat function. Furthermore, a_n vanishes outside of $]x_{n-1}, 1]$ and $a_n(t) = 1$ for $t \geq x_n$. We conclude that the best method S_n uses piecewise linear interpolation. More precisely, the following result holds.

LEMMA 10. *Let $x_0 = 0$ and observe that $f(x_0) = 0$ for $f \in F$. The linear method that is defined by*

$$S_n(f)(t) = f(x_{i-1}) \cdot \frac{x_i - t}{x_i - x_{i-1}} + f(x_i) \cdot \frac{t - x_{i-1}}{x_i - x_{i-1}}$$

for $t \in [x_{i-1}, x_i]$ and

$$S_n(f)(t) = f(x_n)$$

for $t \in [x_n, 1]$ has minimal average error $e_2(S_n, \mathrm{App}_2, w)$ among all linear methods that use the given knots $0 < x_1 < \cdots < x_n \leq 1$.

REMARK 11. For fixed knots x_i the best methods for Int and App_2 are given in Lemma 8 and 10, respectively. Note that these methods are of the same type. For $f \in F$ a function $f^* \in F$ is constructed by linear interpolation of the data $f(0) = 0, f(x_1), \ldots, f(x_n)$ on $[0, x_n]$ and by constant extrapolation of $f(x_n)$ on $[x_n, 1]$. The best methods are $S_n(f) = f^*$ for approximation and $S_n(f) = \mathrm{Int}(f^*)$ for integration. Formally, $S_n(f) = S(f^*)$ for both problems, $S = \mathrm{App}_2$ and $S = \mathrm{Int}$, and f^* is a polynomial spline of degree 1. See Section III.3 for general results on optimality of abstract spline algorithms, which explain this fact.

Furthermore, both methods are local in the sense that f^* is determined on $[x_{i-1}, x_i]$ by the data $f(x_{i-1})$ and $f(x_i)$. This property reflects the quasi-Markov property of the Brownian motion, which states that the restrictions of f on the subsets $[x_{i-1}, x_i]$ and $[0, x_{i-1}] \cup [x_i, 1]$ are independent with respect to w given $f(x_{i-1})$ and $f(x_i)$. See Adler (1981, p. 255).

3.6. Optimal L_2-Approximation. Now we determine optimal knots for piecewise linear interpolation, which is equivalent to determining optimal methods for L_2-approximation with respect to the Wiener measure. Substituting (8) in (7) we get

$$e_2(S_n, \mathrm{App}_2, w)^2 = \int_0^1 (t - a(t)' \cdot b(t)) \, dt$$

for the error of piecewise linear interpolation. If $t \in [x_{i-1}, x_i]$ and $i \geq 2$ then (9) yields

$$t - a(t)' \cdot b(t) = t - a_{i-1}(t) \cdot x_{i-1} - a_i(t) \cdot t = \frac{(t - x_{i-1}) \cdot (x_i - t)}{x_i - x_{i-1}}.$$

If $t \in [0, x_1]$ then

$$t - a(t)' \cdot b(t) = \frac{t \cdot (x_1 - t)}{x_1}$$

due to (10). Finally (11) implies

$$t - a(t)' \cdot b(t) = t - x_n$$

for $t \in [x_n, 1]$. It follows that

$$e_2(S_n, \mathrm{App}_2, w)^2 = \sum_{i=1}^{n} \int_{x_{i-1}}^{x_i} \frac{(t - x_{i-1}) \cdot (x_i - t)}{x_i - x_{i-1}} \, dt + \int_{x_n}^{1} (t - x_n) \, dt$$

$$= \tfrac{1}{6} \sum_{i=1}^{n} (x_i - x_{i-1})^2 + \tfrac{1}{2} (1 - x_n)^2.$$

The latter expression is minimal iff

$$x_i - x_{i-1} = \frac{x_n}{n}$$

and

$$x_n = \frac{3n}{3n + 1}.$$

The minimal value is

$$e_2(S_n, \mathrm{App}_2, w)^2 = (2 \cdot (3n + 1))^{-1}.$$

We have established the following result.

PROPOSITION 12. *Let*

$$x_i = \frac{3i}{3n + 1},$$

and let S_n denote the corresponding piecewise linear interpolation, which is defined in Lemma 10. The linear method S_n is average case optimal with respect to the Wiener measure and $q = 2$. The nth minimal error is

$$e_2(n, \Lambda^{\mathrm{std}}, \mathrm{App}_2, w) = e_2(S_n, \mathrm{App}_2, w) = \frac{1}{\sqrt{2 \cdot (3n + 1)}}.$$

REMARK 13. According to Proposition 4 and Remark 6 the minimal errors $e_2(n, \Lambda^{\mathrm{std}}, \mathrm{Int}, w)$ and $e_2(n, \Lambda^{\mathrm{std}}, \mathrm{App}_2, w)$ differ at least by a factor that is proportional to $n^{-1/2}$. Propositions 9 and 12 show that this estimate is sharp for the Wiener measure, i.e.,

$$e_2(n, \Lambda^{\mathrm{std}}, \mathrm{Int}, w) \asymp n^{-1/2} \cdot e_2(n, \Lambda^{\mathrm{std}}, \mathrm{App}_2, w).$$

For both problems, Int and App_2, the optimal methods use equidistant knots, i.e., $x_i - x_{i-1}$ depends on n but not on i. However, the optimal knots are not symmetric, i.e., $x_1 \neq 1 - x_n$. The absence of symmetry is not surprising, since the covariance kernel of w is not symmetric in this sense, either. In Section 3.7 we mention two measures that are closely related to the Wiener measure and enjoy symmetry.

So far we have only used the mean m and the covariance kernel K of the Wiener measure, while we did not use the fact that w is Gaussian. This observation is generalized in Section III.2. On the other hand, the analysis of L_p-approximation with $p \neq 2$ or q-average errors with $q \neq 2$ relies on the fact that w is Gaussian. See Chapter IV for general results, which hold in particular for the Wiener measure.

3.7. Related Results. Consider the image of w under the transformation $f \mapsto g$ with $g(t) = f(t) - t \cdot f(1)$. This transformation is linear, so that the image measure is Gaussian with zero mean on the class of all function $g \in C([0,1])$ with $g(0) = g(1) = 0$. Its covariance kernel is given by

$$\int_F (f(s) - s \cdot f(1)) \cdot (f(t) - t \cdot f(1)) \, dw(f) = \min(s, t) - s \cdot t$$

$$= \min(s, t) \cdot (1 - \max(s, t)).$$

The associated random function is called the *Brownian bridge*, see Lifshits (1995, p. 31). The optimal knots for integration as well as for L_2-approximation turn out to be $x_i = i/(n+1)$, see Notes and References 3.8.1 and 3.8.2.

Consider now the sum of two independent Brownian motions, one of which with reversed time. The corresponding measure is formally constructed as the image of the product measure $w \otimes w$ on F^2 under the mapping $(f_1, f_2) \mapsto g$ with $g(t) = f_1(t) + f_2(1-t)$. Hereby we get a zero mean Gaussian measure on $C([0,1])$ with covariance kernel

$$\int_{F^2} (f_1(s) + f_2(1-s)) \cdot (f_1(t) + f_2(1-t)) \, d(w \otimes w)(f_1, f_2)$$

$$= \int_F f_1(s) \cdot f_1(t) \, dw(f_1) + \int_F f_2(1-s) \cdot f_2(1-t) \, dw(f_2)$$

$$= \min(s, t) + \min(1-s, 1-t) = 1 - |s - t|.$$

This kernel is sometimes called the linear covariance kernel. The optimal knots for L_2-approximation are symmetric and equidistant. Furthermore, x_1 is the real root of

$$8n \cdot x_1^3 + (9n - 5) \cdot x_1^2 + (3n - 5) \cdot x_1 - 1 = 0,$$

see Notes and References 3.8.2.

3.8. Notes and References. 1. Proposition 9 is due to Suldin (1959). A generalization to stochastic processes with stationary independent increments is given in Samaniego (1976) and Cressie (1978). The analogue to Suldin's result for the Brownian bridge is derived in Yanovich (1988).

2. Proposition 12 is due to Lee (1986). For the Brownian bridge and for the linear covariance kernel the analogue is obtained in Abt (1992) and Müller-Gronbach (1996).

3. For L_p-approximation with $p \neq 2$ optimal methods and corresponding errors with respect to the Wiener measure are obtained in Ritter (1990).

CHAPTER III

Second-Order Results for Linear Problems

We derive average case results that depend on the measure P only through the mean and the covariance kernel, i.e., thorough the first two moments. No further properties of P are used in most of this section. The only exception is Section 3.5, which deals with Gaussian measure.

In Sections 1 and 2 we study reproducing kernel Hilbert spaces and their role in the average case analysis. Optimality properties of spline algorithms are derived in Section 3. Here, optimality refers only to the selection of coefficients of linear methods, while the functionals λ_i are fixed. A spline in a reproducing kernel Hilbert space is an interpolating function with minimal norm. A spline algorithm interpolates the data and then solves the linear problem for the interpolating spline. In Section 4 we discuss kriging or best linear unbiased prediction. Here we have a linear parametric model for the unknown mean of P, and the discrete observations of the random function define a linear regression model with correlated errors.

1. Reproducing Kernel Hilbert Spaces

Hilbert spaces with reproducing kernel play an important role in the average case analysis of linear problems. The kernel is given as the covariance kernel $K : D^2 \to \mathbb{R}$ of the measure P, and the Hilbert space $H(K)$ consists of real-valued functions on D. By a fundamental isomorphism, the average case for a linear problem with a functional S is equivalent to a worst case problem on the unit ball in $H(K)$. Therefore we study properties of these Hilbert spaces, as well as smoothness properties of their elements.

1.1. Construction and Elementary Properties.
The covariance kernel K of a measure P is *nonnegative definite*, since

$$\sum_{i,j=1}^{n} a_i a_j \cdot K(x_i, x_j) = \int_F \left(\sum_{i=1}^{n} (f(x_i) - m(x_i)) \cdot a_i \right)^2 dP(f) \geq 0$$

for all $a_i \in \mathbb{R}$ and $x_i \in D$.

A Hilbert space $H(K)$ of real-valued functions on D is associated with a nonnegative definite function K on D^2. The space is called a *Hilbert space with reproducing kernel K*, and we use $\langle \cdot, \cdot \rangle_K$ and $\| \cdot \|_K$ to denote the scalar product and the norm in $H(K)$, respectively. Furthermore, let

$$B(K) = \{h \in H(K) : \|h\|_K \le 1\}$$

denote the unit ball in $H(K)$. We mention that the results stated in Sections 1.1 and 1.2 hold for an arbitrary nonempty set D.

PROPOSITION 1. *Let $K : D^2 \to \mathbb{R}$ be a nonnegative definite function. Then there exists a uniquely determined Hilbert space $H(K)$ of real-valued functions on D such that*

$$K(\cdot, t) \in H(K)$$

and the reproducing property

$$\langle h, K(\cdot, t) \rangle_K = h(t)$$

hold for all $h \in H(K)$ and $t \in D$.

PROOF. Consider the space $H_0 = \operatorname{span}\{K(\cdot, t) : t \in D\}$ and let $h = \sum_{i=1}^{n} a_i \cdot K(\cdot, x_i)$ and $g = \sum_{j=1}^{\ell} b_j \cdot K(\cdot, y_j)$ be two elements of H_0. By

$$(1) \qquad \langle g, h \rangle_K = \sum_{i=1}^{n} \sum_{j=1}^{\ell} a_i b_j \cdot K(x_i, y_j)$$

we get a well defined symmetric bilinear form on H_0, since

$$\langle g, h \rangle_K = \sum_{i=1}^{n} a_i \cdot g(x_i) = \sum_{j=1}^{\ell} b_j \cdot h(y_j).$$

Moreover, $\langle h, h \rangle_K \ge 0$ since K is nonnegative definite. By definition we have the reproducing property on H_0, and the Cauchy-Schwarz inequality yields $h = 0$ iff $\langle h, h \rangle_K = 0$.

Consider a Cauchy sequence of functions $h_n \in H_0$. Since

$$|h_n(t) - h_\ell(t)| = |\langle h_n - h_\ell, K(\cdot, t) \rangle_K| \le \|h_n - h_\ell\|_K \cdot K(t, t)^{1/2},$$

the pointwise limit

$$h(t) = \lim_{n \to \infty} h_n(t), \qquad t \in D,$$

exists. In particular, if $h = 0$ then $\lim_{n \to \infty} \langle h_n, h_\ell \rangle_K = 0$ for fixed ℓ, and

$$\langle h_n, h_n \rangle_K \le |\langle h_n, h_\ell \rangle_K| + |\langle h_n, h_n - h_\ell \rangle_K| \le |\langle h_n, h_\ell \rangle_K| + \|h_n\|_K \cdot \|h_n - h_\ell\|_K$$

yields $\lim_{n \to \infty} \langle h_n, h_n \rangle_K = 0$.

We define $H(K)$ to consist of the pointwise limits of arbitrary Cauchy sequences in H_0, and we put

$$\langle g, h \rangle_K = \lim_{n \to \infty} \langle g_n, h_n \rangle_K$$

for pointwise limits g and h of Cauchy sequences g_n and h_n in H_0. It is easily checked that $\langle \cdot, \cdot \rangle_K$ is a well defined symmetric bilinear form on $H(K)$. The reproducing property on $H(K)$ follows from

$$\langle h, K(\cdot, t) \rangle_K = \lim_{n \to \infty} \langle h_n, K(\cdot, t) \rangle_K = \lim_{n \to \infty} h_n(t) = h(t),$$

and hereby we get $\langle h, h \rangle_K = 0$ iff $h = 0$.

Note that $\lim_{n \to \infty} \langle g, h_n \rangle_K = \langle g, h \rangle_K$ for any fixed $g \in H(K)$ and

$$\langle h_n - h, h_n - h \rangle_K \leq |\langle h_n - h, h_\ell - h \rangle_K| + |\langle h_n - h, h_n - h_\ell) \rangle_K|$$
$$\leq |\langle h_n - h, h_\ell - h \rangle_K| + \|h_n - h\|_K \cdot \|h_n - h_\ell\|_K.$$

Thus a Cauchy sequence in H_0 tends to its pointwise limit also in the norm $\| \cdot \|_K$. We conclude that H_0 is dense in $H(K)$ and finally that $H(K)$ is a complete space.

Now let H be any Hilbert space of real-valued functions on D with $H_0 \subset H$ and with the reproducing property. Then the scalar product on H_0 is necessarily given by (1). Moreover, H_0 is dense in H, and convergence in H implies pointwise convergence. We conclude that the linear spaces H and $H(K)$ coincide. By continuity, the scalar products coincide, too. $\square$

Clearly the functionals $\delta_t(h) = h(t)$ are bounded on $H(K)$. Conversely, for any Hilbert space H of real-valued functions on D with bounded evaluations δ_t, there exists a uniquely determined nonnegative definite function K such that H is the Hilbert space with reproducing kernel K. Namely, $K(\cdot, t) \in H$ is necessarily given as the representer of δ_t in Riesz's Theorem. Then we have the reproducing property and

$$(2) \qquad \sum_{i,j=1}^{n} a_i a_j \cdot K(x_i, x_j) = \sum_{i,j=1}^{n} a_i a_j \cdot \langle K(\cdot, x_i), K(\cdot, x_j) \rangle_K$$
$$= \left\langle \sum_{i=1}^{n} a_i \cdot K(\cdot, x_i), \sum_{j=1}^{n} a_j \cdot K(\cdot, x_j) \right\rangle_K \geq 0,$$

i.e., K is nonnegative definite. It remains to apply the uniqueness stated in Proposition 1.

In particular, any closed subspace $G \subset H(K)$ possesses a uniquely determined reproducing kernel M, i.e., a nonnegative definite function $M : D^2 \to \mathbb{R}$ with $M(\cdot, t) \in G$ and

$$\langle g, M(\cdot, t) \rangle_M = \langle g, M(\cdot, t) \rangle_K = g(t)$$

for all $t \in D$ and $g \in G$. It is easily verified that K is the sum of the kernels corresponding to G and $G^\perp$.

The sum of any two nonnegative definite kernels is nonnegative, too. The corresponding Hilbert space is given as follows.

LEMMA 2. Let

$$K = K_1 + K_2$$

where $K_1, K_2 : D^2 \to \mathbb{R}$ are nonnegative definite. Then

$$H(K) = H(K_1) + H(K_2),$$

and the norm in $H(K)$ is given by

$$\|h\|_K^2 = \min\{\|h_1\|_{K_1}^2 + \|h_2\|_{K_2}^2 : h = h_1 + h_2, \ h_i \in H(K_i)\}.$$

PROOF. Consider the Hilbert space $H = H(K_1) \times H(K_2)$ with norm given by

$$\|(h_1, h_2)\|_H^2 = \|h_1\|_{K_1}^2 + \|h_2\|_{K_2}^2.$$

Let H' denote the orthogonal complement of the closed subspace

$$H_0 = \{(h, -h) \in H : h \in H(K_1) \cap H(K_2)\}$$

in H. Note that $(h_1', h_2') \mapsto h_1' + h_2'$ defines a linear and one-to-one mapping from H' onto the space $H(K_1) + H(K_2)$. Thus $H(K_1) + H(K_2)$ becomes a Hilbert space by letting

$$\|h_1' + h_2'\|^2 = \|h_1'\|_{K_1}^2 + \|h_2'\|_{K_2}^2.$$

By construction, $\|h\|^2$ is the minimum of $\|h_1\|_{K_1}^2 + \|h_2\|_{K_2}^2$ over $h = h_1 + h_2$ such that $h_i \in H(K_i)$.

Clearly $K(\cdot, t) = K_1(\cdot, t) + K_2(\cdot, t)$ belongs to $H(K_1) + H(K_2)$, and it remains to verify the reproducing property in the latter space. To this end let

$$h = h_1' + h_2', \qquad (h_1', h_2') \in H'.$$

Analogously, let

$$K(\cdot, t) = K_1'(\cdot, t) + K_2'(\cdot, t), \qquad (K_1'(\cdot, t), K_2'(\cdot, t)) \in H'.$$

Then

$$\begin{aligned}
h(t) = h_1'(t) + h_2'(t) &= \langle h_1', K_1(\cdot, t)\rangle_{K_1} + \langle h_2', K_2(\cdot, t)\rangle_{K_2} \\
&= \langle (h_1', h_2'), (K_1(\cdot, t), K_2(\cdot, t))\rangle_H = \langle (h_1', h_2'), (K_1'(\cdot, t), K_2'(\cdot, t))\rangle_H \\
&= \langle h_1', K_1'(\cdot, t)\rangle_{K_1} + \langle h_2', K_2'(\cdot, t)\rangle_{K_2} = \langle h_1' + h_2', K_1'(\cdot, t) + K_2'(\cdot, t)\rangle \\
&= \langle h, K(\cdot, t)\rangle,
\end{aligned}$$

which completes the proof. $\qquad\qquad\qquad\qquad\qquad\qquad\qquad\qquad\qquad\qquad\square$

The nonnegative definite functions on D^2 form a cone. A natural order on this cone is given by letting $K \ll L$ if $L - K$ is nonnegative definite. On the other hand, reproducing kernel Hilbert spaces on a domain D are ordered by inclusion. In the sequel we study the relation between these orderings.

Let K be nonnegative definite and let $c > 0$. The uniqueness part of Proposition 1 immediately implies that

(3) $$H(K) = H(cK)$$

as sets and

$$c \cdot \langle h, h\rangle_{cK} = \langle h, h\rangle_K.$$

Suppose that

$$H(K) \subset H(L)$$

for two kernels K and L on D^2. Then the closed graph theorem implies that the embedding $H(K) \hookrightarrow H(L)$ is continuous. We add that the same argument applies to any pair of Banach spaces $F \subset G$ of functions on D if the functionals δ_t are bounded on F as well as on G.

LEMMA 3. *Let K and L be nonnegative definite on D^2. Then*

$$H(K) \subset H(L)$$

iff

$$K \ll c\,L$$

for some constant $c > 0$.

PROOF. Assume that $H(K) \subset H(L)$. Due to the closed graph theorem,

$$\|h\|_L \le c \cdot \|h\|_K, \qquad h \in H(K),$$

for some $c > 0$. Define

$$\varphi(h) = \sum_{i=1}^{n} h(x_i) \cdot a_i$$

for arbitrary $a_i \in \mathbb{R}$ and $x_i \in D$. Then φ is a bounded linear functional on $H(K)$ as well as on $H(L)$. The respective norms are related by

$$\sum_{i,j=1}^{n} a_i a_j \cdot K(x_i, x_j) = \left\| \sum_{i=1}^{n} a_i \cdot K(\cdot, x_i) \right\|_K^2 = \sup\{\varphi(h)^2 : h \in B(K)\}$$

$$\le c^2 \cdot \sup\{\varphi(h)^2 : h \in B(L)\} = c^2 \cdot \left\| \sum_{i=1}^{n} a_i \cdot L(\cdot, x_i) \right\|_L^2$$

$$= c^2 \cdot \sum_{i,j=1}^{n} a_i a_j \cdot L(x_i, x_j).$$

Thus $K \ll c^2\, L$.

To prove the converse, let $c\,L = K + M$ where $M = c\,L - K$ is nonnegative definite. By Lemma 2 and (3), $H(K) \subset H(c\,L) = H(L)$. □

We obtain nonnegative definite functions on smaller domains by restriction. The corresponding reproducing kernel Hilbert spaces are characterized as follows.

LEMMA 4. *Let $\widehat{D} \subset D$ and*

$$\widehat{K} = K|_{\widehat{D} \times \widehat{D}}$$

with K being nonnegative definite on D^2. Then

$$H(\widehat{K}) = \{h|_{\widehat{D}} : h \in H(K)\}$$

and

$$\|\widehat{h}\|_{\widehat{R}} = \min\{\|h\|_K : h \in H(K),\ h|_{\widehat{D}} = \widehat{h}\}$$

for $\widehat{h} \in H(\widehat{K})$. *In particular,* $\|\widehat{h}\|_{\widehat{R}} = \|h\|_K$ *for* $h = \sum_{i=1}^{n} a_i \cdot K(\cdot, x_i)$ *and* $\widehat{h} = h|_{\widehat{D}}$, *if* $x_i \in \widehat{D}$.

PROOF. Let H_1 denote the orthogonal complement of the closed subspace

$$H_0 = \{h \in H(K) : h(t) = 0 \text{ for } t \in \widehat{D}\}$$

in $H(K)$ and put

$$H = \{h|_{\widehat{D}} : h \in H_1\}.$$

Clearly $h \mapsto h|_{\widehat{D}}$ defines a linear and one-to-one mapping from H_1 onto H, so that H becomes a Hilbert space by letting

$$\langle h|_{\widehat{D}}, h|_{\widehat{D}} \rangle = \langle h, h \rangle_K, \qquad h \in H_1.$$

We claim that $\widehat{K}$ is the reproducing kernel for H. Indeed, for $t \in \widehat{D}$ we have $K(\cdot, t) \in H_1$, which shows

$$\widehat{K}(\cdot, t) = K(\cdot, t)|_{\widehat{D}} \in H.$$

Furthermore, if $h \in H_1$ then

$$\langle h|_{\widehat{D}}, \widehat{K}(\cdot, t) \rangle = \langle h, K(\cdot, t) \rangle_K = h(t).$$

To characterize the norm in $H(\widehat{K})$ let $h = h_0 + h_1$ with $h_i \in H_i$. Then

$$\|h\|_K^2 = \|h_0\|_K^2 + \|h_1\|_K^2 = \|h_0\|_K^2 + \|(h_1)|_{\widehat{D}}\|^2 = \|h_0\|_K^2 + \|h|_{\widehat{D}}\|_{\widehat{R}}^2,$$

since $h|_{\widehat{D}} = (h_1)|_{\widehat{D}}$. Consequently, $\|h|_{\widehat{D}}\|_{\widehat{R}} \le \|h\|_K$ with equality iff $h_0 = 0$. The latter property holds for $h = \sum_{i=1}^{n} a_i \cdot K(\cdot, x_i)$ with $x_i \in \widehat{D}$. $\square$

EXAMPLE 5. Consider the Sobolev space

$$W_2^1([0, 1]) = \{h \in C([0, 1]) : h \text{ absolutely continuous},\ h' \in L_2([0, 1])\},$$

equipped with the scalar product

$$\langle h_1, h_2 \rangle = h_1(0) \cdot h_2(0) + \int_0^1 h_1'(s) \cdot h_2'(s)\, ds.$$

Moreover, let

$$R(s, t) = 1 + \min(s, t), \qquad s, t \in [0, 1].$$

Then $R(\cdot, t) \in W_2^1([0, 1])$ for all $t \in [0, 1]$, and we have the reproducing property

$$\langle h, R(\cdot, t) \rangle = h(0) + \int_0^t h'(s)\, ds = h(t).$$

We conclude that R is nonnegative definite on $[0, 1]^2$, cf. (2). By Proposition 1,

$$W_2^1([0, 1]) = H(R)$$

with coinciding scalar products.

From Example II.1 we know that

$$K(s, t) = \min(s, t), \qquad s, t \in [0, 1],$$

is the covariance kernel of the Wiener measure. Clearly we have $K(\cdot, t) \in H(R)$ and $K(0, t) = 0$. Moreover, for all $h \in H(R)$ with $h(0) = 0$,

$$\langle h, K(\cdot, t) \rangle_R = \langle h, R(\cdot, t) \rangle_R = h(t).$$

We conclude that

$$H(K) = \{h \in W_2^1([0, 1]) : h(0) = 0\},$$

with scalar product given by

$$\langle h_1, h_2 \rangle_K = \int_0^1 h_1'(t) \cdot h_2'(t)\, dt.$$

1.2. The Fundamental Isomorphism. Recall the definition of the probability space $(F, \mathfrak{B}, P)$ where P is defined on the σ-algebra $\mathfrak{B}$ of Borel sets in F, see Section II.1.1. Let $L_2(F, \mathfrak{B}, P)$ denote the space of all square integrable random variables λ with respect to P, which is equipped with the norm

$$\left(\int_F \lambda(f)^2\, dP(f) \right)^{1/2}.$$

Recall that Λ^{std} denotes the set of all Dirac functionals δ_t on F.

As a general assumption we suppose square integrability

$$\int_F f(t)^2\, dP(f) < \infty$$

of all functionals δ_t. Thus $\Lambda^{\mathrm{std}} \subset L_2(F, \mathfrak{B}, P)$ and we may define

$$\Lambda^P = \overline{\operatorname{span} \Lambda^{\mathrm{std}}}$$

with closure understood in $L_2(F, \mathfrak{B}, P)$. By definition, Λ^P is generated by the functionals of the form

$$\lambda(f) = \sum_{i=1}^n f(x_i) \cdot a_i,$$

i.e., by quadrature formulas. Functionals that coincide almost everywhere are identified in Λ^P.

The following result from Loève (1948) is an important tool for the average case analysis of linear problems.

PROPOSITION 6. *Suppose that P is a zero mean measure with covariance kernel K. Then*

$$\mathfrak{I}\delta_t = K(\cdot, t)$$

extends to a Hilbert space isomorphism $\mathfrak{I} : \Lambda^P \to H(K)$. Furthermore

$$(\mathfrak{I}\lambda)(t) = \int_F \lambda(f) \cdot f(t)\, dP(f)$$

for every $\lambda \in \Lambda^P$ and $t \in D$.

PROOF. Let $\lambda(f) = \sum_{i=1}^{n} f(x_i) \cdot a_i$ and $h = \sum_{i=1}^{n} a_i \cdot K(\cdot, x_i)$. Then

$$\|h\|_K^2 = \sum_{i,j=1}^{n} a_i a_j \cdot K(x_i, x_j) = \int_F \lambda(f)^2 \, dP(f).$$

We conclude that $\Im \lambda = h$ yields a well defined isomorphism between span Λ^{std} and span$\{K(\cdot, t) : t \in D\}$. By extension we get the isomorphism between the corresponding Hilbert spaces.

Clearly

$$(\Im \lambda)(t) = \langle \Im \lambda, K(\cdot, t) \rangle_K = \langle \Im \lambda, \Im \delta_t \rangle_K = \int_F \lambda(f) \cdot f(t) \, dP(f)$$

if $\lambda \in \Lambda^P$ and $t \in D$. $\square$

COROLLARY 7. *Let $S, \lambda_1, \ldots, \lambda_n \in \Lambda^P$, $a_i \in \mathbb{R}$, and put $S_n = \sum_{i=1}^{n} a_i \cdot \lambda_i$. Then*

$$\left(\int_F (S(f) - S_n(f))^2 \, dP(f) \right)^{1/2} = \|\Im S - \Im S_n\|_K$$

$$= \sup_{h \in B(K)} |\langle \Im S, h \rangle_K - \langle \Im S_n, h \rangle_K|.$$

The corollary implies that recovery of functionals in Λ^P and in $H(K)$ is equivalent. The q-average error with respect to P and $q = 2$ coincides with the maximal error on the unit ball $B(K)$ in $H(K)$. The linear functionals, which define the problem and the linear method, are transformed via $\Im$. We study important consequences of this equivalence in Sections 2 and 3.

REMARK 8. Consider a stochastic process $(X_t)_{t \in D}$ on an arbitrary probability space $(\Omega, \mathfrak{A}, Q)$, see Section II.1.2. Let E denote the expectation with respect to Q and assume that $E(X_t) = 0$ and $E(X_t^2) < \infty$ for all $t \in D$. Thus $(X_t)_{t \in D}$ is a second-order random function. The covariance kernel $K(s, t) = E(X_s \cdot X_t)$ exists, and the closure of span$\{X_t : t \in D\}$ in $L_2(\Omega, \mathfrak{A}, Q)$ is called the Hilbert space spanned by the process $(X_t)_{t \in D}$. The mapping $X_t \mapsto K(\cdot, t)$ extends to an isomorphism between the Hilbert space spanned by $(X_t)_{t \in D}$ and $H(K)$. The proof is analogous to that of Proposition 6.

Thus we observe that no smoothness assumption on $(X_t)_{t \in D}$ is needed for an analogue to Proposition 6 to hold. Equivalently, we might consider measures P on $\mathbb{R}^D$ instead of $F \subset C^k(D)$.

1.3. Smoothness of Covariance Kernels. To apply Proposition 6 and its corollary, we require some smoothness of the covariance kernel K. Specifically, we assume that there exists an integer $r \in \mathbb{N}_0$ such that the kernel K has continuous partial derivatives $K^{(\alpha_1, \alpha_2)}$ on D^2 for all $\alpha_1, \alpha_2 \in \mathbb{N}_0^d$ with $|\alpha_i| \leq r$. We use the notation

(4) $K \in C^{r,r}(D^2).$

We first show that the smoothness of K on D^2 is completely determined by its smoothness at the diagonal $\{(t,t) \in D^2 : t \in D\}$.

LEMMA 9. *The following properties are equivalent for a nonnegative definite function K on D^2.*

(i) K *is continuous on D^2.*
(ii) K *is continuous at every diagonal point $(t,t) \in D^2$.*
(iii) $t \mapsto K(\cdot, t)$ *is a continuous mapping between D and $H(K)$.*

PROOF. We have

$$\|K(\cdot, s) - K(\cdot, t)\|_K^2 = K(s,s) - 2K(s,t) + K(t,t)$$
$$\leq |K(s,s) - K(t,t)| + 2 \cdot |K(s,t) - K(t,t)|.$$

Hence (iii) follows from (ii).

Furthermore,

$$|K(u,v) - K(s,t)|$$
$$= |\langle K(\cdot, u), K(\cdot, v) - K(\cdot, t)\rangle_K + \langle K(\cdot, t), K(\cdot, u) - K(\cdot, s)\rangle_K|$$
$$\leq \|K(\cdot, u)\|_K \cdot \|K(\cdot, v) - K(\cdot, t)\|_K + \|K(\cdot, t)\|_K \cdot \|K(\cdot, u) - K(\cdot, s)\|_K.$$

Hence (i) follows from (iii). $\square$

By $\mathrm{int}\, D$ we denote the interior of D. Recall that D is the closure of $\mathrm{int}\, D$, according to our general hypothesis. The role of the diagonal is apparent also in the next lemma.

LEMMA 10. *Let $\gamma \in \mathbb{N}_0^d$ with $|\gamma| = 1$, and assume that the limit*

$$\lim_{a,b \to 0} \frac{K(t + a\gamma, t + b\gamma) - K(t + a\gamma, t) - K(t, t + b\gamma) + K(t,t)}{a \cdot b}$$

exists for every $t \in \mathrm{int}\, D$. Then $K^{(\gamma,\gamma)}$ exists and is nonnegative definite on $(\mathrm{int}\, D)^2$.

PROOF. Consider a sequence of real numbers c_ℓ that tends to zero. Put

$$h_\ell = \frac{K(\cdot, t + c_\ell \gamma) - K(\cdot, t)}{c_\ell}$$

for $t \in \mathrm{int}\, D$ and $|c_\ell|$ sufficiently small. Then

$$\langle h_n, h_\ell \rangle_K = \frac{K(t + c_n\gamma, t + c_\ell\gamma) - K(t + c_n\gamma, t) - K(t, t\gamma + c_\ell) + K(t,t)}{c_n \cdot c_\ell}.$$

By the hypothesis of the lemma, we see that $\langle h_n, h_\ell \rangle_K$ converges for n and ℓ tending to ∞. The functions h_ℓ form a Cauchy sequence in $H(K)$ since

$$\lim_{n,\ell \to \infty} \|h_n - h_\ell\|_K^2 = \lim_{n,\ell \to \infty} (\langle h_n, h_n \rangle_K - 2\langle h_n, h_\ell \rangle_K + \langle h_\ell, h_\ell \rangle_K) = 0.$$

The limit is determined pointwise as

$$\lim_{\ell \to \infty} h_\ell(s) = K^{(0,\gamma)}(s,t),$$

whence h_ℓ tends to $K^{(0,\gamma)}(\cdot, t)$ in $H(K)$.

For all $h \in H(K)$,

$$\langle h, K^{(0,\gamma)}(\cdot, t)\rangle_K = \lim_{\ell \to \infty} \langle h, h_\ell\rangle_K = \lim_{\ell \to \infty} \frac{h(t + c_\ell \gamma) - h(t)}{c_\ell} = h^{(\gamma)}(t).$$

In particular, for $s \in \operatorname{int} D$ and $h = K^{(0,\gamma)}(\cdot, s)$, we get the existence of $K^{(\gamma,\gamma)}$ at (s, t), and

$$K^{(\gamma,\gamma)}(s, t) = \lim_{\ell \to \infty} \frac{\psi_\ell(s, t)}{c_\ell^2},$$

where

$$\psi_\ell(s, t) = K(s + c_\ell \gamma, t + c_\ell \gamma) - K(s + c_\ell \gamma, t) - K(s, t + c_\ell \gamma) + K(s, t).$$

Let $a_i \in \mathbb{R}$ and $x_i \in \operatorname{int} D$ for $i = 1, \ldots, n$, and put $a_{i+n} = -a_i$ and $x_{i+n} = x_i + c_\ell \gamma$. Then

$$\sum_{i,j=1}^{n} a_i a_j \cdot K^{(\gamma,\gamma)}(x_i, x_j) = \lim_{\ell \to \infty} \frac{1}{c_\ell^2} \sum_{i,j=1}^{n} a_i a_j \cdot \psi_\ell(x_i, x_j)$$

$$= \lim_{\ell \to \infty} \frac{1}{c_\ell^2} \sum_{i,j=1}^{2n} a_i a_j \cdot K(x_i, x_j) \geq 0,$$

so that $K^{(\gamma,\gamma)}$ is nonnegative definite. $\qquad \square$

Now we give a first description of the smoothness of functions in $H(K)$ in terms of smoothness of K.

PROPOSITION 11. *Suppose that* (4) *holds for a nonnegative definite function* K. *Then*

$$H(K) \subset C^r(D)$$

with a compact embedding. Moreover, $t \mapsto K^{(0,\alpha)}(\cdot, t)$ *defines a continuous mapping between* D *and* $H(K)$ *if* $|\alpha| \leq r$, *and*

$$h^{(\alpha)}(t) = \langle h, K^{(0,\alpha)}(\cdot, t)\rangle_K$$

for $h \in H(K)$.

PROOF. At first assume that $r = 0$ in (4). Recall that D is compact, according to our general hypothesis, and note that

$$|h(s) - h(t)|^2 = \langle h, K(\cdot, s) - K(\cdot, t)\rangle_K^2 \leq \|h\|_K^2 \cdot \|K(\cdot, s) - K(\cdot, t)\|_K^2$$

for $h \in H(K)$. It remains to apply Lemma 9.

Now assume that the lemma holds for some $r \in \mathbb{N}_0$. Let $K \in C^{r+1,r+1}(D^2)$. Moreover, let $\alpha, \beta, \gamma \in \mathbb{N}_0^d$ with $|\beta| = r$, $|\gamma| = 1$, and $\alpha = \beta + \gamma$. Inductively we conclude as follows.

Let

(5) $$h_a = K^{(0,\beta)}(\cdot, t + a\gamma) - K^{(0,\beta)}(\cdot, t)$$

for $t \in \text{int } D$ and $|a|$ sufficiently small. Then $h_a \in H(K)$ and

$$\langle h_a, h_b \rangle_K$$
$$= K^{(\beta,\beta)}(t + a\gamma, t + b\gamma) - K^{(\beta,\beta)}(t + a\gamma, t) - K^{(\beta,\beta)}(t, t + b\gamma) + K^{(\beta,\beta)}(t, t)$$
$$= \int_0^a (K^{(\alpha,\beta)}(t + c_1\gamma, t + b\gamma) - K^{(\alpha,\beta)}(t + c_1\gamma, t))\, dc_1$$
$$= \int_0^a \int_0^b K^{(\alpha,\alpha)}(t + c_1\gamma, t + c_2\gamma)\, dc_2\, dc_1.$$

Hence

$$\lim_{a,b \to 0} \frac{\langle h_a, h_b \rangle_K}{a \cdot b} = K^{(\alpha,\alpha)}(t, t),$$

and we have convergence of $1/a \cdot h_a$ in $H(K)$ as a tends to zero. The limit is given pointwise as

$$\lim_{a \to 0} \frac{h_a(s)}{a} = K^{(0,\alpha)}(s, t).$$

Any function $h \in H(K)$ has partial derivatives of order $r + 1$, since

$$\langle h, K^{(0,\alpha)}(\cdot, t) \rangle_K = \lim_{a \to 0} \frac{\langle h, h_a \rangle_K}{a} = \lim_{a \to 0} \frac{h^{(\beta)}(t + a\gamma) - h^{(\beta)}(t)}{a} = h^{(\alpha)}(t).$$

In particular,

$$\|K^{(0,\alpha)}(\cdot, s) - K^{(0,\alpha)}(\cdot, t)\|_K^2 = K^{(\alpha,\alpha)}(s, t) - 2\, K^{(\alpha,\alpha)}(s, t) + K^{(\alpha,\alpha)}(t, t),$$

which shows that $t \mapsto K^{(0,\alpha)}(\cdot, t)$ is a uniformly continuous mapping between $\text{int } D$ and $H(K)$. Its extension to D is given by $t \mapsto K^{(0,\alpha)}(\cdot, t)$. Using

$$|h^\alpha(s) - h^{(\alpha)}(t)| \leq \|h\|_K \cdot \|K^{(0,\alpha)}(\cdot, s) - K^{(0,\alpha)}(\cdot, t)\|_K$$

for $|\alpha| \leq r + 1$, we see that $H(K) \hookrightarrow C^{r+1}(D)$ is compact. $\square$

COROLLARY 12. *Assume that* (4) *holds for the covariance kernel K of a zero mean measure P on $F \subset C^k(D)$ with $k \leq r$. Then*

$$\Lambda^{\text{all}} \subset \Lambda^P$$

for the dual space Λ^{all} of $C^k(D)$, and

$$\lambda(h) = \langle \Im\lambda, h \rangle_K$$

for $\lambda \in \Lambda^{\text{all}}$ and $h \in H(K)$.

PROOF. By Proposition 11, $H(K) \subset C^r(D) \subset C^k(D)$. Take $\lambda_n \in \text{span } \Lambda^{\text{std}}$ such that

$$\lim_{n \to \infty} \lambda_n(f) = \lambda(f)$$

for every $f \in C^k(D)$. Clearly $\lambda_n(h) = \langle \Im\lambda_n, h \rangle_K$ for $h \in H(K)$. From Proposition 11 we know that the unit ball $B(K)$ is relatively compact in $C^r(D)$. Hence

$$\lim_{n \to \infty} \sup_{h \in B(K)} |\lambda(h) - \lambda_n(h)| = 0,$$

so that $\Im\lambda_n$ and λ_n form Cauchy sequences in $H(K)$ and Λ^P, respectively. It follows that λ_n tends to λ in Λ^P and $\lambda(h) = \langle \Im\lambda, h\rangle_K$. □

Let us discuss some consequences of the previous results. The corollary shows that, under mild assumptions, the class Λ^P contains all linear functionals that are permitted for linear problems. In the Hilbert space $H(K)$ the representer of a bounded linear functional $\lambda \in \Lambda^{\mathrm{all}}$ is given by $\Im\lambda$ and

(6) $$(\Im\lambda)(s) = \langle \Im\lambda, K(\cdot, s)\rangle_K = \lambda(K(\cdot, s)).$$

Furthermore, since Λ^{all} is actually a dense subset of Λ^P,

$$e_2(n, \Lambda^{\mathrm{all}}, S, P) = e_2(n, \Lambda^P, S, P),$$

i.e., Λ^{all} is as powerful as Λ^P for arbitrary linear problems and q-average errors with $q = 2$.

Let $|\alpha|, |\beta| \le k$ and $\lambda(f) = f^{(\alpha)}(t)$. Then $\Im\lambda = K^{(0,\alpha)}(\cdot, t)$ and

(7) $$\int_F f^{(\alpha)}(s) \cdot f^{(\beta)}(t)\, dP(f) = \langle K^{(0,\alpha)}(\cdot, s), K^{(0,\beta)}(\cdot, t)\rangle = K^{(\alpha,\beta)}(s, t).$$

For $\alpha = \beta$ we see that $K^{(\alpha,\alpha)}$ is the covariance kernel of the image of P under $f \mapsto f^{(\alpha)}$.

Suppose that $k < r$ in the assumptions of Corollary 12. As demonstrated in the following section, the derivatives of order $k < |\alpha| \le r$ exist at least in quadratic mean.

1.4. Smoothness in Quadratic Mean. Let K denote the covariance kernel of a zero mean measure P. Consider the canonical stochastic process $(\delta_t)_{t\in D}$ that is associated with P, see Section II.1.2. The smoothness of K does not only determine the smoothness of functions in $H(K)$, but also the smoothness of $(\delta_t)_{t\in D}$ in a certain sense. The respective link is provided by the fundamental isomorphism, and the smoothness is called *smoothness in quadratic mean*. The following results are mainly reformulations of the facts that were derived in Section 1.3.

The process $(\delta_t)_{t\in D}$ is said to be *continuous in quadratic mean* if

$$\lim_{s\to t} \int_F (f(s) - f(t))^2\, dP(f) = 0$$

for all $t \in D$. By definition of the Hilbert space Λ^P, this property is equivalent to continuity of the mapping $t \mapsto \delta_t$ between D and Λ^P. Proposition 6 and Lemma 9 immediately give the following characterization.

LEMMA 13. *The process $(\delta_t)_{t\in D}$ is continuous in quadratic mean iff $K \in C(D^2)$.*

Quadratic mean derivatives are defined as directional derivatives of the mapping $t \mapsto \delta_t$ in the usual way. For $\alpha = (0, \ldots, 0)$ we put $\delta_t^\alpha = \delta_t$. If $\alpha = \beta + \gamma$

where $\beta, \gamma \in \mathbb{N}_0^d$ with $|\gamma| = 1$, then

$$\delta_t^\alpha = \lim_{a \to 0} \frac{1}{a} \cdot (\delta_{t+a\gamma}^\beta - \delta_t^\beta),$$

provided that this limit exists in Λ^P. Continuity of such a derivative means that

$$\lim_{s \to t} \int_F (\delta_s^\alpha(f) - \delta_t^\alpha(f))^2 \, dP(f) = 0.$$

LEMMA 14. *The process $(\delta_t)_{t \in D}$ has continuous derivatives up to order r in quadratic mean iff $K \in C^{r,r}(D^2)$. If either of these properties holds then*

$$\Im \delta_t^\alpha = K^{(0,\alpha)}(\cdot, t)$$

for $|\alpha| \le r$ and $t \in D$.

PROOF. See Lemma 13 for the case $r = 0$. Assume that the lemma holds for some $r \in \mathbb{N}_0$. Inductively we proceed as follows.

If $K \in C^{r+1,r+1}(D^2)$, let $\alpha = \beta + \gamma$ with $|\beta| = r$ and $|\gamma| = 1$, and put

$$\lambda_a = \delta_{t+a\gamma}^\beta - \delta_t^\beta$$

for $t \in \operatorname{int} D$ and $|a|$ sufficiently small. We have

$$\Im \lambda_a = K^{(0,\beta)}(\cdot, t + a\gamma) - K^{(0,\beta)}(\cdot, t) = h_a,$$

see (5). As shown in the proof of Proposition 11, $1/a \cdot h_a$ converges to $K^{(0,\alpha)}(\cdot, t)$ in $H(K)$. Hence δ_t^α exists and $\Im \delta_t^\alpha = K^{(0,\alpha)}(\cdot, t)$. We apply Proposition 11 to get the existence of δ_t^α for $t \in D$ as well as the continuity of the respective mapping.

Conversely, assume that δ_t^α exists and depends continuously on t if $|\alpha| \le r + 1$. Observe that we can use Proposition 11 to see that

$$\int_F \delta_s^\alpha(f) \cdot \delta_t^\beta(f) \, dP(f) = \langle K^{(0,\alpha)}(\cdot, s), K^{(0,\beta)}(\cdot, t) \rangle_K = K^{(\alpha,\beta)}(s, t)$$

for $|\alpha|, |\beta| \le r$. We conclude that $K^{(\alpha,\beta)}(s, t)$ exists in fact for $|\alpha|, |\beta| \le r + 1$ and $s, t \in \operatorname{int} D$. Continuity of of the quadratic mean derivatives allows us to extend these partial derivative of K to D^2. □

The notions of continuity and differentiability in quadratic mean are defined for arbitrary second-order random functions $(X_t)_{t \in D}$. The previous two lemmata extend to this case, cf. Remark 8.

In addition to smoothness in quadratic mean, pathwise smoothness is studied in the literature, see Gihman and Skorohod (1974), and Adler (1981, 1990). In our framework pathwise smoothness of the canonical process $(\delta_t)_{t \in D}$ is expressed by the assumption $F \subset C^k(D)$. Observe that we use $r \in \mathbb{N}_0$ to denote quadratic mean smoothness and that we indicate pathwise smoothness by the parameter $k \in \mathbb{N}_0$. The latter is often of minor importance in the average case analysis, cf. Remark II.2.

EXAMPLE 15. A measure P on $F \subset C^k(D)$ does not necessarily lead to a canonical process that is continuous in quadratic mean. For instance, take a sequence of points $t_n \in D$ that converges to $t \in D$, and take a sequence of functions $f_n \in C^\infty(D)$ with disjoint supports, $f_n(t_n) = 2^{n/2}$, and $f_n(t) = 0$. Assign the probability $2^{-(n+1)}$ to $\pm f_n$ to get a zero mean measure P on $F = C^\infty(D)$. We obtain $K(t_n, t_n) = 1$ but $K(t, t) = 0$ for the covariance kernel of P. Thus continuity of the sample paths does not imply continuity in quadratic mean.

The converse is not true, either. A counterexample is given by the Poisson process, whose covariance kernel coincides with the Brownian motion kernel $K(s, t) = \min(s, t)$.

REMARK 16. Let P denote a zero mean measure on $F \subset C^k(D)$ with covariance kernel K. A sufficient condition for $K \in C^{k,k}(D^2)$ to hold is

$$(8) \qquad \max_{|\alpha| \le k} \int_F \|f^{(\alpha)}\|_\infty^2 \, dP(f) < \infty.$$

In fact, Lebesgue's Theorem yields the existence and continuity of the quadratic mean derivatives $\delta_t^\alpha(f) = f^{(\alpha)}(t)$ up to order k. It remains to apply Lemma 14.

We add that (8) holds in particular for all Gaussian measures, see Section 1.4.

1.5. Notes and References. 1. The results from this section are well known. See Aronszajn (1950), Parzen (1959), Vakhania, Tarieladze, and Chobanyan (1987), Wahba (1990), and Atteia (1992) for a detailed treatment of reproducing kernel Hilbert spaces.

2. Proposition 6 is due to Loève (1948). It is used in numerous papers on linear problems for stochastic processes; among the first are Parzen (1959, 1962) and Hájek (1962).

2. Linear Problems with Values in a Hilbert Space

We apply results from the preceding section to problems with bounded linear operators

$$S : C^k(D) \to G$$

where G is a Hilbert space. Important examples are the integration problem, $S = \mathrm{Int}_\varrho$, and the L_2-approximation problem, $S = \mathrm{App}_{2,\varrho}$. Throughout this section we consider a zero mean measure P on $F \subset C^k(D)$ with covariance kernel $K \in C^{r,r}(D^2)$ for some $r \ge k$, and we study errors defined by second moments, i.e.,

$$q = 2.$$

By Proposition 11, $H(K) \subset C^k(D)$ under these assumptions. Therefore we can study a linear problem, given by S and $\Lambda \subset \Lambda^{\mathrm{all}}$, in the average case setting on F as well as in the worst case setting on the unit ball $B(K)$ of $H(K)$.

For integration the average case is equivalent to the worst case on $B(K)$. For L_2-approximation the average case is related to the worst case for $\mathrm{App}_{2,\varrho}$ and App_∞ on $B(K)$. Moreover, optimal methods for L_2-approximation with $\Lambda = \Lambda^{\mathrm{all}}$ can be derived from the spectral representation of the integral operator with kernel $\varrho(s)^{1/2}\,\varrho(t)^{1/2}\cdot K(s,t)$. These methods are given by the first n terms of a Karhunen-Loève expansion.

2.1. Linear Problems with Values in $\mathbb{R}$. Let S_n denote a linear method for the approximation of a bounded linear functional

$$S : C^k(D) \to \mathbb{R}.$$

Then $S, S_n \in \Lambda^{\mathrm{all}}$ by definition. We have

$$\int_F \lambda(f)\,dP(f) = 0$$

for every $\lambda \in \Lambda^P$, since P has zero mean and Λ^P is the L_2-closure of span Λ^{std}. Furthermore, $\Lambda^{\mathrm{all}} \subset \Lambda^P$, see Corollary 12. Hence

$$e_2(S_n + a_0, S, P)^2$$
$$= \int_F (S(f) - S_n(f))^2\,dP(f) - 2a_0 \int_F (S(f) - S_n(f))\,dP(f) + a_0^2$$
$$= e_2(S_n, S, P)^2 + a_0^2$$

for any affine linear method $S_n + a_0$. It is therefore enough to study linear methods instead of affine linear ones.

Corollary 7 states the equivalence of an average case problem on F, the approximation of S by S_n, and a worst case problem on the unit ball $B(K)$. The latter problem is defined by representers $\Im S$ and $\Im S_n$. By Corollary 12,

$$\langle \Im S, h\rangle_K = S(h), \qquad \langle \Im S_n, h\rangle_K = S_n(h)$$

for every $h \in H(K)$, so that the worst case problem also consists of approximating S by S_n. We conclude that the average error of S_n with respect to P coincides with the maximal error of S_n on $B(K)$.

PROPOSITION 17. *Let S_n denote a linear method, which approximates the linear functional S. Then*

$$e_2(S_n, S, P) = e_{\max}(S_n, S, B(K)) = \|\Im S - \Im S_n\|_K.$$

For the integration problem we have

$$S(f) = \mathrm{Int}_\varrho(f) = \int_D f(t)\cdot\varrho(t)\,dt$$

with $\varrho \in L_1(D)$. We mainly analyze quadrature formulas S_n that use function values only, but sometimes derivatives are permitted as well. Letting

$$S_n(f) = \sum_{i=1}^{n} f^{(\alpha_i)}(x_i)\cdot a_i$$

with $|\alpha_i| \leq k$, we obtain

$$e_2(S_n, \mathrm{Int}_\varrho, P) = e_{\max}(S_n, \mathrm{Int}_\varrho, B(K)) = \left\| \xi - \sum_{i=1}^n a_i \cdot K^{(0,\alpha_i)}(\cdot, x_i) \right\|_K$$

where $\xi = \mathfrak{I} \, \mathrm{Int}_\varrho$. By (6),

$$(9) \qquad\qquad \xi(t) = \int_D K(s,t) \cdot \varrho(s) \, ds.$$

EXAMPLE 18. According to Proposition II.9, a trapezoidal rule with knots $x_i = 2i/(2n+1)$ is average case optimal for the problem Int with respect to the Wiener measure. From Example 5 and Proposition 17 we conclude that the same formula is worst case optimal on the unit ball

$$B(K) = \{ h \in W_2^1([0,1]) : h(0) = 0, \; \|h'\|_2 \leq 1 \}$$

where $K(s,t) = \min(s,t)$.

2.2. The General Case. Let S_n denote a linear method for the approximation of an operator S with values in a Hilbert space G. Assume, without loss of generality, that G is separable and let $(g_j)_j$ denote an orthonormal basis of G. Put

$$S^j(f) = \langle S(f), g_j \rangle_G, \qquad S_n^j(f) = \langle S_n(f), g_j \rangle_G,$$

to obtain

$$e_2(S_n + a_0, S, P)^2 = \sum_j \int_F \left(S^j(f) - S_n^j(f) - \langle a_0, g_j \rangle_G \right)^2 dP(f)$$

$$= \sum_j e_2(S_n^j + \langle a_0, g_j \rangle, S^j, P)^2$$

for any affine method $S_n + a_0$. From the previous section we conclude that $a_0 = 0$ is optimal. It is therefore enough to study linear methods.

We have made the same observation without restrictions on G and q, but only for symmetric measures P in Remark II.3. In other cases affine methods may be better than linear methods.

The previous representation

$$e_2(S_n, S, P)^2 = \sum_j e_2(S_n^j, S^j, P)^2$$

allows us to apply the fundamental isomorphism. Hence

$$e_2(S_n, S, P)^2 = \sum_j e_{\max}(S_n^j, S^j, B(K))^2 = \sum_j \|\mathfrak{I} S^j - \mathfrak{I} S_n^j\|_K^2,$$

see Proposition 17, so that the average error $e_2(S_n, S, P)$ is expressed as a series of maximal errors on the unit ball in $H(K)$.

Furthermore, it turns out that $e_2(S_n, S, P)$ depends on the zero mean measure P only through its covariance kernel K. In the particular case of the Wiener measure w and the integration and L_2-approximation problem this fact was already observed in Section II.3.

2.3. L_2-Approximation. For L_2-approximation we use methods

$$S_n(f) = \sum_{i=1}^{n} \lambda_i(f) \cdot a_i, \qquad a_i\, \varrho^{1/2} \in L_2(D).$$

Note that the particular cases of approximation based on functions values, derivatives, and Fourier or wavelet coefficients are covered.

It turns out that the maximal errors of S_n on $B(K)$ for L_2- and L_∞-approximation yield bounds for the average error $e_2(S_n, \mathrm{App}_{2,\varrho}, P)$.

PROPOSITION 19. *Any linear method S_n for L_2-approximation satisfies*

$$e_{\max}(S_n, \mathrm{App}_{2,\varrho}, B(K)) \le e_2(S_n, \mathrm{App}_{2,\varrho}, P) \le e_{\max}(S_n, \mathrm{App}_\infty, B(K)) \cdot \|\varrho\|_1^{1/2}.$$

PROOF. Fubini's Theorem and Proposition 17, applied to $S = \delta_t$, yield

$$e_2(S_n, \mathrm{App}_{2,\varrho}, P)^2 = \int_D \int_F (f(t) - S_n(f)(t))^2\, dP(f) \cdot \varrho(t)\, dt$$

$$= \int_D \sup_{h \in B(K)} |h(t) - S_n(h)(t)|^2 \cdot \varrho(t)\, dt.$$

Thus

$$e_2(S_n, \mathrm{App}_{2,\varrho}, P)^2 \le \sup_{h \in B(K)} \|h - S_n(h)\|_\infty^2 \cdot \|\varrho\|_1$$

and

$$e_2(S_n, \mathrm{App}_{2,\varrho}, P)^2 \ge \sup_{h \in B(K)} \int_D |h(t) - S_n(h)(t)|^2 \cdot \varrho(t)\, dt,$$

as claimed. $\qquad\square$

REMARK 20. L_2-approximation in the average case setting and L_∞-approximation in the worst case setting can be regarded as approximating a manifold in a Hilbert space by some manifold in a finite dimensional subspace. The manifold is given by $(\delta_t)_{t \in D}$ or $(K(\cdot, t))_{t \in D}$ and the subspace is spanned by $\lambda_1, \ldots, \lambda_n$ or $\mathfrak{I}\lambda_1, \ldots, \mathfrak{I}\lambda_n$. Clearly Proposition 19 can be generalized to other curves, for instance those which correspond to differentiation, i.e., to the recovery of derivatives.

REMARK 21. In terms of minimal errors, Proposition 19 says

$$e_{\max}(n, \Lambda, \mathrm{App}_{2,\varrho}, B(K)) \le e_2(n, \Lambda, \mathrm{App}_{2,\varrho}, P)$$

$$\le e_{\max}(n, \Lambda, \mathrm{App}_\infty, B(K)) \cdot \|\varrho\|_1^{1/2}$$

for any $\Lambda \subset \Lambda^{\mathrm{all}}$. As already mentioned in Remark II.6, minimal maximal errors for integration and L_2-approximation often do not differ much in their asymptotic behavior for $\Lambda = \Lambda^{\mathrm{std}}$. In such a case Proposition 4 shows that the lower bound in the previous estimate is not sharp. The upper bound turns out to be sharp, modulo constants, in many cases. See, for instance Example 28, which deals with the Wiener measure.

It would be interesting to know general conditions for covariance kernels K and classes Λ that imply $e_2(n, \Lambda, \mathrm{App}_{2,\varrho}, P) \asymp e_{\max}(n, \Lambda, \mathrm{App}_\infty, B(K))$.

Assume that the weight function ϱ is continuous and positive on D, and consider the covariance kernel

$$R(s,t) = \varrho(s)^{1/2} \cdot \varrho(t)^{1/2} \cdot K(s,t).$$

This kernel is continuous and corresponds to the image of P under the transformation $f \mapsto \varrho^{1/2} f$. Using the same notation for the kernel and the integral operator, we define

$$(R\eta)(s) = \int_D R(s,t) \cdot \eta(t)\, dt$$

on the space $L_2(D)$.

Let $\mu_1 \geq \mu_2 \geq \cdots > 0$ denote the nonzero eigenvalues of R, repeated according to their multiplicity, and let $\xi_1, \xi_2, \ldots$ denote a corresponding orthonormal system of eigenfunctions. We may assume continuity of the eigenfunctions.

In the following, sums $\sum_j$ are either finite or countable according to the rank of R, and $\langle \cdot, \cdot \rangle_2$ denotes the scalar product in $L_2(D)$. Due to Mercer's Theorem, R has the representation

$$R(s,t) = \sum_j \mu_j \cdot \xi_j(s)\, \xi_j(t),$$

being uniformly and absolutely convergent. Thus

$$\int_D R(t,t)\, dt = \sum_j \mu_j.$$

LEMMA 22. *Let Z denote the closed subspace of $L_2(D)$ that is generated by $(\xi_j)_j$, and put*

$$\langle h_1, h_2 \rangle = \sum_j \mu_j^{-1} \cdot \langle h_1, \xi_j \rangle_2 \langle h_2, \xi_j \rangle_2.$$

Then

$$H(R) = \{ h \in C(D) \cap Z : \langle h, h \rangle < \infty \}$$

and the scalar product in $H(R)$ is given by $\langle \cdot, \cdot \rangle$. Furthermore,

$$R^{1/2}\eta = \sum_j \mu_j^{1/2} \langle \eta, \xi_j \rangle_2 \cdot \xi_j$$

defines an isomorphism between Z and $H(R)$. In particular, the functions $\mu_j^{1/2}\xi_j$ form an orthonormal basis of $H(R)$.

PROOF. Consider the Hilbert space

$$Z_0 = \{h \in Z : \langle h, h \rangle < \infty\},$$

equipped with the scalar product $\langle \cdot, \cdot \rangle$. For every $h \in Z_0$,

$$\sup_{t \in D} \sum_j |\langle h, \xi \rangle_2 \cdot \xi_j(t)| \leq \left(\sum_j \mu_j^{-1} \cdot \langle h, \xi_j \rangle_2^2 \right)^{1/2} \cdot \sup_{t \in D} \left(\sum_j \mu_j \cdot \xi_j(t)^2 \right)^{1/2}$$

$$= \langle h, h \rangle^{1/2} \cdot \sup_{t \in D} R(t, t)^{1/2},$$

so that $\sum_j \langle h, \xi_j \rangle_2 \cdot \xi_j$ converges uniformly to h. Thus Z_0 forms a Hilbert space of continuous function on D.

It is easily checked that

$$R(\cdot, t) = \sum_j \mu_j \, \xi_j(t) \cdot \xi_j \in Z_0.$$

Moreover,

$$\langle h, R(\cdot, t) \rangle = \sum_j \mu_j^{-1} \cdot \langle R(\cdot, t), \xi_j \rangle_2 \, \langle h, \xi_j \rangle_2 = \sum_j \langle h, \xi_j \rangle_2 \cdot \xi_j(t) = h(t),$$

and Proposition 1 yields the characterization of $H(R)$, as claimed. Now it is straightforward to verify the remaining assertions. $\qquad \square$

LEMMA 23. *An isomorphism between* $H(K)$ *and* $H(R)$ *is defined by* $h \mapsto \varrho^{1/2} h$.

PROOF. Clearly $h \mapsto \varrho^{1/2} h$ is a well defined mapping between $\mathrm{span}\{K(\cdot, t) : t \in D\}$ and $\mathrm{span}\{R(\cdot, t) : t \in D\}$. Since

$$\|\varrho^{1/2}(\cdot) \, K(\cdot, t)\|_R^2 = \|\varrho^{-1/2}(t) \, R(\cdot, t)\|_R^2 = K(t, t) = \|K(\cdot, t)\|_K^2,$$

this mapping extends to an isomorphism T between $H(K)$ and $H(R)$. We have $Th = \varrho^{1/2} h$ for every $h \in H(K)$, since

$$Th(t) = \lim_{k \to \infty} Th_k(t) = \varrho(t)^{1/2} h(t)$$

if the sequence of functions $h_k \in \mathrm{span}\{K(\cdot, t) : t \in D\}$ tends to h. $\qquad \square$

The following proposition determines the minimal errors for L_2-approximation with $q = 2$ if arbitrary functionals from Λ^{all} are permitted.

PROPOSITION 24. *Suppose that* $\varrho > 0$ *is continuous. Any linear method* S_n *for* L_2-*approximation satisfies*

$$e_2(S_n, \mathrm{App}_{2,\varrho}, P) \geq \left(\sum_{j > n} \mu_j \right)^{1/2}.$$

Equality holds for

$$S_n(f) = \sum_{i=1}^n \langle f, \varrho^{1/2} \cdot \xi_i \rangle_2 \cdot \xi_i / \varrho^{1/2}$$

with $n \leq \mathrm{rank}\, R$.

PROOF. Let $S_n(f) = \sum_{i=1}^{n} \lambda_i(f) \cdot a_i$ with linearly independent functionals $\lambda_i \in \Lambda^{\text{all}}$, and let $\varphi_1, \ldots, \varphi_n$ denote an orthonormal basis of $\text{span}\{\Im\lambda_1, \ldots, \Im\lambda_n\}$ in $H(K)$. Then we get

$$e_2(S_n, \text{App}_{2,\varrho}, P)^2 = \int_D \left\| K(\cdot, t) - \sum_{i=1}^{n} a_i(t) \cdot \Im\lambda_i \right\|_K^2 \cdot \varrho(t)\, dt$$

$$\geq \int_D \left\| K(\cdot, t) - \sum_{i=1}^{n} \langle K(\cdot, t), \varphi_i \rangle_K \cdot \varphi_i \right\|_K^2 \cdot \varrho(t)\, dt$$

$$= \int_D R(t, t)\, dt - \sum_{i=1}^{n} \int_D \varphi_i(t)^2 \cdot \varrho(t)\, dt,$$

using Fubini's Theorem and Proposition 17.

The functions $\varphi_i \varrho^{1/2}$ are orthonormal in $H(R)$, see Lemma 23. Thus, $\varphi_i \varrho^{1/2} = R^{1/2}\eta_i$ for orthonormal functions $\eta_i \in L_2(D)$, see Lemma 22, and

$$\langle \varphi_i \varrho^{1/2}, \varphi_i \varrho^{1/2} \rangle_2 = \langle R\eta_i, \eta_i \rangle_2.$$

We conclude that

$$e_2(S_n, \text{App}_{2,\varrho}, P)^2 \geq \sum_j \mu_j - \sum_{i=1}^{n} \langle R\eta_i, \eta_i \rangle_2 \geq \sum_{j>n} \mu_j$$

because of the minimax property of the eigenvalues of R.

The second estimate becomes an equality for $\eta_i = \xi_i$, and in this case

$$\langle \varphi_i, h \rangle_K = \langle R^{1/2}\xi_i, h\varrho^{1/2} \rangle_R = \mu_i^{-1/2} \int_D h(s)\varrho(s)^{1/2} \cdot \xi_i(s)\, ds.$$

Taking

$$\lambda_i(f) = \mu_i^{-1/2} \int_D f(s)\varrho(s)^{1/2} \cdot \xi_i(s)\, ds$$

we get $\Im\lambda_i = \varphi_i$, see Corollary 12. For

$$a_i(t) = \varphi_i(t) = \frac{\mu_i^{1/2} \cdot \xi_i(t)}{\varrho^{1/2}(t)}$$

we also obtain equality in the first estimate. $\qquad\square$

REMARK 25. Consider the method S_n that is optimal for L_2-approximation with $\Lambda = \Lambda^{\text{all}}$ according to Proposition 24. Let Q denote the image of the measure P with respect to $f \mapsto \varrho^{1/2} f \in C(D)$. Then

$$(10) \qquad e_2(S_n, \text{App}_{2,\varrho}, P)^2 = \int_{C(D)} \left\| g - \sum_{i=1}^{n} \int_D g(s) \cdot \xi_i(s)\, ds \cdot \xi_i \right\|_2^2 dQ(g).$$

Hence it is optimal to approximate g by the first n terms of the series

$$(11) \qquad \sum_j \langle g, \xi_j \rangle_2 \cdot \xi_j,$$

which is known as the *Karhunen-Loève expansion* of the random function g.

The functions ξ_j are orthonormal in $L_2(D)$, and the integrals $\int_D g(s)\cdot\xi_i(s)\,ds$ are uncorrelated random variables with variance μ_i since

$$\int_{C(D)} \langle g,\xi_i\rangle_2 \cdot \langle g,\xi_k\rangle_2 \,dQ(g) = \langle R\xi_i,\xi_k\rangle_2 = \mu_i \cdot \delta_{i,k}.$$

Here $\delta_{i,k}$ is the Kronecker symbol. In addition to convergence in quadratic mean and L_2-norm, see (10), the series (11) is also pointwise convergent in quadratic mean.

REMARK 26. There is a worst case analogue to Proposition 24, see Traub and Woźniakowski (1980), Micchelli and Rivlin (1985), and Traub, Wasilkowski, and Woźniakowski (1988, Section 4.5.3). Namely,

$$e_{\max}(n, H(K)^*, \mathrm{App}_{2,\varrho}, B(K))$$
$$= \inf_{h_1,\ldots,h_n \in H(K)} \sup\{\|h\|_{2,\varrho} : h \in B(K),\ \langle h, h_i\rangle_K = 0 \text{ for } i = 1,\ldots,n\}$$
$$= \inf_{h_1,\ldots,h_n \in H(R)} \sup\{\|h\|_2 : h \in B(R),\ \langle h, h_i\rangle_R = 0 \text{ for } i = 1,\ldots,n\}$$
$$= \mu_{n+1}^{1/2}$$

for the minimal error of methods that use n bounded linear functionals on $H(K)$. Note that the last equality is the minimax property of the eigenvalues of R. We add that the same method S_n is optimal in the worst case and average case setting.

REMARK 27. Proposition 24 basically solves the problem of L_2-approximation if weighted integrals may be used to recover f. However, we are mainly interested in linear methods that are based on function values (samples) of f, i.e., in the case $\Lambda = \Lambda^{\mathrm{std}}$ instead of $\Lambda = \Lambda^{\mathrm{all}}$. We briefly compare the power of these two classes for L_2-approximation.

Results from Müller-Gronbach (1996) show that sampling is optimal in Λ^{all} only under rather strict assumptions on the covariance kernel K of P and on ϱ. Usually the lower bound $b_n = (\sum_{j>n}\mu_j)^{1/2}$ cannot be reached by methods that use n function values. Hence we may ask whether at least

$$e_2(n, \Lambda^{\mathrm{all}}, \mathrm{App}_{2,\varrho}, P) \asymp e_2(n, \Lambda^{\mathrm{std}}, \mathrm{App}_{2,\varrho}, P).$$

It would be interesting to know general conditions that guarantee *order optimality of sampling* in this sense.

The following approach is discussed in Woźniakowski (1992): approximate the first k terms in the Karhunen-Loève expansion (11) by means of quadrature formulas that use the same set of knots $x_1,\ldots,x_n$ for all j. If, for instance, $b_k = O(k^{-\beta})$ then the choice $k \asymp n^{1/(2\beta+1)}$ yields a method S_n with error $O(n^{-1/(2+1/\beta)})$. This is important in particular for small β. The proof is non-constructive and similar to the one of Proposition II.4.

(Order) optimality of sampling is also studied in the worst case setting. See Micchelli and Rivlin (1985) for results and further references. Worst case and average case results on this topic are related. For instance, if

$$e_{\max}(n, H(K)^*, \mathrm{App}_{2,\varrho}, B(K)) \asymp e_{\max}(n, \Lambda^{\mathrm{std}}, \mathrm{App}_{2,\varrho}, B(K))$$

in the worst case setting and

$$n \cdot \mu_n = O\Big(\sum_{j>n} \mu_j\Big)$$

then sampling is also order optimal in the average case setting.

We sketch the proof in the case $\varrho = 1$. By assumption, there are linear methods S_n for L_2-approximation that use n function values and yield

$$e_{\max}(S_n, \mathrm{App}_2, B(K)) = O(\mu_{n+1}^{1/2}),$$

see Remark 26. Define

$$T_n(f) = \sum_{k=1}^{n} \langle S_n(f), \xi_k \rangle_2 \cdot \xi_k$$

to obtain

$$e_2(T_n, \mathrm{App}_2, P)^2 = \sum_j \int_F \langle f - T_n(f), \xi_j \rangle_2^2 \, dP(f)$$

$$= \sum_{j=1}^{n} \sup\{\langle h - S_n(h), \xi_j \rangle_2^2 : h \in B(K)\} + \sum_{j>n} \mu_j$$

$$\leq n \cdot \mu_{n+1} + \sum_{j>n} \mu_j = O\Big(\sum_{j>n} \mu_j\Big).$$

EXAMPLE 28. Let us look at L_2-approximation with $\varrho = 1$ in the average case setting with respect to the Wiener measure w. Proposition II.12 determines optimal methods that use function values only, and the minimal errors are

$$e_2(n, \Lambda^{\mathrm{std}}, \mathrm{App}_2, w) = (2 \cdot (3n + 1))^{-1/2} \approx (6n)^{-1/2}.$$

Recall that $K(s, t) = \min(s, t)$ for the Wiener measure. The corresponding unit ball $B(K)$ consists of all function $h \in W_2^1([0, 1])$ with $h(0) = 0$ and $\|h'\|_2 \leq 1$, see Example 5. In the worst case setting,

$$e_{\max}(n, \Lambda^{\mathrm{std}}, \mathrm{App}_2, B(K)) \asymp n^{-1}$$

and

$$e_{\max}(n, \Lambda^{\mathrm{std}}, \mathrm{App}_\infty, B(K)) \asymp n^{-1/2},$$

see Pinkus (1985). Thus the upper bound from Remark 21 is sharp, up to a multiplicative constant.

Consider the eigenvalue problem

$$\int_0^1 \min(s, t) \cdot \xi(t) \, dt = \mu \cdot \xi(s)$$

for the integral operator with the kernel K. Differentiating twice we obtain an equivalent boundary value problem

$$-\xi(s) = \mu \cdot \xi''(s), \qquad \xi(0) = \xi'(1) = 0.$$

It follows that the eigenvalues and normalized eigenfunctions are given by

$$\mu_j = \left(\left(j - \tfrac{1}{2}\right)\pi\right)^{-2}$$

and

$$\xi_j(t) = \sqrt{2}\,\sin\left(\left(j - \tfrac{1}{2}\right)\pi \cdot t\right).$$

Proposition 24 shows that

$$S_n(f) = \sum_{i=1}^{n} \int_0^1 f(t) \cdot \xi_i(t)\,dt \cdot \xi_i$$

is the optimal linear method for L_2-approximation using n functionals from Λ^{all}. For this method we have

$$e_2(S_n, \mathrm{App}_2, w_0) = e_2(n, \Lambda^{\mathrm{all}}, \mathrm{App}_2, w_0)$$

$$= \left(\frac{1}{2} - \frac{4}{\pi^2}\sum_{i=1}^{n}(2i-1)^{-2}\right)^{1/2} \approx (\pi^2 n)^{-1/2}.$$

Thus sampling is order optimal, and for large n the minimal errors for the classes Λ^{std} and Λ^{all} differ roughly by the factor $\pi/\sqrt{6}$. See Richter (1992) for errors of further linear methods with respect to the Wiener measure.

2.4. Notes and References. 1. Proposition 17 is due to Sacks and Ylvisaker (1970b), who study the integration problem.

2. Proposition 19 is from Woźniakowski (1992).

3. Proposition 24 is from Micchelli and Wahba (1981). The result is generalized in Wasilkowski and Woźniakowski (1986) to arbitrary linear problems with a Hilbert space G. The eigenvalues μ_j and eigenvectors ξ_i are replaced by the corresponding quantities of the covariance operator of the image measure SP on G.

3. Splines and Their Optimality Properties

In this section we introduce splines in reproducing kernel Hilbert spaces and study their optimality for linear problems. Optimality holds for linear problems with values in a Hilbert space or with respect to Gaussian measures.

3.1. Splines in Reproducing Kernel Hilbert Spaces. Let K denote a nonnegative definite function on a set D^2. We consider a finite number of bounded linear functionals on $H(K)$ that are represented by linearly independent functions $\xi_1, \ldots, \xi_n \in H(K)$. Clearly one can interpolate arbitrary data $y_1, \ldots, y_n \in \mathbb{R}$ in the sense that there exist functions $h \in H(K)$ with

$$(12) \qquad \langle \xi_i, h \rangle_K = y_i, \qquad i = 1, \ldots, n.$$

In particular, there exists a unique function $h \in H(K)$ with (12) that has minimal norm $\|h\|_K$ among all functions with property (12). This function is called the (abstract) *spline* in $H(K)$ that interpolates the data $y_1, \ldots, y_n$ for given $\xi_1, \ldots, \xi_n$.

We are mainly interested in splines that interpolate function values. In this case $\xi_i = K(\cdot, x_i)$ for given knots $x_i \in D$, and the spline has minimal norm among all functions $h \in H(K)$ with $h(x_i) = y_i$.

LEMMA 29. *Let* $\xi_1, \ldots, \xi_n \in H(K)$ *be linearly independent. Moreover, let* $y_1, \ldots, y_n \in \mathbb{R}$. *The following properties are equivalent for* $h \in H(K)$.

(i) h *is the spline that interpolates the data* $y_1, \ldots, y_n$ *for given* $\xi_1, \ldots, \xi_n$.

(ii) h *satisfies* (12) *and*

$$h \in \operatorname{span}\{\xi_1, \ldots, \xi_n\}.$$

(iii) *For every* $t \in D$,

$$h(t) = (y_1, \ldots, y_n) \cdot \Sigma^{-1} \cdot (\xi_1(t), \ldots, \xi_n(t))',$$

where

$$\Sigma = (\langle \xi_i, \xi_j \rangle_K)_{1 \le i,j \le n}$$

is the Gram matrix corresponding to $\xi_1, \ldots \xi_n$.

PROOF. The equivalence of (i) and (ii) follows from

$$\{h \in H(K) : \langle \xi_i, h \rangle_K = 0 \text{ for } i = 1, \ldots, n\}^{\perp} = \operatorname{span}\{\xi_1, \ldots, \xi_n\}.$$

Elementary linear algebra yields the equivalence of (ii) and (iii).   □

The following fact is easily verified.

LEMMA 30. *Let* $\xi_1, \ldots, \xi_n \in H(K)$ *be linearly independent. The covariance kernel* M *of the subspace* $\operatorname{span}\{\xi_1, \ldots, \xi_n\}$ *of* $H(K)$ *is given by*

$$M(s,t) = (\xi_1(s), \ldots, \xi_n(s)) \cdot \Sigma^{-1} \cdot (\xi_1(t), \ldots, \xi_n(t))'.$$

3.2. Optimality of K-Spline Algorithms on Hilbert Spaces. In the remaining part of Section 3 we consider a linear problem with operator $S : C^k(D) \to G$ and a zero mean measure P on $F \subset C^k(D)$. Let K denote the covariance kernel of P and assume, for simplicity, that $K \in C^{r,r}(D^2)$ for some $r \ge k$.

Suppose we are given fixed functionals $\lambda_1, \ldots, \lambda_n \in \Lambda^{\text{all}}$, e.g., $\lambda_i(f) = f(x_i)$. How do we choose the coefficients $a_i \in G$ to obtain a small average error $e_q(S_n, S, P)$ for

$$S_n(f) = \sum_{i=1}^{n} \lambda_i(f) \cdot a_i?$$

Without loss of generality we assume that $\lambda_1, \ldots, \lambda_n$ are linearly independent, since otherwise we may ignore some data $\lambda_i(f)$ without increasing the average error.

In two important cases the best linear method is given by the following simple scheme. For $f \in F$ let h denote the spline in $H(K)$ that interpolates the data $\lambda_1(f), \ldots, \lambda_n(f)$ for given

$$\xi_i = \Im \lambda_i.$$

Here $\Im$ denotes the fundamental isomorphism between Λ^P and $H(K)$. The interpolation property (12) for h reads

$$\lambda_i(h) = \lambda_i(f), \qquad i = 1, \ldots, n,$$

see Corollary 12. The K-spline algorithm is defined by

$$S_n(f) = S(h),$$

i.e., we apply the operator S to the spline that interpolates the data. Clearly S_n is a linear method using the functionals λ_i, see Lemma 29. Because of this lemma, we consider the subspace

$$V = \mathrm{span}\{\Im \lambda_i : i = 1, \ldots, n\}$$

of $H(K)$. In particular, if

(13) $$\lambda_i(f) = f^{(\alpha_i)}(x_i), \qquad x_i \in D, \ |\alpha_i| \leq k,$$

then

$$V = \mathrm{span}\{K^{(0,\alpha_i)}(\cdot, x_i) : i = 1, \ldots, n\}.$$

We will show optimality of the K-spline algorithm in particular for linear functionals S and average errors with $q = 2$. In this case the error $e_2(S_n, S, P)$ of any linear method is defined as the distance between S and S_n in the Hilbert space Λ^P. Hence the best linear method is given as the orthogonal projection of S onto the subspace that is generated by the functionals λ_i. By means of the fundamental isomorphism, orthogonality is equivalent to exactness on the subspace V, and this yields the following result.

PROPOSITION 31. *Consider a linear problem with a Hilbert space G. Let K denote the covariance kernel of the zero mean measure P, take $q = 2$, and assume that functionals $\lambda_1, \ldots, \lambda_n \in \Lambda^{\mathrm{all}}$ are given. Then the K-spline algorithm has minimal error $e_2(S_n, S, P)$ among all linear methods S_n that use the functionals λ_i.*

PROOF. We use the facts and notation from Section 2.2. Clearly $\Im S_n^j \in V$ for every linear method S_n. Moreover, S_n minimizes the error $e_2(S_n, S, P)$ iff $\Im S_n^j$ is the orthogonal projection of $\Im S^j$ onto V for all j. The latter property is equivalent to

$$0 = \langle \Im S^j - \Im S_n^j, h \rangle_K = S^j(h) - S_n^j(h)$$

for every $h \in V$ and every j, see Corollary 12. Therefore S_n has minimal error among all linear methods that use the functionals λ_i iff

$$S_n(h) = S(h)$$

for all $h \in V$. By definition, the latter property holds for the K-spline algorithm.

$\square$

The K-spline algorithm S_n vanishes on $V^\perp$, as does every linear method that uses the functionals λ_i. For $S \in \Lambda^{\text{all}}$, i.e., for linear problems with values in $\mathbb{R}$, we therefore have

$$(14) \qquad e_2(S_n, S, P) = e_{\max}(S_n, S, B(K))$$
$$= \sup\{|S(h)| : h \in B(K), \ \lambda_i(h) = 0 \text{ for } i = 1, \ldots, n\}.$$

Proposition 31 extends to linear problems with arbitrary normed spaces G, if P has certain symmetry properties. For instance, this holds true for Gaussian measures P, and even nonlinear methods may be allowed. See Sections 3.5 and VII.2.2. However, it does not extend to the non-Hilbert case without additional assumptions.

EXAMPLE 32. We give an example of discrete L_∞-approximation where the K-spline algorithm does not give the best linear method. This example is used in Novak and Ritter (1989) in a slightly different context.

Let $X = G = \mathbb{R}^3$ be equipped with the supremum norm, and consider the uniform distribution P on $F = \{\pm(3, -1, 1), \pm(0, 2, 1), \pm(0, -2, -3)\}$. Suppose we want to approximate $S(f) = f$, knowing only $\lambda(f) = f_1 + f_2 - f_3$. Note that $\lambda(F) = \{\pm 1\}$. Best linear approximation in each coordinate leads to the spline algorithm $S_1(f) = \lambda(f) \cdot (1, -\frac{1}{3}, -\frac{1}{2})$. We compute $e_2(S_1, S, P)^2 = 149/27$. However, $e_2(\widetilde{S}_1, S, P) = 2$ holds for the method $\widetilde{S}_1(f) = \lambda(f) \cdot (1, 0, -1)$.

3.3. Optimal Coefficients for Integration. Lemma 29 yields explicit formulas for the coefficients $a = (a_1, \ldots, a_n)'$ of the K-spline algorithm. For instance, let S be a bounded linear functional, and let

$$(15) \qquad \Sigma = (\lambda_i(\Im \lambda_j))_{1 \le i, j \le n}$$

denote the covariance matrix of the functionals λ_i. Furthermore, let

$$(16) \qquad b = (S(\Im \lambda_i))_{1 \le i \le n}$$

denote the vector of covariances between S and λ_i. The coefficients of the K-spline algorithm are given as solutions of the normal equation

$$(17) \qquad \Sigma \cdot a = b.$$

By Proposition 31, the error

$$(18) \qquad e_2(S_n, S, P) = (\|\Im S\|_K^2 - b' \cdot \Sigma^{-1} \cdot b)^{1/2}$$

of the K-spline algorithm is minimal among all quadrature formulas that are based on the functionals λ_i.

If function or derivative values are available for integration, i.e., in the case (13), we get

$$\Sigma = (K^{(\alpha_i, \alpha_j)}(x_i, x_j))_{1 \le i, j \le n},$$

and

$$b = (\xi^{(\alpha_i)}(x_i))_{1 \le i \le n},$$

where $\xi = \mathfrak{I}S$ is the representer of $S = \mathrm{Int}_\varrho$ on $H(K)$, see (9). The variance of Int_ϱ is given by

$$\int_F \mathrm{Int}_\varrho(f)^2 \, dP(f) = \|\xi\|_K^2 = \int_{D^2} K(s,t) \cdot \varrho(s) \, \varrho(t) \, d(s,t).$$

3.4. Optimal Coefficients for L_2-Approximation. For this problem the operator $S = \mathrm{App}_{2,\varrho}$ is an embedding and the coefficients $a = (a_1, \ldots, a_n)$ of linear methods S_n are real-valued functions on D. The K-spline algorithm is optimal and its coefficients solve the normal equations

$$\Sigma \cdot a(t) = b(t), \qquad t \in D.$$

Here Σ denotes the covariance matrix (15) and

$$b(t) = (\mathfrak{I}\lambda_i(t))_{1 \le i \le n},$$

which is $b(t) = (K(t, x_i))_{1 \le i \le n}$ if the data consist of function values $\lambda_i(f) = f(x_i)$. The error of the K-spline algorithms satisfies

$$(19) \quad e_2(S_n, \mathrm{App}_{2,\varrho}, P)^2 = \int_D \int_F \left(f(t) - \sum_{i=1}^n \lambda_i(f) \cdot a_i(t) \right)^2 dP(f) \cdot \varrho(t) \, dt$$

$$= \int_D (K(t,t) - b(t)' \cdot \Sigma^{-1} \cdot b(t)) \cdot \varrho(t) \, dt$$

$$= \int_D \sup\{h(t)^2 : h \in B(K), \, \lambda_i(h) = 0 \text{ for } i = 1, \ldots, n\} \cdot \varrho(t) \, dt.$$

Here we have used (14) and (18) with $S = \delta_t$. Analogous statements hold for differentiation of f if the error is measured in some weighted L_2-norm.

EXAMPLE 33. Consider the Wiener measure w. In this case $K(s,t) = \min(s,t)$, and $H(K)$ consists of all function $h \in W_2^1([0,1]$ with $h(0) = 0$, see Example 5. The spline in $H(K)$ that interpolates function values $y_1, \ldots, y_n$ at knots $0 < x_1 < \cdots < x_n \le 1$ is the piecewise linear function that satisfies $h(x_i) = y_i$, $h(0) = 0$, and that is constant on $[x_n, 1]$. Thus h is a polynomial spline of degree one. Polynomial splines of higher order play an important role in the sequel.

For integration and L_2-approximation optimality of the corresponding K-spline algorithms was previously shown in Section II.3.

3.5. Optimality of K-Spline Algorithms for Gaussian Measures. If P is Gaussian then Proposition 31 holds for q-average errors with $1 \le q < \infty$ and without any further assumption on the range G of the linear problem. In Section VII.2.2 we will see that the conclusion remains true even if arbitrary nonlinear methods are allowed that use given functionals λ_i. This optimality of K-spline algorithms is a consequence of symmetry properties, which hold in particular for Gaussian measures.

PROPOSITION 34. *Let $1 \leq q < \infty$. Consider an arbitrary linear problem and a zero mean Gaussian measure P on $F \subset C^k(D)$. Let K denote the covariance kernel of P and assume that functionals $\lambda_1, \ldots, \lambda_n \in \Lambda^{\text{all}}$ are given. Then the K-spline algorithm has minimal error $e_q(S_n, S, P)$ among all linear methods S_n that use the functionals λ_i.*

PROOF. Let Q denote a zero mean Gaussian measure on $C^k(D)$, and consider a bounded linear functional λ on $C^k(D)$. By Remark 16, we may apply Corollary 12 to conclude that $\lambda \in \Lambda^Q$ and

$$\int_F \lambda(f) \cdot f(t) \, dQ(f) = \langle \Im\lambda, L(\cdot, t) \rangle_L = \lambda(L(\cdot, t)).$$

Here L denotes the covariance kernel of Q. Hence λ vanishes on $H(L)$ iff $\lambda(f) = 0$ with probability one.

By definition, the support of a measure Q on $C^k(D)$ is the smallest closed subset A of $C^k(D)$ such that $Q(A) = 1$. Since Q is Gaussian with zero mean, its support is a linear subspace and given as the closure of $H(L)$ in $C^k(D)$, see Vakhania (1975).

Put $V = \text{span}\{\Im\lambda_i : i = 1, \ldots, n\}$. Let K_0 and K_1 denote the reproducing kernels of the subspaces $V^\perp$ and V, respectively, of $H(K)$. Furthermore, let P_i denote the zero mean Gaussian measure on F with covariance kernel K_i. Since $K = K_0 + K_1$ we conclude that P is the convolution of P_0 and P_1.

Note that λ_i vanishes on $V^\perp$ by definition. Therefore

$$\int_F \|S(f) - S_n(f)\|_G^q \, dP(f)$$

$$= \int_F \int_F \|S(f_0 + f_1) - S_n(f_0 + f_1)\|_G^q \, dP_0(f_0) \, dP_1(f_1)$$

$$= \int_F \int_F \|S(f_0) - (S_n(f_1) - S(f_1))\|_G^q \, dP_0(f_0) \, dP_1(f_1).$$

Because of the symmetry of P_0 the inner integral is minimized if $S_n(f_1) = S(f_1)$, see Remark II.3. Thus $e_q(S_n, S, P)$ is minimal if

$$P_1(\{f \in F : S_n(f) = S(f)\}) = 1.$$

Since V is the support of P_1, minimality of $e_q(S_n, S, P)$ is equivalent to

$$V \subset \{f \in F : S_n(f) = S(f)\}).$$

This criterion, which also appeared in the proof of Proposition 31, is satisfied for the K-spline algorithm S_n. $\square$

For L_p-approximation the K-spline algorithm simply takes the interpolating spline as the reconstruction of f from data $\lambda_i(f)$. This method is optimal for every $1 \leq p \leq \infty$, every weight function ϱ, and every $1 \leq q < \infty$, if average errors with respect to Gaussian measures are studied.

REMARK 35. Consider a linear problem with $\Lambda = \Lambda^{\text{std}}$. Searching optimal methods requires minimization of the error with respect to the coefficients and knots of linear methods. For fixed knots the best coefficients are determined under rather general assumptions in Propositions 31 and 34. The selection of the knots is left as the main problem.

Determining an optimal quadrature formula is equivalent to the following problem of best approximation for the representer $\xi(t) = \text{Int}_\varrho(K(\cdot, t))$ of Int_ϱ in $H(K)$: minimize the distance

$$d(\xi, V_n) = \inf_{h \in V_n} \|\xi - h\|_K,$$

where V_n runs through all at most n-dimensional subspaces spanned by functions $K(\cdot, s)$.

Determining an optimal linear method for L_2-approximation is equivalent to the following problem in $H(K)$: minimize the average distance

$$\int_D d(K(\cdot, t), V_n)^2 \cdot \varrho(t)\, dt = \int_D \inf_{h \in V_n} \|K(\cdot, t) - h\|_K^2 \cdot \varrho(t)\, dt$$

over all subspaces V_n. Both problems are very difficult and have only been solved for some particular covariance kernels K and weight functions ϱ.

3.6. Notes and References. 1. See, e.g., Atteia (1992) for a study of splines in reproducing kernel Hilbert spaces.

2. Proposition 31 is due to Kimeldorf and Wahba (1970a, 1970b) for $S = \delta_t$ and generalized to linear problems with a Hilbert space G in Micchelli and Rivlin (1985) and Wasilkowski and Woźniakowski (1986). See Wahba (1990, Sections 2.4, 2.5) and Mardia, Kent, Goodall, and Little (1996) for extensions to conditionally positive kernels and intrinsic random functions.

3. Proposition 34 is due to Lee and Wasilkowski (1986), see also Notes and References VII.2.6.3.

4. Optimality properties of splines are known in various settings. Results concerning the worst case optimality of spline algorithms are summarized in Traub et al. (1988). For various problems with noisy data $\lambda_i(f) + \varepsilon_i$, smoothing splines are known to be optimal. See Kimeldorf and Wahba (1970a), Eubank (1988), Wahba (1990), Plaskota (1996), and also Section V.1.

5. We focus on asymptotically optimal or order optimal methods and error bounds, and we do not study the questions of existence, uniqueness, and characterization of optimal knots in detail. In the worst case setting these questions are often related to determining splines with minimal norm. See Zhensykbaev (1983), Micchelli and Rivlin (1985), Traub et al. (1988), Korneichuk (1991), Bojanov, Hakopian, and Sahakian (1993). For average case results we refer to Sacks and Ylvisaker (1966, 1970b), Eubank, Smith, and Smith (1981, 1982), and Müller-Gronbach (1996).

4. Kriging

The results from the previous sections hold for zero mean measures. Now we drop the assumption $m = 0$ for the mean of P, and we distinguish three cases. In the nonparametric case only some smoothness properties of m are known. In the linear parametric case we know a finite dimensional space of functions that contains m. Finally m might be nonzero, but known. In the latter case we are basically dealing with the zero mean case, see Section 4.1. The nonparametric case will be addressed in Section IV.4.2. Here we focus on the linear parametric case, and we discuss best linear unbiased prediction, which is known as universal kriging in geostatistics, see Section 4.2. In case of a single unknown parameter, the design problem for parameter estimation is equivalent to the design problem for integration, see Section 4.3.

4.1. Simple Kriging. Suppose that functionals $S, \lambda_1, \ldots, \lambda_n \in \Lambda^{\text{all}}$ are given. We wish to approximate S by an affine linear method

$$S_n(f) = a_0 + \sum_{i=1}^{n} \lambda_i(f) \cdot a_i.$$

We study q-average errors with $q = 2$, and we assume that the covariance kernel K of P is known. Furthermore, we assume

$$\int_F S(f) \, dP(f) = S(m)$$

and

$$\int_F \lambda_i(f) \, dP(f) = \lambda_i(m).$$

These relations obviously hold in the most important example, when we use function values $\lambda_i(f) = f(x_i)$ to approximate the value $S(f) = f(x_0)$ at an untried point. Moreover, the relations hold if P is a measure on $C^k(D)$, $m \in C^k(D)$, and if (4) holds for some $r \geq k$, see Section 1.3.

Let Σ denote the covariance matrix of the functionals λ_i and let b denote the vector of covariances between S and λ_i, see (15) and (16). For simplicity we assume that Σ is invertible. In the previous example, $\Sigma = (K(x_i, x_j))_{1 \leq i,j \leq n}$ and $b = (K(x_0, x_i))_{1 \leq i \leq n}$.

If $m = 0$ then the average errors depend on the measure P only through K, and $a_0 = 0$ is the best choice of a_0, see Section 2.1. In general these errors only depend on m and K, and the best choice of a_0 is given by

$$a_0 = S(m) - \sum_{i=1}^{n} \lambda_i(m) \cdot a_i.$$

With this choice, we obtain

$$S_n(f) = S(m) + \sum_{i=1}^{n} \lambda_i(f - m) \cdot a_i$$

as well as unbiasedness of S_n, i.e.,

$$\int_F S_n(f)\,dP(f) = \int_F S(f)\,dP(f).$$

Furthermore,

$$e_2(S_n, S, P)^2 = \int_F (S(f) - S_n(f))^2\,dP(f)$$

$$= \int_F (S(f - m) - \sum_{i=1}^n \lambda_i(f - m) \cdot a_i)^2\,dP(f).$$

The right-hand side is the squared average error of a linear method with respect to a zero mean measure with covariance kernel K. The optimal coefficients $a_1, \ldots, a_n$ are therefore determined by the normal equation $\Sigma \cdot a = b$, see (17), and $e_2(S_n, S, P)$ is then given by (18). With the optimal choice of coefficients, S_n is called the *best affine linear predictor* for S with respect to the squared-error loss, and the term *simple kriging* is used for this method in geostatistics. In our terms simple kriging means application of the K-spline algorithm with the centered data $\lambda_i(f) - \lambda_i(m)$ and addition of $S(m)$. Observe that simple kriging relies on knowing the mean m or at least the values $S(m), \lambda_1(m), \ldots, \lambda_n(m)$.

4.2. Universal Kriging. Now we turn to the linear parametric case, where the mean is of the form

$$m = \sum_{k=1}^\ell \beta_k \cdot f_k$$

with linearly independent and known functions $f_k : D \to \mathbb{R}$ and unknown parameters $\beta_k \in \mathbb{R}$.

We only consider methods S_n that are unbiased for every $\beta \in \mathbb{R}^\ell$. This property is equivalent to

$$a_0 = 0$$

and

$$X' \cdot a = s$$

for $a = (a_1, \ldots, a_n)'$, where

$$X = (\lambda_i(f_k))_{1 \le i \le n,\ 1 \le k \le \ell}$$

and

$$s = (S(f_k))_{1 \le k \le \ell}.$$

For simplicity we assume that X has rank ℓ.

As a consequence of unbiasedness for every β, the error $e_2(S_n, S, P)$ does not depend on the unknown mean m, since

$$\int_F (S(f) - S_n(f))^2\,dP(f) = \int_F (S(f - m) - S_n(f - m))^2\,dP(f).$$

It therefore makes sense to minimize the error in the class of all linear unbiased methods. Every such method that minimizes the error is called a *best linear unbiased predictor*, and the term *universal kriging* is used in geostatistics. Existence and uniqueness follow from our assumptions on Σ and X.

PROPOSITION 36. *The best linear unbiased predictor for S using $\lambda_1, \ldots, \lambda_n$ is given by*

$$S_n(f) = s' \cdot \widehat{\beta}(f) + (N(f) - X \cdot \widehat{\beta}(f))' \cdot \Sigma^{-1} \cdot b$$

where

$$N(f) = (\lambda_1(f), \ldots, \lambda_n(f))'$$

and

$$\widehat{\beta}(f) = (X' \cdot \Sigma^{-1} \cdot X)^{-1} \cdot X' \cdot \Sigma^{-1} \cdot N(f).$$

PROOF. Let $a^* = \Sigma^{-1} \cdot b$ and

$$(20) \qquad S_n^\beta(f) = s' \cdot \beta + (N(f) - X \cdot \beta)' \cdot a^*.$$

From Section 4.1 we know that S_n^β is the best affine linear predictor for S and S_n^β is unbiased for the particular β.

Let $S_n(f) = a' \cdot N(f)$ denote a linear unbiased method. Clearly

$$\int_F (S(f) - S_n(f))^2 \, dP(f) = c_1 + 2 c_2 + c_3,$$

where

$$c_1 = \int_F (S(f) - S_n^\beta(f))^2 \, dP(f),$$

$$c_2 = \int_F (S(f) - S_n^\beta(f)) \cdot (S_n^\beta(f) - S_n(f)) \, dP(f),$$

$$c_3 = \int_F (S_n^\beta(f) - S_n(f))^2 \, dP(f).$$

Using the unbiasedness of S_n and S_n^β and the optimality property of the coefficients a_i^*, we get

$$c_2 = \int_F \left(S(f - m) - \sum_{i=1}^n \lambda_i(f - m) \cdot a_i^*\right) \cdot \sum_{i=1}^n \lambda_i(f - m) \cdot (a_i^* - a_i) \, dP(f) = 0.$$

It remains to minimize c_3.

We have

$$S_n^\beta(f) - S_n(f) = (a^* - a)' \cdot N(f) - (X' \cdot a^* - s)' \cdot \beta$$

and

$$\int_F (a^* - a)' \cdot N(f) \, dP(f) = (X' \cdot a^* - s)' \cdot \beta,$$

so that $(a^* - a)' \cdot N$ is a linear unbiased estimator for $(X' \cdot a^* - s)' \cdot \beta$ and c_3 is its variance. The variance is minimized iff

$$(a^* - a)' \cdot N(f) = (X' \cdot a^* - s)' \cdot \widehat{\beta}(f),$$

see Christensen (1987, Chapter 2), and this relation yields the best linear unbiased predictor as claimed. □

REMARK 37. Let $e(f) = N(f) - X \cdot \beta$. Writing the data $N(f)$ in the form

(21) $$X \cdot \beta + e(f)$$

we obtain a *linear regression model* with correlated errors. The random vector e has zero mean and covariance matrix Σ with respect to P.

In the previous proof we considered two different tasks, prediction of the unknown random quantity $S(f)$ and estimation of the deterministic vector β. In both cases we considered unbiased linear methods and errors in mean square sense.

The *best linear unbiased estimator* for β is given by $\widehat{\beta}$ as defined in Proposition 36, and $\widehat{\beta}$ is also called the *generalized least squares estimator*. It turns out that the best linear unbiased predictor for S is obtained in the following way. Since β is not known, we cannot use the best linear predictor S_n^β for S, see (20). However, substituting β by $\widehat{\beta}$ in (20) we get the best linear unbiased predictor for S.

We add that e describes the systematic departure of the data from the mean $X \cdot \beta$. Measurement errors are incorporated by an additive white noise. See also Chapter V for the noisy data case.

4.3. Integration and Regression Design. Due to Proposition 17, every quadrature formula S_n satisfies $e_2(S_n, \mathrm{Int}_\varrho, P) = e_{\max}(S_n, \mathrm{Int}_\varrho, B(K))$. Thus the average case and the worst case for integration are equivalent. Concerning the selection of knots, integration is also equivalent to a regression design problem.

Consider the linear regression model (21) and assume $\lambda_i(f) = f(x_i)$ with pairwise different $x_i \in D$. We wish to estimate the unknown vector $\beta \in \mathbb{R}^n$ from the data $X \cdot \beta + e(f)$, using the best linear unbiased estimator.

Experimental design theory deals with the optimal selection of the sampling design $x_1, \ldots, x_n$. Optimality is defined with respect to various functionals of the covariance matrix of the best linear unbiased estimator, e.g., its determinant, trace, or maximal eigenvalue. The classical theory is devoted to the case where Σ is a multiple of the unit matrix. Here we are dealing with correlated errors, and $\Sigma = (K(x_i, x_j))_{1 \leq i, j \leq n}$ depends on the covariance kernel of the random function f as well as on the sampling design.

Assume that $\ell = 1$, so that a single parameter $\beta = \beta_1 \in \mathbb{R}$ has to be estimated, and optimality is defined in terms of the variance of the best linear unbiased estimator. Moreover, assume that Σ is invertible and

$$X = (f_1(x_1), \ldots, f_1(x_n))' \neq 0.$$

For any sampling design the best linear unbiased estimator is

$$\widehat{\beta}(f) = \frac{1}{X' \cdot \Sigma^{-1} \cdot X} \cdot X' \cdot \Sigma^{-1} \cdot N(f),$$

see Remark 37. The design problem consists of finding an n-point design such that the variance

$$\int_F (\beta - \widehat{\beta}(f))^2 \, dP(f) = \frac{1}{X' \cdot \Sigma^{-1} \cdot X}$$

of $\widehat{\beta}$ is minimal.

Let $\varrho \in C(D)$. Assume that f_1 is of the form

$$f_1(t) = \int_D K(s,t) \cdot \varrho(s) \, ds,$$

and consider the integration problem Int_ϱ with weight function ϱ, where average errors with respect to P and $q = 2$ are used. The error of the K-spline algorithm S_n is given by

$$e_2(S_n, \mathrm{Int}_\varrho, P) = (\|\mathfrak{I}S\|_K^2 - X' \cdot \Sigma^{-1} \cdot X)^{1/2},$$

see (18). Both, the average error of the K-spline algorithm as well as the variance of the best linear unbiased estimator are decreasing functions of $X' \cdot \Sigma^{-1} \cdot X$. Thus the design problem for estimation of β and the problem of finding optimal knots for Int_ϱ are equivalent. In both cases $X' \cdot \Sigma^{-1} \cdot X$ has to be maximized over all n-point designs.

4.4. Notes and References. 1. Proposition 36 is well known, see, e.g., Christensen (1987, Theorem 12.2.3) and Christensen (1991, Chapter VI).

2. We refer to Cressie (1993) for a detailed analysis of kriging and its application in geostatistics. See also Christensen (1991, Chapter VI) and Hjort and Omre (1994). The method is named after the mining engineer D. G. Krige.

3. The result from Section 4.3 is due to see Sacks and Ylvisaker (1970b).

4. Designs for regression problems with correlated error are also studied in Cambanis (1985) and Näther (1985). See Pilz (1983) and O'Hagan (1992) for a Bayesian approach concerning the parameter β. The design problem is also related to a signal detection problem, see Cambanis and Masry (1983) and Cambanis (1985) for details, results, and references.

5. See Fedorov (1972) and Pukelsheim (1993) for optimal design theory in the case of uncorrelated observations.

CHAPTER IV

Integration and Approximation
of Univariate Functions

Most of the known average case results for concrete problems are for integration and approximation of univariate functions. For several classes $F \subset C^r([0,1])$ and measures P on F asymptotically or order optimal methods and corresponding error bounds are known. On the other hand, one is interested in results that do not depend on the particularly chosen P but hold for a class of measures. Such classes are often defined by properties of the covariance kernel K of P. The latter approach is taken in this chapter.

Sacks-Ylvisaker regularity conditions for K are introduced in Section 1 and applied to integration and approximation in Sections 2 and 3. These regularity conditions yield sharp bounds for minimal errors. In Section 5 we discuss Hölder conditions and locally stationary or stationary random functions. The first two of these conditions are too weak to determine the asymptotical behavior of minimal errors. It is unknown whether sharp bounds can be derived for stationary random functions.

Optimality of quadrature formulas or linear methods for approximation refers to the selection of coefficients and knots. The main problem is the choice of the knots. We need not completely know the covariance kernel and mean of P in order to find good knots. Instead, it suffices to know the local smoothness of the random function. We discuss this issue in Section 6, which is also a summary of main results from this chapter. In Section 4 we demonstrate that partial knowledge concerning the mean and covariance kernel also allows us to find good coefficients.

1. Sacks-Ylvisaker Regularity Conditions

In a series of papers, Sacks and Ylvisaker (1966, 1968, 1970a, 1970b) introduce regularity conditions for covariance kernels K and study the integration problem Int_ϱ. We show that these conditions determine the reproducing kernel Hilbert space $H(K)$ uniquely up to a finite dimensional subspace of polynomials. We give estimates for the norm on $H(K)$ and for the eigenvalues of the integral operator with kernel K. Further analysis of the integration and approximation problems is based on these conclusions.

1.1. Definition and Examples. We denote one-sided limits at the diagonal in $[0, 1]^2$ in the following way. Let

$$\Omega_+ = \{(s,t) \in {]0, 1[}^2 : s > t\}, \qquad \Omega_- = \{(s,t) \in {]0, 1[}^2 : s < t\},$$

and let cl A denote the closure of a set A. Suppose that L is a continuous function on $\Omega_+ \cup \Omega_-$ such that $L|_{\Omega_j}$ is continuously extendible to cl Ω_j for $j \in \{+, -\}$. By L_j we denote the extension of L to $[0, 1]^2$ that is continuous on cl Ω_j and on $[0, 1]^2 \setminus$ cl Ω_j.

A covariance kernel K on $[0, 1]^2$ satisfies *Sacks-Ylvisaker conditions* of order $r \in \mathbb{N}_0$ if the following conditions hold.

(A) $K \in C^{r,r}([0, 1]^2)$, the partial derivatives of $L = K^{(r,r)}$ up to order two are continuous on $\Omega_+ \cup \Omega_-$ and continuously extendible to cl Ω_+ as well as to cl Ω_-.

(B)

$$L_-^{(1,0)}(s, s) - L_+^{(1,0)}(s, s) = 1, \qquad 0 \le s \le 1.$$

(C) $L_+^{(2,0)}(s, \cdot) \in H(L)$ for all $0 \le s \le 1$ and

$$\sup_{0 \le s \le 1} \|L_+^{(2,0)}(s, \cdot)\|_L < \infty.$$

(D)

$$K^{(r,k)}(\cdot, 0) = 0, \qquad k = 0, \ldots, r - 1.$$

We also say that a zero mean measure P on $F \subset C^r([0, 1])$ satisfies Sacks-Ylvisaker conditions of order r if its covariance kernel has properties (A), (B), (C), and (D). Note that (D) is void if $r = 0$.

REMARK 1. The diagonal is distinguished in $[0, 1]^2$ because continuity or differentiability of K on $[0, 1]^2$ already follows from continuity or differentiability at all points (s, s) with $0 \le s \le 1$. We also note that $L = K^{(r,r)}$ is the covariance kernel of the image of P under $f \mapsto f^{(r)}$. See Section III.1.3.

Property (A) yields

$$(1) \qquad \int_F (f^{(r)}(s) - f^{(r)}(t))^2 \, dP(f) = L(s, s) - 2 L(s, t) + L(t, t)$$

$$\le 2 \cdot \sup_{s,t \in \Omega_+ \cup \Omega_-} |L^{(1,0)}(s,t)| \cdot |s - t|.$$

Hence the random function $f^{(r)}$ is Hölder continuous with exponent $\frac{1}{2}$ in quadratic mean.

If K is any nonnegative definite function on $[0, 1]^2$ that satisfies Sacks-Ylvisaker conditions, then (1) implies that a Gaussian measure on $C^r([0, 1])$ with covariance kernel K exists, see Adler (1981, Theorem 3.4.1). Hence it is sufficient to find such functions K in order to have examples of measures that satisfy Sacks-Ylvisaker conditions.

Moreover, we basically have to deal with the case of order zero. Namely, let $\theta_0, \ldots, \theta_{r-1} \in \mathbb{R}$ where $r \geq 1$. Define

$$(2) \qquad (V_r g)(t) = \sum_{k=0}^{r-1} \frac{\theta_k}{k!} \cdot t^k + (T_r g)(t)$$

for $g \in C([0,1])$ where T_r is the operator of r-fold integration,

$$(3) \qquad (T_r g)(t) = \int_0^t g(u) \cdot \frac{(t-u)^{r-1}}{(r-1)!} \, du.$$

Suppose that g is a random function whose covariance kernel satisfies Sacks-Ylvisaker conditions of order zero and $\theta_0, \ldots, \theta_{r-1}$ are zero mean random variables with finite second moments. Then V_r defines a random function whose covariance kernel obviously satisfies conditions (A), (B), and (C) with the chosen r. Moreover, every such instance arises this way.

Condition (D) states that $f^{(r)}(t)$ and $f^{(k)}(0)$ are uncorrelated for $0 \leq t \leq 1$ and $k = 0, \ldots, r-1$ since

$$K^{(r,k)}(t,0) = \int_F f^{(r)}(t) \cdot f^{(k)}(0) \, dP(f),$$

see (III.7). Thus we get examples for Sacks-Ylvisaker conditions of order $r \geq 1$ by r-fold integration with uncorrelated boundary conditions.

In particular, (D) holds if $f^{(k)}(0) = 0$ for $k = 0, \ldots, r-1$ and almost every f. Equivalently, $K^{(0,k)}(\cdot, 0) = 0$. We add that (D) is not needed for some of the results that are presented in the sequel.

REMARK 2. The smoothness of a random function f is restricted by (B), which implies that the derivative $f^{(r+1)}$ does not exist in quadratic mean. Furthermore, by (A) and (B),

$$\lim_{s \to t} \frac{1}{|s-t|} \cdot \int_F (f^{(r)}(s) - f^{(r)}(t))^2 \, dP(f) = 1,$$

i.e., $f^{(r)}$ has the local Hölder constant one for Hölder exponent $\frac{1}{2}$ at every point $t \in [0,1]$.

For $r = 0$, Sacks and Ylvisaker (1966, 1968) and Su and Cambanis (1993) require conditions (A) and (C) together with

(B') *The function*

$$\alpha(s) = L_-^{(1,0)}(s,s) - L_+^{(1,0)}(s,s)$$

is continuous, nonnegative, and not identically zero on $[0,1]$.

We show that (B') is not more general than (B), up to normalization.

By (A) and (C), the functions $L_+^{(2,0)}$ and $L_+^{(1,1)}$ are continuous on $[0,1]^2$, and

$$\alpha(s+\delta) - \alpha(s)$$
$$= L_-^{(1,0)}(s+\delta, s+\delta) - L_-^{(1,0)}(s,s) - L_+^{(1,0)}(s+\delta, s+\delta) + L_+^{(1,0)}(s,s)$$
$$= \left(L_-^{(2,0)}(\varepsilon_1, s+\delta) + L_-^{(1,1)}(s,\varepsilon_2) - L_+^{(1,1)}(s+\delta, \varepsilon_3) - L_+^{(2,0)}(\varepsilon_4, s)\right) \cdot \delta$$
$$= \left(L_+^{(2,0)}(\varepsilon_1, s+\delta) - L_+^{(2,0)}(\varepsilon_4, s) + L_+^{(1,1)}(s,\varepsilon_2) - L_+^{(1,1)}(s+\delta, \varepsilon_3)\right) \cdot \delta,$$

where $0 < s < \varepsilon_k < s+\delta < 1$. Hence $\alpha'(s) = 0$ and α is constant.

For convenience we use the normalization $\alpha = 1$ in these notes. A different value $\alpha > 0$ yields additional multiplicative constants $\alpha^{1/2}$ in results for integration and approximation.

Functions α are also permitted by the following authors. Müller-Gronbach (1996) considers generalized conditions (A) and (B') together with (C) in the case $r = 0$. Benhenni and Cambanis (1992a, 1992b) consider (A), (B), and a modified condition (C) with regularity $r \geq 0$. However, examples with nonconstant functions α seem to be unknown.

A random function is called *(wide sense) stationary* if its covariances $K(s,t)$ depend only on $|s-t|$, see Adler (1981). If K satisfies (A) and L corresponds to a stationary random function, then the difference α is a nonnegative constant. This follows from

$$\alpha(s) = \lim_{\delta \to 0+} \frac{L(s,s) - L(s-\delta, s)}{\delta} - \lim_{\delta \to 0+} \frac{L(s+\delta, s) - L(s,s)}{\delta}$$
$$= \lim_{\delta \to 0+} \frac{2 \cdot (L(0,0) - L(\delta, 0))}{\delta}$$

if $0 < s < 1$, together with the Cauchy-Schwarz inequality $L(s,t)^2 \leq L(s,s) \cdot L(t,t)$. The latter holds for every covariance kernel.

Properties (A), (B), and (C) are further discussed in Sacks and Ylvisaker (1966).

EXAMPLE 3. Various examples of kernels satisfying Sacks-Ylvisaker conditions are known. For $r = 0$, these conditions are satisfied in particular if K is any of the following kernels,

$$Q_0(s,t) = \min(s,t),$$
$$P_0(s,t) = \min(s,t) - st,$$
(4) $$R_0(s,t) = 1 + \min(s,t),$$
$$K_1(s,t) = \tfrac{1}{2}(1 - |s-t|),$$
$$K_2(s,t) = \tfrac{1}{2}\exp(-|s-t|).$$

Zero mean Gaussian measures P on classes $F \subset C([0,1])$ with these covariance kernels are given as follows. For Q_0 we get the Wiener measure w on the class $\{f \in C([0,1]) : f(0) = 0\}$, see Example II.1. The zero mean Gaussian measure on $\{f \in C([0,1]) : f(0) = f(1) = 0\}$ with kernel P_0 corresponds to the

Brownian bridge, see Section II.3.7. The kernels R_0 and K_1 correspond to the sum $(f_1(t)+f_2(1-t))\cdot 2^{-1/2}$ where f_1 and f_2 are independent and f_1 is distributed according to w. For R_0 the function f_2 is constant and normally distributed with variance one, while for K_1 the function f_2 is also distributed according to w. See Section II.3.7. The kernel K_2 corresponds to an *Ornstein-Uhlenbeck process*, see Lifshits (1995, p. 39). For R_0, K_1, or K_2 we take $F = C([0,1])$. More complicated examples are given in Sacks and Ylvisaker (1966) and Benhenni and Cambanis (1992a, 1992b).

The Hilbert spaces $H(K)$ are known in the examples above. We have $H(K_i) = W_2^1([0,1])$ with the norms

$$\|h\|_{K_1}^2 = \|h'\|_2^2 + (h(0) + h(1))^2$$

and

$$\|h\|_{K_2}^2 = \|h'\|_2^2 + \|h\|_2^2 + h(0)^2 + h(1)^2,$$

see Mitchell, Morris, and Ylvisaker (1990). Here $W_2^1([0,1])$ denotes the Sobolev space of functions with square integrable derivative on $[0,1]$, see Example III.5. The Hilbert spaces corresponding to P_0, Q_0, and R_0 are discussed in the sequel.

Conditions (A) and (B), with $r = 0$, obviously hold for the kernels above. Condition (C) holds, since $K_+^{(2,0)} = K$ for $K = K_2$ and $K_+^{(2,0)} = 0$ in the other cases. Here (C) can be verified without knowing $H(K)$ and its norm explicitly.

Now we consider functions of higher regularity. Taking $\theta_0 = \cdots = \theta_{r-1} = 0$ in (2) and considering the Wiener measure $w = w_0$, we obtain the *r-fold integrated Wiener measure* $w_r = T_r w_0$ on the class

$$F = \{f \in C^r([0,1]) : f^{(k)}(0) = 0 \text{ for } k = 0,\ldots,r\}.$$

Clearly w_r is a zero mean Gaussian measure, and its covariance kernel Q_r is given by

$$Q_r(s,t) = \int_0^1 \frac{(s-u)_+^r \cdot (t-u)_+^r}{(r!)^2} \, du,$$

where $z_+ = z$ if $z > 0$ and $z_+ = 0$ otherwise. Let

$$W_2^{r+1}([0,1]) = \{h \in C^r([0,1]) : h^{(r)} \text{ absolutely continuous}, h^{(r+1)} \in L_2([0,1])\}.$$

The Hilbert space with reproducing kernel Q_r is given by

$$H(Q_r) = \{h \in W_2^{r+1}([0,1]) : h^{(k)}(0) = 0 \text{ for } k = 0,\ldots,r\},$$

and

$$\|h\|_{Q_r} = \|h^{(r+1)}\|_2.$$

Suppose that g is a random function with covariance kernel R_0. Moreover, let $g, \theta_0, \ldots, \theta_{r-1}$ be uncorrelated, with θ_k having variance $\sigma^2 > 0$. Then (2) yields a measure on $C^r([0,1])$ with covariance kernel

$$R_r(s,t) = \sigma^2 \cdot \sum_{k=0}^r \frac{s^k \cdot t^k}{(k!)^2} + Q_r(s,t).$$

The reproducing kernel Hilbert space is the *Sobolev space*

$$H(R_r) = W_2^{r+1}([0,1])$$

equipped with the norm

$$\|h\|_{R_r}^2 = \frac{1}{\sigma^2} \cdot \sum_{k=0}^{r} h^{(k)}(0)^2 + \|h^{(r+1)}\|_2^2.$$

The limiting case $\sigma^2 \to \infty$ yields an improper prior. See Wahba (1978a, 1990).

Observe that the kernels K_1 and K_2 correspond to stationary random functions, since $K_i(s,t)$ depends only on $|s-t|$. Mitchell *et al.* (1990) and Lasinger (1993) study whether stationarity can be preserved by r-fold integration using suitable stochastic boundary conditions. They give a positive answer in particular for the kernels K_1 and K_2. See Remark 9 for a further discussion.

The importance of splines in $H(K)$ is demonstrated in Section III.3. The kernels Q_r, R_r, and P_0 yield polynomial splines of degree $2r+1$, and $K = K_2$ yields exponential splines.

1.2. Reproducing Kernel Hilbert Spaces. In the following we study the reproducing kernel Hilbert spaces under Sacks-Ylvisaker conditions. Let P_r denote the kernel that corresponds to the closed subspace of all functions from $H(R_r) = W_2^{r+1}([0,1])$ vanishing at the boundary, i.e.,

$$H(P_r) = \{h \in H(R_r) : h^{(k)}(0) = h^{(k)}(1) = 0 \text{ for } k = 0,\ldots,r\}$$

with the norm

$$\|h\|_{P_r} = \|h^{(r+1)}\|_2.$$

We note that

(5) $$P_r(s,t) = Q_r(s,t) - \gamma(s)' \cdot \Gamma^{-1} \cdot \gamma(t)$$

where

$$\gamma(u) = (Q_r^{(0,k)}(u,1))_{0 \le k \le r} \quad \text{and} \quad \Gamma = (Q_r^{(k,j)}(1,1))_{0 \le k,j \le r},$$

see Lemmata III.2 and III.30. In particular, P_0 is given as in (4). Although P_r does not satisfy condition (C) if $r \ge 1$, this kernel plays an important role in the analysis of reproducing kernel Hilbert spaces under Sacks-Ylvisaker conditions. Furthermore, results for integration and approximation that hold under Sacks-Ylvisaker conditions are valid for P_r as well.

The spaces $H(P_r)$, $H(Q_r)$ and $H(R_r)$ differ only by some polynomials. That is,

$$H(P_r) \oplus \mathbb{P}_{2r+1} = H(R_r) \quad \text{and} \quad H(Q_r) \oplus \mathbb{P}_r = H(R_r),$$

where $\mathbb{P}_i$ is the space of polynomials of degree at most i. We show that the Hilbert space $H(K)$ with K satisfying Sacks-Ylvisaker conditions of order r is closely related to the spaces $H(P_r)$ and $H(R_r)$ and may also differ from the space $H(R_r)$ only by some polynomials. Moreover, we compare the (semi)norms $\|h\|_K$ and $\|h^{(r+1)}\|_2$ for functions h that vanish at the knots

$$0 \le x_1 < \cdots < x_n \le 1.$$

Put

$$\delta = \max_{i=1,\ldots,n+1} (x_i - x_{i-1})$$

where $x_0 = 0$ and $x_{n+1} = 1$. The following estimate can be verified by induction.

LEMMA 4. *If* $n \ge r+1$ *and* $h \in W_2^{r+1}([0,1])$ *with* $h(x_i) = 0$ *for* $i = 1,\ldots,n$, *then*

$$\|h^{(k)}\|_2 \le \frac{(r+1)!}{k!} \cdot \delta^{r-k+1} \cdot \|h^{(r+1)}\|_2, \qquad k = 0,\ldots,r.$$

LEMMA 5. *Let* K *satisfy Sacks-Ylvisaker conditions of order* $r = 0$. *Then*

$$H(P_0) = \{h \in H(K) : h(0) = h(1) = 0\}$$

and

$$H(K) \subset H(R_0)$$

as a closed subspace. Moreover, there exists a positive constant c, *depending only on* K, *with the following property. For every* $h \in H(K)$ *such that* $h(x_i) = 0$ *for* $i = 1,\ldots,n$, *we have*

$$\left(\int_{x_1}^{x_n} h'(s)^2 \, ds \right)^{1/2} \le (1 + c\,\delta) \cdot \|h\|_K.$$

If $x_1 = 0$ *and* $x_n = 1$, *then also*

$$(1 + c\,\delta)^{-1} \cdot \|h\|_K \le \|h'\|_2.$$

PROOF. Put $\widehat{D} = [x_1, x_n]$ and let $\widehat{K}$ denote the restriction of K to $\widehat{D} \times \widehat{D}$. Moreover, let

$$h = \sum_{j=1}^m b_j K(\cdot, t_j)$$

with distinct $t_j \in \widehat{D}$ such that $h(x_i) = 0$ for $i = 1,\ldots,n$. Put $\widehat{h} = h|_{\widehat{D}}$. In the sequel we use Lemma III.4 concerning reproducing kernel Hilbert spaces and restrictions.

Condition (A) implies $h \in W_2^1([0,1])$, and integration by parts yields

$$\int_{x_1}^{x_n} \widehat{h}'(s)^2 \, ds = \sum_{j,k=1}^m b_j b_k \cdot \Big(K(x_n, t_j)\, K_+^{(1,0)}(x_n, t_k) - K(x_1, t_j)\, K_-^{(1,0)}(x_1, t_k)$$

$$+ K(t_j, t_k)\big(K_-^{(1,0)}(t_k, t_k) - K_+^{(1,0)}(t_k, t_k) \big)$$

$$- \int_{x_1}^{x_n} K(s, t_j)\, K_+^{(2,0)}(s, t_k) \, ds \Big).$$

Observe that $0 = \widehat{h}(x_1) = \widehat{h}(x_n) = \sum_{j=1}^{m} b_j \cdot K(x_1, t_j) = \sum_{j=1}^{m} b_j \cdot K(x_n, t_j)$. Using (B), we have

$$\sum_{j,k=1}^{m} b_j b_k \cdot K(t_j, t_k) \left(K_-^{(1,0)}(t_k, t_k) - K_+^{(1,0)}(t_k, t_k) \right)$$

$$= \sum_{j,k=1}^{m} b_j b_k \cdot K(t_j, t_k) = \|\widehat{h}\|_{\widehat{R}}^2.$$

Thus

$$\|\widehat{h}\|_{\widehat{R}}^2 - \int_{x_1}^{x_n} \widehat{h}'(s)^2 \, ds = c_1,$$

where

$$c_1 = \sum_{j,k=1}^{m} b_j b_k \cdot \int_{x_1}^{x_n} K(s, t_j) K_+^{(2,0)}(s, t_k) \, ds = \sum_{k=1}^{m} b_k \cdot \int_{x_1}^{x_n} \widehat{h}(s) K_+^{(2,0)}(s, t_k) \, ds.$$

For $x_1 \leq s \leq x_n$ let $\widehat{g}_s$ denote the restriction of $K_+^{(2,0)}(s, \cdot)$ to $[x_1, x_n]$. From (C) we get $\widehat{g}_s \in H(\widehat{K})$. Hence

$$\sum_{k=1}^{m} b_k \cdot K_+^{(2,0)}(s, t_k) = \sum_{k=1}^{m} b_k \cdot \langle \widehat{g}_s, \widehat{K}(\cdot, t_k) \rangle_{\widehat{R}} = \langle \widehat{g}_s, \widehat{h} \rangle_{\widehat{R}}$$

and

$$|c_1| = \left| \int_{x_1}^{x_n} \widehat{h}(s) \langle \widehat{g}_s, \widehat{h} \rangle_{\widehat{R}} \, ds \right|$$

$$\leq c \cdot \|\widehat{h}\|_{\widehat{R}} \cdot \int_{x_1}^{x_n} |\widehat{h}(s)| \, ds \leq c \delta \cdot \|\widehat{h}\|_{\widehat{R}} \cdot \left(\int_{x_1}^{x_n} \widehat{h}'(s)^2 \, ds \right)^{1/2},$$

with

$$c = \sup_{x_1 \leq s \leq x_n} \|\widehat{g}_s\|_{\widehat{R}} \leq \sup_{0 \leq s \leq 1} \|K_+^{(2,0)}(s, \cdot)\|_K < \infty,$$

see (C) and Lemma 4. Therefore we obtain

$$\left| \|\widehat{h}\|_{\widehat{R}}^2 - \int_{x_1}^{x_n} \widehat{h}'(s)^2 \, ds \right| \leq c \delta \cdot \|\widehat{h}\|_{\widehat{R}} \cdot \left(\int_{x_1}^{x_n} \widehat{h}'(s)^2 \, ds \right)^{1/2}.$$

This implies

$$(6) \qquad (1 + c \delta)^{-1} \cdot \|\widehat{h}\|_{\widehat{R}} \leq \left(\int_{x_1}^{x_n} \widehat{h}'(s)^2 \, ds \right)^{1/2} \leq (1 + c \delta) \cdot \|\widehat{h}\|_{\widehat{R}}.$$

Let M denote the reproducing kernel of the closed subspace

$$\{ h \in H(K) : h(x_i) = 0 \text{ for } i = 1, \dots, n \} = \text{span}\{ K(\cdot, x_i) : i = 1, \dots, n \}^\perp$$

of $H(K)$, and let $\widehat{M}$ denote the restriction of M to $\widehat{D} \times \widehat{D}$. We conclude that (6) holds for every $\widehat{h} \in H(\widehat{M})$, since $H(\widehat{M}) \cap \text{span}\{ \widehat{K}(\cdot, t) : t \in \widehat{D} \}$ is dense in $H(\widehat{M})$. In particular, $H(\widehat{M}) \subset W_2^1(\widehat{D})$. Since $\|h|_{[x_1, x_n]}\|_{\widehat{R}} \leq \|h\|_K$, every function $h \in H(M)$ satisfies the first estimate from Lemma 5. Furthermore, the second estimate follows if $x_1 = 0$ and $x_n = 1$.

In the sequel we only use the previous results with $x_1 = 0$ and $x_2 = 1$. By (6),

$$H(M) = \{h \in H(K) : h(0) = h(1) = 0\}$$

is a closed subspace of $H(P_0)$. It remains to show $H(K) \subset H(R_0)$ as a closed subspace and $H(M) = H(P_0)$.

To prove that $H(K) \subset H(R_0)$, we proceed as follows. Consider the subspace

$$H(M)^{\perp} = \mathrm{span}\{K(\cdot, 0), K(\cdot, 1)\}$$

of $H(K)$. Its reproducing kernel $K - M$ is given as a linear combination of

$$K(s, 0)K(t, 0), \ K(s, 0)K(t, 1), \ K(s, 1)K(t, 0), \ K(s, 1)K(t, 1),$$

see Lemma III.30. From (A),

$$H(M)^{\perp} \subset C^2([0, 1]) \subset H(R_0)$$

and

(7) $$K - M \in C^{2,2}([0, 1]^2).$$

Hence $H(K) = H(M) \oplus H(M)^{\perp} \subset H(R_0)$, and $H(K)$ is closed in $H(R_0)$ as the sum of the finite dimensional space $H(M)^{\perp}$ and $H(M)$, which is already known to be closed. We add that M satisfies (A) and (B) with $r = 0$.

Consider the integral operator $T : C([0, 1]) \to H(M)$ given by

$$(T\varphi)(t) = \int_0^1 M(s, t)\, \varphi(s)\, ds,$$

and let

$$g(t) = \int_0^1 K(s, t)\, \varphi(s)\, ds = (T\varphi)(t) + \int_0^1 (K - M)(s, t)\, \varphi(s)\, ds.$$

If $T\varphi = 0$ then $g \in \mathrm{span}\{K(\cdot, 0), K(\cdot, 1)\}$. Theorem 3.1 of Sacks and Ylvisaker (1966), applied to the kernel K, yields $\varphi = 0$. Hence the operator T is one-to-one.

To prove that $H(M) = H(P_0)$, it is enough to show that the only $h \in H(P_0)$ with $\langle h, T\varphi \rangle_{R_0} = 0$ for all $\varphi \in C([0, 1])$ is $h = 0$. Observe that

(8) $$(T\varphi)''(t) = -\varphi(t) + \int_0^1 N(s, t)\, \varphi(s)\, ds = (U\varphi)(t) - \varphi(t)$$

with $N = M_+^{(0,2)}$ and

$$(U\varphi)(t) = \int_0^1 N(s, t)\, \varphi(s)\, ds.$$

Moreover, (A) and (C) imply $K_+^{(0,2)}(t, t) = K_-^{(0,2)}(t, t)$ for every $0 \le t \le 1$, and therefore $K_+^{(0,2)} \in C([0, 1]^2)$ by (A). See Sacks and Ylvisaker (1966, p. 75) for these facts. Using (7) we obtain $N \in C([0, 1]^2)$, and therefore U is a compact operator on $C([0,1])$. Since T is injective and $T\varphi$ vanishes at zero and one, $(T\varphi)'' = 0$ implies $\varphi = 0$. Thus, 1 is not an eigenvalue of U.

Assume now that $h \in H(P_0)$ with $\langle h, T\varphi \rangle_{R_0} = 0$ for all $\varphi \in C([0,1])$. The Fredholm alternative and (8) imply the existence of φ such that $(T\varphi)'' = -h$. Since

$$0 = \int_0^1 h'(t)(T\varphi)'(t)\,dt = -\int_0^1 h(t)(T\varphi)''(t)\,dt = \|h\|_2^2$$

we have $h = 0$, which completes the proof of $H(M) = H(P_0)$. $\square$

Now we extend Lemma 5 to the case of regularity $r > 0$. Recall that $\mathbb{P}_i$ denotes the space of polynomials of degree at most i, and put $\mathbb{P}_{-1} = \{0\}$.

PROPOSITION 6. *Let K satisfy conditions (A), (B), and (C) with $r \in \mathbb{N}_0$. Then*

$$H(P_r) \subset H(K) + \mathbb{P}_{r-1},$$

and

$$H(K) \subset H(R_r)$$

as a closed subspace. Moreover, there exists a positive constant c, depending only on K, with the following property. For every $h \in H(K)$ and $n > 2r + 1$ such that $h(x_i) = 0$ for $i = 1, \dots, n$, we have

$$\left(\int_{x_{r+1}}^{x_{n-r}} h^{(r+1)}(s)^2\,ds\right)^{1/2} \leq (1 + c\,\delta)\cdot\|h\|_K.$$

PROOF. Because of Lemma 5, it is enough to consider the case $r > 0$. Since $K \in C^{r,r}([0,1]^2)$, we have $H(K) \subset C^r([0,1])$ and $h^{(k)}(t) = \langle h, K^{(0,k)}(\cdot, t)\rangle_K$ for every $h \in H(K)$, $k = 0, \dots, r$, and $0 \leq t \leq 1$, see Proposition III.11. Clearly $L = K^{(r,r)}$ satisfies Sacks-Ylvisaker conditions of order $r = 0$. Define the kernel $V(s,t) = h^{(r)}(s)\cdot h^{(r)}(t)$ with $h \in H(K)$. Then

$$\sum_{i,j=1}^{k} b_i b_j \cdot V(t_i, t_j) = \left\langle h, \sum_{i=1}^{k} b_i \cdot K^{(0,r)}(\cdot, t_i)\right\rangle_K^2 \leq \|h\|_K^2 \cdot \sum_{i,j=1}^{k} b_i b_j \cdot L(t_i, t_j)$$

for all $b_i \in \mathbb{R}$ and $0 \leq t_i \leq 1$. This implies $H(V) \subset H(L)$, see Lemma III.3. Since $H(V) = \operatorname{span}\{h^{(r)}\}$ we have $h^{(r)} \in H(L)$, and

$$Uh = h^{(r)}$$

defines a linear operator from $H(K)$ to $H(L)$. The continuity of this mapping follows from the closed graph theorem. By Lemma 5, $H(L) \subset H(R_0)$ and therefore

$$H(K) \subset H(R_r).$$

Let $U^* : H(L) \to H(K)$ denote the adjoint of U. We show that

$$U(U^* g) = g$$

for every $g \in H(L)$. It is enough to verify this relation for $g = L(\cdot, t)$ with $0 \le t \le 1$. For $h \in H(K)$ we have

$$\langle h, U^* L(\cdot, t) \rangle_K = \langle h^{(r)}, L(\cdot, t) \rangle_L = h^{(r)}(t) = \langle h, K^{(0,r)}(\cdot, t) \rangle_K.$$

Hence,

$$(9) \qquad\qquad U^* L(\cdot, t) = K^{(0,r)}(\cdot, t)$$

and $U(U^* L(\cdot, t)) = K^{(r,r)}(\cdot, t) = L(\cdot, t)$, as claimed. This yields that U is surjective, and therefore Lemma 5 gives

$$H(P_0) \subset H(L) = \{h^{(r)} : h \in H(K)\}.$$

If $g \in H(P_r)$ then $g^{(r)} \in H(P_0)$ and there exists a function $h \in H(K)$ such that $h^{(r)} = g^{(r)}$. This implies $g = h + p$ for some polynomial p of degree at most $r - 1$, i.e.,

$$H(P_r) \subset H(K) + \mathbb{P}_{r-1}.$$

Clearly $\ker U \subset \mathbb{P}_{r-1}$, and

$$\operatorname{ran} U^* = \{h \in H(K) : h = U^* U h\}$$

is easy to verify. Hence, $\operatorname{ran} U^*$ is closed in $H(K)$ and

$$H(K) = \operatorname{ran} U^* \oplus \ker U.$$

To conclude that $H(K)$ is closed in $H(R_r)$ it remains to show that $\operatorname{ran} U^*$ is also closed in $H(R_r)$. Let $h_n \in \operatorname{ran} U^*$ and $h \in H(R_r)$ with $\lim_{n\to\infty} \|h_n - h\|_{R_r} = 0$. Then $h_n^{(r)} \in H(L)$ and $\lim_{n\to\infty} \|h_n^{(r)} - h^{(r)}\|_{R_0} = 0$. Since $H(L)$ is closed in $H(R_0)$ by Lemma 5, we get $h^{(r)} \in H(L)$, and the equivalence of the norms $\|\cdot\|_L$ and $\|\cdot\|_{R_0}$ on $H(L)$ yields $\lim_{n\to\infty} \|h_n^{(r)} - h^{(r)}\|_L = 0$. The continuity of U^* and of the embedding $H(K) \hookrightarrow H(R_r)$, together with $U^* h_n^{(r)} = h_n$, imply that $\lim_{n\to\infty} \|h_n - U^* h^{(r)}\|_{R_r} = 0$. Therefore $h = U^* h^{(r)} \in \operatorname{ran} U^*$.

It remains to establish the norm estimate. If h_1 denotes the orthogonal projection of $h \in H(K)$ onto $\operatorname{ran} U^*$ then

$$\|Uh\|_L^2 = \|U h_1\|_L^2 = \langle U^* U h_1, h_1 \rangle_K = \|h_1\|_K^2 \le \|h\|_K^2.$$

Suppose that $h(x_i) = 0$ for $i = 1, \dots, n$. Then $h^{(r)} \in H(L)$ has zeros $0 < z_1 < \cdots < z_{n-r} < 1$, with $z_1 \le x_{r+1}$ and $z_{n-r} \ge x_{n-r}$ as well as $\max_i (z_i - z_{i-1}) \le (r+1)\delta$. Here $z_0 = 0$ and $z_{n-r+1} = 1$. We get

$$\left(\int_{x_{r+1}}^{x_{n-r}} h^{(r+1)}(s)^2 \, ds \right)^{1/2} \le \left(\int_{z_1}^{z_{n-r}} h^{(r+1)}(s)^2 \, ds \right)^{1/2}$$

$$\le (1 + c(r+1)\delta) \cdot \|h^{(r)}\|_L \le (1 + c(r+1)\delta) \cdot \|h\|_K$$

from Lemma 5. $\qquad\qquad\qquad\qquad\qquad\qquad\qquad\qquad\qquad\qquad\qquad\square$

Suppose that conditions (A), (B), and (C) hold with $r \in \mathbb{N}_0$. If $r = 0$ then the Hilbert space $H(K)$ lies between $H(P_0)$ and $H(R_0)$, see Lemma 5. For $r \geq 1$ Proposition 6 yields the analogous inclusions for the sum $H(K) + \mathbb{P}_{r-1}$. We give an example that demonstrates that $\mathbb{P}_{r-1}$ is essential and $H(P_r)$ does not have to be a subset of $H(K)$ alone. This fact is relevant in the analysis of the integration problem.

EXAMPLE 7. Consider an arbitrary measure P on $C^r([0,1])$ whose kernel M satisfies Sacks-Ylvisaker conditions with regularity $r \geq 1$. Define

$$g(t) = f(t) - \int_0^1 f(u)\, du$$

for $f \in C^r([0,1])$, and consider the image of P under this transformation. Its covariance kernel K equals

$$K(s,t) = M(s,t) - \int_0^1 M(s,u)\, du - \int_0^1 M(t,u)\, du + \int_0^1 \int_0^1 M(u,v)\, du\, dv$$

and hence $K^{(r,r)}(s,t) = M^{(r,r)}(s,t)$. Thus, K satisfies the conditions (A), (B), and (C) with the given r.

Observe that $\int_0^1 K(s,t)\, dt = 0$ for all $0 \leq s \leq 1$. Hence, $H(K)$ consists of functions with zero integral. Since this is not true for $H(P_r)$, we have

$$H(P_r) \not\subset H(K).$$

This illustrates that, in general, the presence of $\mathbb{P}_{r-1}$ in Proposition 6 is necessary.

To guarantee that $H(P_r)$ is a subset of $H(K)$ we have introduced condition (D) in the case $r \geq 1$.

PROPOSITION 8. *Let K satisfy Sacks-Ylvisaker conditions of order $r \in \mathbb{N}_0$. Then*

$$H(P_r) \subset H(K) \subset H(R_r).$$

Moreover, there exists a positive constant c, depending only on K, with the following property. For all $h \in H(P_r)$ and $n \geq r$ such that $h(x_i) = 0$ for $i = 1, \ldots, n$, we have

$$(1 + c\,\delta)^{-1} \cdot \|h\|_K \leq \|h^{(r+1)}\|_2.$$

PROOF. Note that $H(K) \subset H(R_r)$ by Proposition 6. Let $L = K^{(r,r)}$. From (9) and (D) we get

$$(U^* L(\cdot, t))^{(k)}(0) = K^{(r,k)}(t,0) = 0$$

for $k = 0, \ldots, r-1$. Thus U^* is given as r-fold integration, i.e., $U^* = T_r$, see (3). Let $h \in H(P_r)$. Lemma 5 implies $h^{(r)} \in H(L)$, and therefore $h = U^* h^{(r)} \in H(K)$. Furthermore, $\|h\|_K = \|h^{(r)}\|_L$. The norm estimate now follows as in the proof of Proposition 6, using the second estimate from Lemma 5.   □

REMARK 9. Suppose that K corresponds to a stationary random function and satisfies (A), (B), and (C) for some $r \geq 1$. Then (D) cannot hold, since stationarity together with (D) implies that K is constant. However, from Lasinger (1993, Theorems 2, 4) we conclude that $\mathbb{P}_{r-1} \subset H(K)$ if the variances $K^{(k,k)}(0,0)$ are sufficiently large. Consequently, $H(P_r) \subset H(K) \subset H(R_r)$. See Lasinger (1993) and Wahba (1990, p. 22) for examples that also fulfill the norm estimate from Proposition 8.

1.3. Decay of Eigenvalues. A continuous covariance kernel K on $[0,1]^2$ defines a compact and symmetric operator

$$(K\eta)(s) = \int_0^1 K(s,t) \cdot \eta(t) \, dt$$

on $L_2([0,1])$. As in the previous chapter, we use K to denote the kernel as well as the integral operator. Proposition III.24 demonstrates the role of the nonzero eigenvalues $\mu_1 \geq \mu_2 \geq \cdots > 0$ of K for L_2-approximation.

From Example III.28 we know that

$$\mu_n = \frac{1}{\left(\left(n - \frac{1}{2}\right)\pi\right)^2}$$

with eigenfunctions $\xi_n(t) = \sqrt{2}\sin\left(\left(n - \frac{1}{2}\right)\pi \cdot t\right)$ for the kernel $K = Q_0$. The eigenvalues corresponding to the kernels K_1 and K_2 in (4) are known implicitly, see Su and Cambanis (1993) for details.

We show that Sacks-Ylvisaker conditions determine the eigenvalues, up to strong asymptotic equivalence.

PROPOSITION 10. *Let K satisfy Sacks-Ylvisaker conditions of order $r \in \mathbb{N}_0$. Then*

$$\mu_n \approx (\pi n)^{-(2r+2)}.$$

PROOF. According to the minimax property of the eigenvalues,

$$\mu_{n+1}^{1/2} = \inf_{h_1,\ldots,h_n \in H(K)} \sup\{\|h\|_2 : h \in B(K), \, \langle h, h_i \rangle_K = 0 \text{ for } i = 1,\ldots,n\}.$$

We show that $\mu_{n+1}^{1/2}$ coincides with the Kolmogorov n-width

$$d_n(B(K), L_2([0,1])) = \inf_{U_n} \sup_{h \in B(K)} \inf_{g \in U_n} \|h - g\|_2$$

of $B(K)$ in the space $L_2([0,1])$. Here U_n varies over all n-dimensional subspaces of $L_2([0,1])$.

The embedding $H(K) \hookrightarrow L_2([0,1])$ is continuous and $L_2([0,1])$ is a Hilbert space. Therefore $\mu_{n+1}^{1/2} \leq d_n(B(K), L_2([0,1]))$. On the other hand, let $\xi_1, \xi_2, \ldots$ denote an orthonormal system of eigenfunctions corresponding to the eigenvalues $\mu_1 \geq \mu_2 \geq \ldots$. From Lemma III.22 we know that the functions $\mu_i^{1/2} \cdot \xi_i$ form an orthonormal basis of $H(K)$ and $\langle h, \xi_i \rangle_2 = \mu_i \cdot \langle h, \xi_i \rangle_K$ for $h \in H(K)$. Thus

$$\left\| h - \sum_{i=1}^{n} \langle h, \xi_i \rangle_2 \cdot \xi_i \right\|_2^2 = \sum_{i=n+1}^{\infty} \mu_i \cdot \langle h, \mu_i^{1/2} \xi_i \rangle_K^2,$$

which implies

$$d_n(B(K), L_2([0,1]))^2 \leq \sup_{h \in B(K)} \sum_{i=n+1}^{\infty} \mu_i \cdot \langle h, \mu_i^{1/2} \xi_i \rangle_K^2 = \mu_{n+1}.$$

In the sequel we write $\mu_i = \mu_i(K)$ to express the dependence of the eigenvalues on the kernel K. At first we study the case $K = P_r$. Let

(10) $$B_r = \{ h \in W_2^{r+1}([0,1]) : \|h^{(r+1)}\|_2 \leq 1 \}$$

denote the unit ball in the Sobolev space $W_2^{r+1}([0,1])$ with respect to the seminorm $\|h^{(r+1)}\|_2$. The theory of n-widths originated with the work of Kolmogorov (1936), who determined the n-widths $d_n(B_r, L_2([0,1]))$ of B_r in $L_2([0,1])$. These widths satisfy

$$d_n(B_r, L_2([0,1])) \approx (\pi n)^{-(r+1)},$$

see also Lorentz, v. Golitschek, and Makovoz (1996, Prop. 13.8.2). Clearly

$$d_n(B_r, L_2([0,1])) \approx d_n(B(P_r), L_2([0,1]))$$

since

$$B(P_r) = \{ h \in B_r : h^{(k)}(0) = h^{(k)}(1) = 0 \text{ for } k = 0, \dots, r \},$$

and therefore $\mu_n(P_r) \approx (\pi n)^{-(2r+2)}$.

Now we turn to the general case. Let

$$X_k = \{ h \in H(P_r) : h(i/k) = 0 \text{ for } i = 1, \dots, k-1 \}$$

and let $\xi_1, \xi_2, \dots$ denote the eigenfunctions corresponding to the eigenvalues $\mu_1(P_r) \geq \mu_2(P_r) \geq \dots$. The minimax property and Propositions 6 and 8 yield

$$\mu_{n+k+2r+1}(K) \leq \sup\{\|h\|_2^2 : h \in B(K) \cap X_k, \ \langle h, \xi_\ell \rangle_{P_r} = 0 \text{ for } \ell = 1, \dots, n-1\}$$
$$\leq (1 + c/k) \cdot \sup\{\|h\|_2^2 : h \in B(P_r) \cap X_k, \ \langle h, \xi_\ell \rangle_{P_r} = 0 \text{ for } \ell = 1, \dots, n-1\}$$
$$\leq (1 + c/k) \cdot \mu_n(P_r).$$

On the other hand

$$\mu_{n+k-1}(P_r) \leq (1 + c/k) \cdot \mu_n(K).$$

We conclude that $\mu_n(P_r) \approx \mu_n(K)$. $\qquad\square$

1.4. Notes and References. 1. Sacks and Ylvisaker (1966, 1968, 1970a, 1970b) have introduced the smoothness conditions (A)–(C) together with the boundary condition $K^{(0,k)}(\cdot, 0) = 0$ for $k = 0, \ldots, r - 1$ instead of (D). These conditions, or slight modifications thereof, are used by several authors. A partial list of papers includes Benhenni and Cambanis (1992a, 1992b), Su and Cambanis (1993), Istas and Laredo (1994, 1997), Müller-Gronbach (1996), Ritter, Wasilkowski, and Woźniakowski (1995), Ritter (1996a), Müller-Gronbach and Ritter (1996).

2. Propositions 6 and 8 are due to Ritter *et al.* (1995), where the inclusions for the set $H(K)$ are derived, and Ritter (1996a), where the norm estimates are established.

3. Proposition 10 improves a result from Ritter *et al.* (1995), where the weak order of eigenvalues is obtained.

2. Integration under Sacks-Ylvisaker Conditions

We study quadrature formulas

$$S_n(f) = \sum_{i=1}^{n} f^{(\alpha_i)}(x_i) \cdot a_i$$

to approximate integrals

$$\mathrm{Int}_\varrho(f) = \int_0^1 f(t) \cdot \varrho(t) \, dt,$$

where

$$\varrho \in C([0,1]), \qquad \varrho > 0,$$

throughout this section. The latter assumptions can be weakened as indicated in Notes and References 2.4.2. We are mainly interested in the case $\alpha_i = 0$, which corresponds to the class $\Lambda = \Lambda^{\mathrm{std}}$ of permissible functionals. We obtain asymptotically optimal formulas S_n under Sacks-Ylvisaker conditions of order $r \in \mathbb{N}_0$. The errors $e_2(S_n, \mathrm{Int}_\varrho, P)$ of such formulas differ from the nth minimal error $e_2(n, \Lambda^{\mathrm{std}}, \mathrm{Int}_\varrho, P)$ by an arbitrarily small multiplicative constant if n is sufficiently large. The corresponding knots are given as a regular sequence, i.e., as quantiles of a suitably chosen density on $[0, 1]$. We also discuss integration using Hermite data.

Consider average errors $e_2(S_n, \mathrm{Int}_\varrho, P)$ that are defined by second moments, and assume that P has mean zero. According to Proposition III.17, average case results for the problem Int_ϱ depend on P only through its covariance kernel K. The average error coincides with the maximal error on the unit ball $B(K)$ in the Hilbert space with reproducing kernel K, i.e.,

$$e_2(S_n, \mathrm{Int}_\varrho, P) = e_{\max}(S_n, \mathrm{Int}_\varrho, B(K)).$$

From Section 1 we know that $B(K)$ is closely related to the unit ball B_r in the Sobolev space $W_2^{r+1}([0,1])$, if K satisfies Sacks-Ylvisaker conditions. Therefore we can use worst case results on B_r to obtain average case results in this case.

EXAMPLE 11. The r-fold integrated Wiener measure w_r serves as a principal example for Sacks-Ylvisaker conditions. The corresponding reproducing kernel Hilbert space $H(Q_r)$ consists of all functions $h \in W_2^{r+1}([0,1])$ that vanish together with all derivatives at $t = 0$. The norm on $H(Q_r)$ is given by $\|h^{(r+1)}\|_2$. See Example 3. Assume $0 < x_1 < \cdots < x_n \le 1$ in the sequel.

According to Proposition III.17, the error of a quadrature formula S_n that uses the values $f(x_i)$ and coefficients a_i satisfies

$$e_2(S_n, \mathrm{Int}_\varrho, w_r)^2 = \|\xi - \sum_{i=1}^n a_i \cdot Q_r(\cdot, x_i)\|_{Q_r}^2$$

$$= \int_0^1 \left(\int_0^1 \frac{(s-t)_+^r}{r!} \cdot \varrho(s)\, ds - \sum_{i=1}^n a_i \cdot \frac{(x_i - t)_+^r}{r!} \right)^2 dt$$

where $\xi(t) = \mathrm{Int}_\varrho(Q_r(\cdot, t))$. In particular,

$$e_2(S_n, \mathrm{Int}, w_r)^2 = \int_0^1 \left(\frac{(1-t)^{r+1}}{(r+1)!} - \sum_{i=1}^n a_i \cdot \frac{(x_i - t)_+^r}{r!} \right)^2 dt$$

for $\varrho = 1$. For $r = 0$ and $\varrho = 1$, explicit minimization of the error is possible and leads to optimal quadrature formulas, see Section II.3.

Minimization of $e_2(S_n, \mathrm{Int}_\varrho, w_r)$ is a spline approximation problem in L_2-norm with free knots x_i. The general solution of this problem, even in the case $\varrho = 1$, is not known, see Traub, Wasilkowski, and Woźniakowski (1988, Section 5.2.2). Therefore asymptotically optimal quadrature formulas are studied.

REMARK 12. More specific assumptions on P are needed to study q-average errors of quadrature formulas with $q \ne 2$. For zero mean Gaussian measures results are basically independent of q for the following reason. Let

$$\nu_q^q = (2\pi)^{-1/2} \int_{\mathbb{R}} |t|^q \cdot \exp\left(-\tfrac{1}{2}t^2\right) dt$$

denote the qth absolute moment of the standard normal distribution, and assume that P is a zero mean Gaussian measure on $C^k(D)$ for any set $D \subset \mathbb{R}^d$. Then

$$e_q(S_n, \mathrm{Int}_\varrho, P) = \nu_q \cdot e_2(S_n, \mathrm{Int}_\varrho, P),$$

since $\mathrm{Int}_\varrho(f) - S_n(f)$ defines a linear functional on $C^k(D)$, which is normally distributed with respect to the Gaussian measure P.

2.1. Regular Sequences. Let $\psi \ge 0$ be integrable on $[0,1]$. A sequence of knots

$$0 = x_{1,n} < \cdots < x_{n,n} = 1$$

is called a *regular sequence* generated by ψ if

$$\int_0^{x_{i,n}} \psi(t)\, dt = \frac{i-1}{n-1} \cdot \int_0^1 \psi(t)\, dt,$$

see Sacks and Ylvisaker (1970a). Thus $x_{1,n}, \ldots, x_{n,n}$ are quantiles of the density ψ.

Clearly $\psi = 1$ gives equidistant knots. The density $\psi(t) = (t\,(1-t))^{-1/2}/\pi$ of the arcsine distribution yields the zeros $x_{2,n}, \ldots, x_{n-1,n}$ of the $(n-2)$nd Chebyshev polynomial of the second kind. The interpolatory quadrature formulas that use this regular sequence form the Clenshaw-Curtis method, see Braß (1977, p. 117). The Gauß quadrature, however, is not based on a regular sequence.

The class of continuous and positive densities is of particular interest. If $\psi \in C([0,1])$ and $\psi > 0$, then the regular sequence is uniquely determined and also quasi-uniform, i.e.,

(11) $$\max_{i=2,\ldots,n} (x_{i,n} - x_{i-1,n}) \asymp \min_{i=2,\ldots,n} (x_{i,n} - x_{i-1,n}) \asymp n^{-1}.$$

Moreover, $x_{2i-1,2n-1} = x_{i,n}$, so that the function values can be used again if the number of knots increases from n to $2n-1$.

Regular sequences are convenient for analysis and implementation. Hence it is important to know whether regular sequences lead to asymptotically optimal methods. More precisely, we address the following question. Do there exist a density ψ as well as coefficients $a_{i,n}$ such that the corresponding quadrature formulas

$$S_n(f) = \sum_{i=1}^n f(x_{i,n}) \cdot a_{i,n}$$

satisfy

$$e_2(S_n, \mathrm{Int}_\varrho, P) \approx e_2(n, \Lambda^{\mathrm{std}}, \mathrm{Int}_\varrho, P)?$$

Of course, the same question arises in the worst case setting, as well as for other numerical problems for univariate functions, such as the approximation problem. So far answers are available only for a few problems. For several problems, however, the asymptotically best choice of a density ψ is known. Results and further references are given in the sequel.

It would be interesting to know general properties of the operator S and the functions class F or the measure P, respectively, that imply a positive answer to the above question.

2.2. Worst Case Analysis. We report on worst case results on the unit ball B_r in the Sobolev space $W_2^{r+1}([0,1])$ with respect to the seminorm $\|h^{(r+1)}\|_2$, see (10). It is well known that optimal quadrature formulas and optimal recovery on B_r are based on interpolation with natural splines. Let $0 = x_1 < \cdots < x_n = 1$

and let

$$\mathbb{S}_r(x_1,\dots,x_n) = \{s \in C^{r-1}([0,1]) : s|_{[x_{i-1},x_i]} \in \mathbb{P}_r\}$$

denote the space of polynomial splines of degree r with simple knots x_i. A *natural spline* of degree $2r+1$ is a function $s \in \mathbb{S}_{2r+1}(x_1,\dots,x_n)$ that satisfies the boundary conditions $s^{(k)}(0) = s^{(k)}(1) = 0$ for $k = r+1,\dots,2r$. We denote the corresponding space by $\mathbb{N}_{2r+1}(x_1,\dots,x_n)$. Clearly $\mathbb{P}_r \subset \mathbb{N}_{2r+1}(x_1,\dots,x_n)$.

Assume that $n \geq r+1$ in the following. Given $h \in W_2^{r+1}([0,1])$ there exists a uniquely determined $s \in \mathbb{N}_{2r+1}(x_1,\dots,x_n)$ such that $s(x_i) = h(x_i)$ for $i = 1,\dots,n$. The natural spline s has minimal seminorm $\|s^{(r+1)}\|_2$ among all functions from $W_2^{r+1}([0,1])$ that interpolate h at the knots x_i.

Natural spline interpolation defines quadrature formulas S_n via

$$S_n(h) = \sum_{i=1}^n h(x_i) \cdot \mathrm{Int}_\varrho(s_i) = \mathrm{Int}_\varrho(s),$$

where $s_i \in \mathbb{N}_{2r+1}(x_1,\dots,x_n)$ with $s_i(x_j) = \delta_{i,j}$. This formula, which is based on interpolation with $\mathbb{N}_{2r+1}(x_1,\dots,x_n)$, has minimal error $e_{\max}(S_n, \mathrm{Int}_\varrho, B_r)$ among all quadrature formulas that use the knots x_i. This follows from the fact that the set of all functions $h \in B_r$ with $h(x_i) = y_i$ is symmetric with respect to the spline $s \in \mathbb{N}_{2r+1}(x_1,\dots,x_n)$ with $s(x_i) = y_i$. See Traub *et al.* (1988, Corollary 4.5.7.1).

The worst case analysis on B_r will lead to upper bounds in the average case setting. Matching lower bounds are derived from worst case results on the unit ball $B(P_r)$ in the reproducing kernel Hilbert space $H(P_r)$. Recall that $B(P_r)$ consists of all functions from B_r with vanishing derivatives of order $0,\dots,r$ at the boundary. Instead of spline interpolation with natural boundary conditions, we now use interpolating splines from $\mathbb{S}_{2r+1}(x_1,\dots,x_n)$ with Hermite boundary conditions zero. Hereby we get the P_r-spline algorithm, which has minimal error on $B(P_r)$ among all quadrature formulas that use given knots x_i.

For $F = B_r$ or $F = B(P_r)$ let $S_n^{\psi,F}$ denote the best formula in the previous sense that uses knots $x_i = x_{i,n}$ from the regular sequence generated by the density ψ. Clearly $S_n^{\psi,F}$ depends on n, ψ, and F as well as on ϱ.

We will see that the quality of such a method $S_n^{\psi,F}$ is determined by

$$J_{\varrho,r}(\psi) = \left(\int_0^1 \varrho(t)^2 \cdot \psi(t)^{-(2r+2)} \, dt \right)^{1/2} \cdot \left(\int_0^1 \psi(t) \, dt \right)^{r+1}.$$

LEMMA 13.

$$\inf_\psi J_{\varrho,r}(\psi) = J_{\varrho,r}(\varrho^{2/(2r+3)}) = \left(\int_0^1 \varrho(t)^{2/(2r+3)} \, dt \right)^{(2r+3)/2},$$

where ψ varies over all positive densities ψ on $[0,1]$.

PROOF. The Hölder inequality yields

$$\int_0^1 \varrho(t)^{2/(2r+3)}\,dt \leq \left(\int_0^1 \left(\varrho(t)^{2/(2r+3)} \cdot \psi(t)^{-(2r+2)/(2r+3)}\right)^{2r+3} dt\right)^{1/(2r+3)}$$
$$\cdot \left(\int_0^1 \psi(t)\,dt\right)^{(2r+2)/(2r+3)}$$

for every positive and continuous function ψ.   □

Let

$$(12) \qquad\qquad c_r^2 = \frac{|b_{2r+2}|}{(2r+2)!},$$

with b_k denoting the kth Bernoulli number. In particular, $c_0^2 = \frac{1}{12}$, $c_1^2 = \frac{1}{720}$, and c_r tends to zero exponentially fast.

PROPOSITION 14. *Let $F = B(P_r)$ or $F = B_r$. Then*

$$e_{\max}(S_n^{\psi,F}, \mathrm{Int}_\varrho, F) \approx c_r \cdot J_{\varrho,r}(\psi) \cdot n^{-(r+1)}$$

for every continuous and positive density ψ. The minimal errors satisfy

$$e_{\max}(n, \Lambda^{\mathrm{std}}, \mathrm{Int}_\varrho, F) \approx c_r \cdot J_{\varrho,r}(\varrho^{2/(2r+3)}) \cdot n^{-(r+1)},$$

and therefore the regular sequence generated by $\varrho^{2/(2r+3)}$ leads to asymptotically optimal methods.

PROOF. We outline the proof. At first we consider the case $F = B_r$. Let $S_n(h) = \sum_{i=1}^n h(x_i) \cdot a_i$ be an arbitrary quadrature formula with finite error $e_{\max}(S_n, \mathrm{Int}_\varrho, B_r)$. Assume $0 = x_1 < \cdots < x_n = 1$. Since S_n must be exact on $\mathbb{P}_r$, the Peano Kernel Theorem yields

$$e_{\max}(S_n, \mathrm{Int}_\varrho, B_r) = \|K_n\|_2,$$

where

$$K_n(u) = (-1)^{r+1}/r! \cdot \left(\int_0^1 (u-t)_+^r \cdot \varrho(t)\,dt - \sum_{i=1}^n a_i \cdot (u - x_i)_+^r\right).$$

This kernel is called a (generalized) monospline since, in the case $\varrho = 1$, we have $(-1)^{r+1} K_n(u) = u^{r+1}/(r+1)! + s$ with $s \in \mathbb{S}_r(x_1, \dots, x_n)$. It is easy to see that

$$(13) \qquad\qquad K_n^{(k)}(0) = K_n^{(k)}(1) = 0, \qquad k = 0, \dots, r-1.$$

Conversely, every monospline K_n that satisfies these boundary conditions is the Peano kernel of a quadrature formula S_n. Hence, finding optimal quadrature formulas is equivalent to finding monosplines with minimal norm under the constraints (13). See, e.g., Zhensykbaev (1983) and Bojanov, Hakopian, and Sahakian (1993).

Assume now that $S_n = S_n^{\psi,B_r}$, and let K_n^ψ denote the corresponding Peano kernel. Barrow and Smith (1979, Theorem 1) determine the asymptotic order of $\|K_n^\psi\|_2$ and obtain

$$e_{\max}(S_n^{\psi,B_r}, \mathrm{Int}_\varrho, B_r) \approx c_r \cdot J_{\varrho,r}(\psi) \cdot n^{-(r+1)}.$$

As these authors note, this asymptotic behavior is essentially due to Schoenberg (1974) in the case $\varrho = \psi = 1$. From Lemma 13 the asymptotic upper bound for the minimal errors $e_{\max}(n, \Lambda^{\text{std}}, \text{Int}_\varrho, F)$ follows for $F = B_r$ and $F = B(P_r)$.

It remains to establish the asymptotic lower bounds in the case $F = B(P_r)$. By Proposition III.17,

$$e_{\max}(S_n, \text{Int}_\varrho, B(P_r)) = \left\| \xi - \sum_{i=1}^n a_i \cdot P_r(\cdot, x_i) \right\|_{P_r}$$

$$= \left\| \xi^{(r+1)} - \sum_{i=1}^n a_i \cdot P_r^{(r+1,0)}(\cdot, x_i) \right\|_2$$

for every quadrature formula S_n, where $\xi(t) = \text{Int}_\varrho(P_r(\cdot, t))$. From Sacks and Ylvisaker (1970a, Lemma 1) we conclude $\xi \in C^{2r+2}([0, 1])$, where

$$\xi^{(2r+2)} = (-1)^{r+1} \cdot \varrho.$$

Using the representation (5) for P_r and

$$Q_r^{(r+1,0)}(s, t) = \frac{(t - s)_+^r}{r!},$$

$$Q_r^{(r+1,k)}(s, 1) = \frac{(1 - s)^{r-k}}{(r - k)!}$$

we get

$$\mathbb{S}_r(x_1, \ldots, x_n) = \text{span}\{P_r^{(r+1,0)}(\cdot, x_i) : i = 1, \ldots, n\} \oplus \mathbb{P}_r$$

as an orthogonal sum in $L_2([0, 1])$. Moreover, $\xi^{(r+1)}$ is orthogonal to $\mathbb{P}_r$, and therefore

$$e_{\max}(S_n, \text{Int}_\varrho, B(P_r)) = \inf\{\|\xi^{(r+1)} - s\|_2 : s \in \mathbb{S}_r(x_1, \ldots, x_n)\}$$

$$= d(\xi^{(r+1)}, \mathbb{S}_r(x_1, \ldots, x_n))$$

if S_n is the P_r-spline algorithm that uses the knots x_i. See Eubank, Smith, and Smith (1981, Section 5). Finding an optimal method is therefore equivalent to a spline approximation problem in $L_2([0, 1])$ with free knots.

Let $g \in C^{r+1}([0, 1])$. Barrow and Smith (1978, Theorem 2) show that

$$d(g, \mathbb{S}_r(x_{1,n}, \ldots, x_{n,n})) \approx c_r \cdot \left(\int_0^1 g^{(r+1)}(t)^2 \cdot \psi(t)^{-(2r+2)} dt \right)^{1/2} \cdot n^{-(r+1)}$$

if $x_{i,n}$ is the regular sequence generated by the continuous and positive probability density ψ, and

$$\inf\{d(g, \mathbb{S}_r(x_1, \ldots, x_n)) : 0 = x_1 < \cdots < x_n = 1\}$$

$$\approx c_r \cdot \left(\int_0^1 g^{(r+1)}(t)^{2/(2r+3)} dt \right)^{(2r+3)/2} \cdot n^{-(r+1)}.$$

It remains to take $g = \xi^{(r+1)}$. $\square$

2.3. Average Case Analysis. Now we turn to the average case setting with a zero mean measure P on $F \subset C^r([0,1])$. Let K denote the covariance kernel of P. Moreover, let $S_n^{\psi,K}$ denote the K-spline algorithm that uses knots $x_i = x_{i,n}$ from the regular sequence generated by the density ψ. Clearly this quadrature formula depends on n, ψ and K, as well as on ϱ. Recall the definition (12) of the constant c_r.

PROPOSITION 15. *Let P be a zero mean measure with covariance kernel K that satisfies Sacks-Ylvisaker conditions of order $r \in \mathbb{N}_0$. Then*

$$e_2(S_n^{\psi,K}, \text{Int}_\varrho, P) \approx c_r \cdot J_{\varrho,r}(\psi) \cdot n^{-(r+1)}$$

for every continuous and positive density ψ. The minimal errors satisfy

$$e_2(n, \Lambda^{\text{std}}, \text{Int}_\varrho, P) \approx c_r \cdot J_{\varrho,r}(\varrho^{2/(2r+3)}) \cdot n^{-(r+1)},$$

and therefore the regular sequence generated by $\varrho^{2/(2r+3)}$ leads to asymptotically optimal methods.

PROOF. Let $S_n^{\psi,K}$ and $x_{i,n}$ be defined as above. By c we denote different positive constants, which only depend on K, ϱ, and ψ.

Fix $0 < a < \frac{1}{2}$, and let $h \in B(K)$ with $h(x_{i,n}) = 0$. Note that $\|h^{(r+1)}\|_2 \le c$ by Proposition 6, and $\|h\|_2 \le c \cdot n^{-(r+1)}$ by Lemma 4, since the knots $x_{i,n}$ are quasi-uniform, see (11). Therefore

(14)
$$\left| \int_0^a h(t) \cdot \varrho(t)\, dt \right| \le c \cdot \|\varrho \cdot 1_{[0,a]}\|_2 \cdot n^{-(r+1)},$$
$$\left| \int_{1-a}^1 h(t) \cdot \varrho(t)\, dt \right| \le c \cdot \|\varrho \cdot 1_{[1-a,1]}\|_2 \cdot n^{-(r+1)}.$$

If n is sufficiently large, then

$$\left(\int_{a/2}^{1-a/2} h^{(r+1)}(s)^2\, ds \right)^{1/2} \le 1 + \frac{c}{n}$$

according to Proposition 6. Take $g \in C^{r+1}(\mathbb{R})$ with $g = 0$ on $[0, \frac{1}{2}a] \cup [1 - \frac{1}{2}a, 1]$, $g = 1$ on $[a, 1-a]$, and $0 < g < 1$ otherwise. Using Lemma 4 we obtain

$$\|(g \cdot h)^{(r+1)}\|_2$$
$$\le \left(\int_{a/2}^{1-a/2} h^{(r+1)}(s)^2\, ds \right)^{1/2} + c \cdot \sup_{k=1,\dots,r+1} \left(\|g^{(k)}\|_\infty \cdot \|h^{(r+1-k)}\|_2 \right)$$
$$\le A_n$$

with

$$A_n = 1 + \frac{c}{n} \cdot \left(1 + \sup_{k=1,\dots,r+1} \|g^{(k)}\|_\infty \right).$$

Define $\zeta = \varrho \cdot 1_{[a, 1-a]}$, and note that $\mathrm{Int}_\zeta(h) = \mathrm{Int}_\zeta(gh)$. We apply Proposition 14 with ϱ replaced by ζ to obtain

$$\limsup_{n \to \infty}\left(n^{r+1} \cdot \sup\{|\mathrm{Int}_\zeta(h)| : h \in B(K),\ h(x_{i,n}) = 0 \text{ for } i = 1, \ldots, n\}\right)$$

$$\leq \limsup_{n \to \infty}\left(n^{r+1} \cdot A_n \cdot \sup\{|\mathrm{Int}_\zeta(h)| : h \in B_r,\ h(x_{i,n}) = 0 \text{ for } i = 1, \ldots, n\}\right)$$

$$\leq c_r \cdot J_{\zeta,r}(\psi) \leq c_r \cdot J_{\varrho,r}(\psi).$$

Together with Proposition III.17, (14), and (III.14), this implies

$$\limsup_{n \to \infty}\left(n^{r+1} \cdot e_2(S_n^{\psi, K}, \mathrm{Int}_\varrho, P)\right)$$

$$\leq c_r \cdot J_{\varrho,r}(\psi) + c \cdot \left(\|\varrho \cdot 1_{[0,a]}\|_2 + \|\varrho \cdot 1_{[1-a,1]}\|_2\right).$$

Letting a tend to zero, the asymptotic upper bounds for $e_2(S_n^{\psi, K}, \mathrm{Int}_\varrho, P)$ as well as $e_2(n, \Lambda^{\mathrm{std}}, \mathrm{Int}_\varrho, P)$ follow.

To establish the asymptotic lower bounds we consider a quadrature formula S_n using knots $0 \leq x_1 < \cdots < x_n \leq 1$. Put $\delta_n = \max_i(x_i - x_{i-1})$, where $x_0 = 0$ and $x_{n+1} = 1$. By Propositions III.17 and 8,

$$e_2(S_n, \mathrm{Int}_\varrho, P)$$

$$\geq \sup\{|\mathrm{Int}_\varrho(h)| : h \in B(K) \cap H(P_r),\ h(x_i) = 0 \text{ for } i = 1, \ldots, n\}$$

$$\geq (1 + c\,\delta_n)^{-1} \cdot \sup\{|\mathrm{Int}_\varrho(h)| : h \in B(P_r),\ h(x_i) = 0 \text{ for } i = 1, \ldots, n\}$$

if $n \geq r$. In the case $S_n = S_n^{\psi, K}$ we conclude

$$\liminf_{n \to \infty}\left(n^{r+1} \cdot e_2(S_n^{\psi, K}, \mathrm{Int}_\varrho, P)\right) \geq c_r \cdot J_{\varrho,r}(\psi)$$

from Proposition 14 and (III.14), since $\lim_n \delta_n = 0$ by (11).

Now take an asymptotically optimal sequence of quadrature formulas S_n. Without loss of generality we may assume that $\lim_n \delta_n = 0$. Then

$$\liminf_{n \to \infty}\left(n^{r+1} \cdot e_2(n, \Lambda^{\mathrm{std}}, \mathrm{Int}_\varrho, P)\right) \geq \liminf_{n \to \infty}\left(n^{r+1} \cdot e_{\max}(S_n, \mathrm{Int}_\varrho, B(P_r))\right)$$

$$\geq c_r \cdot J_{\varrho,r}(\varrho^{2/(2r+3)})$$

follows again from Proposition 14 and (III.14).   □

REMARK 16. An upper bound

$$e_2(n, \Lambda^{\mathrm{std}}, \mathrm{Int}_\varrho, P) = O(n^{-(r+1)})$$

follows from condition (A), see Sacks and Ylvisaker (1970b, Prop. 4). Condition (B) restricts the smoothness of the random functions f, but (B) alone is not sufficient to show that this bound is sharp. Take, for instance, a measure P with kernel

$$K(s,t) = \min(s,t) - \left(s - \tfrac{1}{2}s^2 + t - \tfrac{1}{2}t^2\right) + \tfrac{1}{3}.$$

In this case $\mathrm{Int}(h) = 0$ for every $h \in H(K)$, and therefore $e_2(n, \Lambda^{\mathrm{std}}, \mathrm{Int}, P) = 0$ for every $n \in \mathbb{N}$. However, K satisfies conditions (A) and (B) with $r = 0$. Cf. Example 7. We add that $K_+^{(2,0)}(s,t) = 1 \notin H(K)$.

While (A), (B), and (C) lead to sharp bounds for regularity $r = 0$, Example 7 shows that this is not true for $r > 0$. Therefore we also use the boundary condition (D) for $r > 0$.

REMARK 17. The last two propositions deal with quadrature formulas that use function values only. Now we briefly discuss formulas

$$S_n(f) = \sum_{i=1}^{n} f^{(\alpha_i)}(x_i) \cdot a_i$$

that use derivatives of orders $\alpha_i \in \{0, \ldots, r\}$. It seems fair to compare quadrature formulas by their errors and by the total number n of functionals used in the formula.

Zhensykbaev (1983, Theorem 5) shows that derivatives do not help in the worst case for classes F of the form

$$F = \{f \in B_r : f^{(k)}(0) = 0 \text{ for } k \in \Gamma_0, \ f^{(k)}(1) = 0 \text{ for } k \in \Gamma_1\}$$

with $\Gamma_0, \Gamma_1 \subset \{0, \ldots, r\}$. The minimal error $e_{\max}(S_n, \text{Int}_\varrho, F)$ in the class of all methods S_n that may use derivatives is attained by a method that uses only function values. It turns out that derivatives do not help at least asymptotically under the assumptions of Proposition 15: the minimal errors in the two classes of formulas with and without derivatives are strongly equivalent. The proof uses Zhensykbaev's result for $F = B(P_r)$ and is similar to the proof of the lower bounds in Proposition 15.

Sacks and Ylvisaker (1970a, 1970b) study quadrature formulas that use *Hermite data* of order r, i.e.,

$$S_n(f) = \sum_{i=1}^{\ell} \sum_{k=0}^{r} f^{(k)}(x_i) \cdot a_{i,k}$$

with

$$n = \ell \cdot (r+1).$$

They obtain an analogue to Proposition 15. Namely, consider the K-spline algorithms that use Hermite data at knots $x_{1,\ell} < \cdots < x_{\ell,\ell}$ from a regular sequence generated by a continuous and positive density ψ. The errors of these quadrature formulas are strongly equivalent to

$$c_r^H \cdot J_{\varrho,r}(\psi) \cdot \ell^{-(r+1)},$$

where

$$c_r^H = \frac{(r+1)!}{((2r+2)! \cdot (2r+3)!)^{1/2}}.$$

The minimal errors in the class of all linear methods that use Hermite data at ℓ knots are strongly equivalent to

$$c_r^H \cdot J_{\varrho,r}(\varrho^{2/(2r+3)}) \cdot \ell^{-(r+1)}.$$

Thus $\psi = \varrho^{2/(2r+3)}$ turns out to be optimal again. Clearly $c_0 = c_0^H$, but note that $c_1 = c_1^H$, too. Since

$$\lim_{r \to \infty} \frac{c_r^H \cdot \ell^{-(r+1)}}{c_r \cdot n^{-(r+1)}} = \lim_{r \to \infty} \frac{(r+1)^{r+1} \cdot c_r^H}{c_r} = \infty,$$

we see that Hermite data are very disadvantageous if n and r are large. Moreover, it may be impractical to use high order derivatives for the integration of smooth functions. The analysis of Hermite data of order r allows to use the Markov property and leads to problems of best approximation by piecewise polynomials with free knots.

REMARK 18. Wahba (1971, 1974) and Hájek and Kimeldorf (1974) construct covariance kernels from differential operators in the following way. Let $\mathfrak{D}g = g'$ and let

$$\mathfrak{L} = \sum_{i=0}^{r+1} \beta_i \cdot \mathfrak{D}^i$$

denote the linear differential operator of order $r + 1$ with coefficient functions $\beta_i \in C^i([0,1])$ where $|\beta_{r+1}| > 0$. Furthermore, let G denote the Green's function for the initial value problem $\mathfrak{L}f = u$ and $f^{(i)}(0) = 0$ for $i = 0, \ldots, r$. Hence $f(s) = \int_0^1 G(s,t) \cdot u(t)\, dt$ is the solution of the initial value problem. Clearly

$$(15) \qquad\qquad K(s,t) = \int_0^1 G(s,u) \cdot G(t,u)\, du$$

defines a nonnegative definite function, and it turns out that its reproducing kernel Hilbert space is

$$H(K) = H(Q_r) = \{h \in W_2^{r+1}([0,1]) : h^{(k)}(0) = 0 \text{ for } k = 0, \ldots, r\},$$

equipped with the norm

$$\|h\|_K = \|\mathfrak{L}h\|_2.$$

The solution of the stochastic differential equation $\mathfrak{L}f = u$, with u being white noise, is a random function with covariance kernel K. Hence K corresponds to an autoregressive random function of order $r + 1$.

For instance, if $\mathfrak{L} = \beta_{r+1} \cdot \mathfrak{D}^{r+1}$ then

$$K(s,t) = \int_0^1 \frac{(s-u)_+^r \cdot (t-u)_+^r}{(r! \cdot \beta_{r+1}(u))^2}\, du,$$

and

$$K^{(r,r)}(s,t) = \min(T(s), T(t))$$

where $T' = \beta_{r+1}^{-2}$ and $T(0) = 0$. While $K \in C^{r,r}([0,1]^2)$ satisfies (A) and (D) in the Sacks-Ylvisaker conditions, (B) and (C) are only fulfilled if β_{r+1} is constant. The kernel $K^{(r,r)}$ corresponds to Brownian motion under the time transformation T, and K corresponds to the r-fold integral of this random function.

Wahba (1971), in a particular case, and Hájek and Kimeldorf (1974), in the general case, study the problem Int_ϱ when Hermite data are available. They obtain an analogue to Proposition 15, similar to the one described in Remark 17. Namely, consider the regular sequence $x_{1,\ell} < \cdots < x_{\ell,\ell}$ that is generated by a continuous and positive density ψ. The errors of K-spline algorithms that use Hermite data at these knots are strongly equivalent to

$$c_r^H \cdot J_{\zeta,r}(\psi) \cdot \ell^{-(r+1)},$$

where

(16) $$\zeta = \varrho/\beta_{r+1}.$$

The minimal errors in the class of all linear methods that use Hermite data at ℓ knots are strongly equivalent to

$$c_r^H \cdot J_{\zeta,r}\left(|\zeta|^{2/(2r+3)}\right) \cdot \ell^{-(r+1)}.$$

A generalization of these results is due to Wahba (1974). She considers kernels K such that the corresponding zero mean Gaussian measure on $C^r([0,1])$ is equivalent to the zero mean Gaussian measure with a kernel given by (15). For two kernels of the form (15) this equivalence holds iff the leading coefficients β_{r+1} of the differential operators $\mathfrak{L}$ coincide. Another generalization is due to Wittwer (1976), where vector processes are studied.

Our proof technique for the problem Int_ϱ when only function values are available also applies to covariance kernels K of the form (15). Let $h \in H(K)$, and note that the norms $\|h\|_K$, $\|h\|_{Q_r} = \|h^{(r+1)}\|_2$, and $\|\beta_{r+1} \cdot h\|_{Q_r}$ are equivalent. Assume that h vanishes at the knots $0 < x_1 < \cdots < x_n \leq 1$ and put $\delta = \max_i(x_i - x_{i-1})$ where $x_0 = 0$ and $x_{n+1} = 1$. From Lemma 4 we get

$$\left| \|h\|_K - \|\beta_{r+1} \cdot h\|_{Q_r} \right| \leq \|\mathfrak{L}h - (\beta_{r+1} \cdot h)^{(r+1)}\|_2 \leq c \cdot \delta \cdot \|h^{(r+1)}\|_2$$

with a constant c, which depends only on $\mathfrak{L}$. Hence

$$(1 + c\,\delta)^{-1} \cdot \|\beta_{r+1} \cdot h\|_{Q_r} \leq \|h\|_K \leq (1 + c\,\delta) \cdot \|\beta_{r+1} \cdot h\|_{Q_r}$$

for some constant c, and therefore

$$(1 + c\,\delta)^{-1} \cdot \sup\{|\text{Int}_\varrho(h)| : \beta_{r+1} \cdot h \in B(Q_r),\ h(x_i) = 0 \text{ for } i = 1,\ldots,n\}$$
$$\leq \sup\{|\text{Int}_\varrho(h)| : h \in B(K),\ h(x_i) = 0 \text{ for } i = 1,\ldots,n\}$$
$$\leq (1 + c\,\delta) \cdot \sup\{|\text{Int}_\varrho(h)| : \beta_{r+1} \cdot h \in B(Q_r),\ h(x_i) = 0 \text{ for } i = 1,\ldots,n\}.$$

Clearly $\text{Int}_\varrho(h) = \text{Int}_\zeta(\beta_{r+1} \cdot h)$ with ζ given by (16), and applying the proof of Proposition 15 we get

$$e_2(S_n^{\psi,K}, \text{Int}_\varrho, P) \approx c_r \cdot J_{\zeta,r}(\psi) \cdot n^{-(r+1)}$$

and

$$e_2(n, \Lambda^{\text{std}}, \text{Int}_\varrho, P) \approx c_r \cdot J_{\zeta,r}\left(|\zeta|^{2/(2r+3)}\right) \cdot n^{-(r+1)},$$

where P is a zero mean measure with covariance kernel K.

Under the assumptions of the last two propositions, regular sequences lead to asymptotically optimal quadrature formulas. This result also holds for quadrature formulas that use Hermite data, see Remarks 17 and 18. In the latter case, ϱ must be replaced by ϱ/β_{r+1}, see (16).

2.4. Notes and References. 1. Proposition 14 is due to Barrow and Smith (1979) for $F = B_r$ and Eubank *et al.* (1981) for $F = B(P_r)$. The spaces $F = B(Q_r)$ and $F = \{h \in B(Q_r) : h^{(k)}(1) = 0 \text{ for } k = \ell, \ldots, r\}$ with $1 \leq \ell \leq r$ are considered in Eubank, Smith, and Smith (1982).

2. Proposition 15 is due to Sacks and Ylvisaker (1966, 1970a) for $r = 0$ and 1. For arbitrary r the asymptotic analysis for regular sequences is due to Benhenni and Cambanis (1992a) under slightly different smoothness assumptions; the result on the minimal errors is due to Ritter (1996a). Sacks and Ylvisaker (1966, 1970b) have even obtained the asymptotic optimality of the density $|\varrho|^{2/(2r+3)}$ for every continuous ϱ if $r = 0$ and, under mild assumptions on the zeros of ϱ, if $r = 1$. In general, $\varrho > 0$ can be replaced by a condition that restricts the behavior of ϱ in the neighborhood of the zeros. The case $\varrho \in L_2([0, 1])$ is considered in Barrow and Smith (1979).

Proposition 15, with $r \geq 2$, confirms a conjecture by Eubank *et al.* (1982, Remark 1). Further support to the conjecture was given by Benhenni and Cambanis (1992a).

3. The results for autoregressive processes, see Remark 18, are due to Wahba (1971, 1974) and Hájek and Kimeldorf (1974) for Hermite data and due to Ritter (1996a) if only function evaluations are allowed.

4. Processes whose derivatives have stationary independent increments are considered in Temirgaliev (1988), the particular case w_1 is considered in Voronin and Skalyga (1984).

5. See Lee and Wasilkowski (1986) and Gao (1993) for further results for the integration problem involving r-fold integrated Wiener measures.

3. Approximation under Sacks-Ylvisaker Conditions

We approximate functions f from a class $F \subset C^r([0, 1])$ with respect to a weighted L_p-norm. The weight function ϱ is assumed to be piecewise continuous and positive. For simplicity we deal with a single piece, allowing zeros or singularities of ϱ only at $t = 0$ and $t = 1$. Hence

$$\varrho \in L_1([0, 1]), \qquad \varrho > 0 \text{ and continuous on }]0, 1[,$$

throughout this section.

We are mainly interested in linear methods that use function values only, i.e., the class of permissible functionals is given by $\Lambda = \Lambda^{\text{std}}$. In addition, we consider the Hermite data case and the case $\Lambda = \Lambda^{\text{all}}$, where, in particular, weighted integrals may be used to recover functions $f \in F$.

Results and methods of proof are different for finite p and for $p = \infty$. We assume Sacks-Ylvisaker conditions of order r for the covariance kernel K of the zero mean measure P on F. While this is sufficient in the case $p = 2$, we have to be more specific for $p \neq 2$. In the latter case we consider Gaussian measures P. Detailed proofs are presented for L_2-approximation; in the other cases we give basic ideas.

Consider average errors that are defined by second moments, $q = 2$, and fix functionals $\lambda_1, \ldots, \lambda_n \in \Lambda^{\mathrm{all}}$. The corresponding K-spline algorithm S_n^K minimizes the error and yields

(17) $\quad e_2(S_n^K, \mathrm{App}_{2,\varrho}, P)^2$

$$= \int_0^1 \sup\{h(t)^2 : h \in B(K),\ \lambda_i(h) = 0 \text{ for } i = 1, \ldots, n\} \cdot \varrho(t)\, dt,$$

see (III.19). Here $B(K)$ denotes the unit ball in the Hilbert space with reproducing kernel K. Note that for all $t \in [0,1]$ the extremal function is a spline in $H(K)$.

3.1. L_2-Approximation with Hermite Data.

A linear method that uses Hermite data of order r is of the form

$$S_n(f) = \sum_{i=1}^{\ell} \sum_{k=0}^{r} f^{(k)}(x_i) \cdot a_{i,k},$$

where

$$n = \ell \cdot (r+1)$$

and ℓ denotes the number of knots. In particular, we study K-spline algorithms that use Hermite data of order r at knots $x_i = x_{i,\ell}$ from regular sequences generated by densities ψ. Such methods are denoted by $S_n^{\psi, K, r}$. We require

(18) $\quad \psi \in L_1([0,1]), \qquad \psi$ continuous on $]0,1[, \qquad \inf_{0 < t < 1} \psi(t) > 0,$

which is slightly stronger than our assumption on ϱ.

Analogously to the functional $J_{\varrho,r}$ we define

$$I_{\varrho,r}(\psi) = \left(\int_0^1 \varrho(t) \cdot \psi(t)^{-(2r+1)}\, dt \right)^{1/2} \cdot \left(\int_0^1 \psi(t)\, dt \right)^{r+1/2}$$

and obtain

(19) $\quad \inf\{I_{\varrho,r}(\psi) : \psi \in L_1([0,1]),\ \psi > 0 \text{ and continuous on }]0,1[\}$

$$= I_{\varrho,r}(\varrho^{1/(2r+2)}) = \left(\int_0^1 \varrho(t)^{1/(2r+2)}\, dt \right)^{r+1},$$

cf. Lemma 13. Furthermore, we put

$$d_r^H = \left(\frac{2r+1}{(4r+3)!} \right)^{1/2} \cdot \frac{(2r)!}{r!}$$

and we use the notation

$$h[x] = (h(x), h'(x), \ldots, h^{(r)}(x)).$$

PROPOSITION 19. *Let P be a zero mean measure with covariance kernel K that satisfies Sacks-Ylvisaker conditions of order $r \in \mathbb{N}_0$. Then*

$$e_2(S_n^{\psi,K,r}, \mathrm{App}_{2,\varrho}, P) \approx d_r^H \cdot I_{\varrho,r}(\psi) \cdot \ell^{-(r+1/2)}$$

for every density ψ that satisfies (18). The minimal errors in the class of all linear methods that use Hermite data of order r at ℓ knots are strongly equivalent to

$$d_r^H \cdot I_{\varrho,r}(\varrho^{1/(2r+2)}) \cdot \ell^{-(r+1/2)}.$$

PROOF. Let $S_n^{K,r}$ be a K-spline algorithm that uses Hermite data of order r at knots

$$0 = x_{1,\ell} < \cdots < x_{\ell,\ell} = 1,$$

put $X_\ell = \{x_{1,\ell}, \ldots, x_{\ell,\ell}\}$, and define

$$\delta(X_\ell) = \max_{i=2,\ldots,\ell} (x_{i,\ell} - x_{i-1,\ell}).$$

Note that

$$e(S_n^{K,r}, \mathrm{App}_{2,\varrho}, P)^2 = \int_0^1 \sup\{h(t)^2 : h \in B(K), \ h[x] = 0 \text{ for } x \in X_\ell\} \cdot \varrho(t)\, dt.$$

From Propositions 6 and 8 we get $H(P_r) \subset H(K) \subset H(R_r)$, and therefore the error $e_2(S_n^{K,r}, \mathrm{App}_{2,\varrho}, P)$ tends to zero iff $\delta(X_\ell)$ does. Thus we assume $\lim_{\ell \to \infty} \delta(X_\ell) = 0$ in the following.

The estimates for $\|h\|_K$ and $\|h^{(r+1)}\|_2$ in Propositions 6 and 8 imply

$$e(S_n^{K,r}, \mathrm{App}_{2,\varrho}, P)^2 \approx \int_0^1 \sup\{h(t)^2 : h \in B_r, \ h[x] = 0 \text{ for } x \in X_\ell\} \cdot \varrho(t)\, dt,$$

where B_r is the unit ball in $W_2^{r+1}([0,1])$ with respect to the seminorm $\|h^{(r+1)}\|_2$.

The solutions of the latter extremal problems are known. For every $t \in \,]x_{i-1,\ell}, x_{i,\ell}[$ the extremal function is a polynomial spline of degree $2r+1$ with support $[x_{i-1,\ell}, x_{i,\ell}]$, and the extremal value is

$$(20) \qquad \sup\{h(t)^2 : h \in B_r, \ h[x] = 0 \text{ for } x \in X_\ell\}$$

$$= (x_{i,\ell} - x_{i-1,\ell})^{2r+1} \cdot P_r\left(\frac{t - x_{i-1,\ell}}{x_{i,\ell} - x_{i-1,\ell}}, \frac{t - x_{i-1,\ell}}{x_{i,\ell} - x_{i-1,\ell}}\right)$$

$$= \frac{1}{(2r+1) \cdot (r!)^2} \cdot \left(\frac{(t - x_{i-1,\ell}) \cdot (x_{i,\ell} - t)}{x_{i,\ell} - x_{i-1,\ell}}\right)^{2r+1},$$

see Speckman (1979). Therefore

$$A_\ell(a,b) = \int_a^b \sup\{h(t)^2 : h \in B_r, \ h[x] = 0 \text{ for } x \in X_\ell\} \cdot \varrho(t)\, dt$$

satisfies

$$A_\ell(0,1) = \frac{1}{(2r+1)\cdot(r!)^2} \cdot \sum_{i=2}^{\ell} \int_{x_{i-1,\ell}}^{x_{i,\ell}} \left(\frac{(t-x_{i-1,\ell})\cdot(x_{i,\ell}-t)}{x_{i,\ell}-x_{i-1,\ell}} \right)^{2r+1} \cdot \varrho(t)\, dt$$

$$= \frac{\int_0^1 (t\cdot(1-t))^{2r+1}\, dt}{(2r+1)\cdot(r!)^2} \cdot \sum_{i=2}^{\ell} \varrho(\tau_{i,\ell}) \cdot (x_{i,\ell}-x_{i-1,\ell})^{2r+2}$$

$$= \left(d_r^H\right)^2 \cdot \sum_{i=2}^{\ell} \varrho(\tau_{i,\ell}) \cdot (x_{i,\ell}-x_{i-1,\ell})^{2r+2},$$

where $\tau_{i,\ell} \in \,]x_{i-1}, x_i[$.

We determine an asymptotic lower bound for the error of $S_n^{K,r}$. Fix $m \in \mathbb{N}$ and let $c_{j,m}$ denote the infimum of ϱ on $[(j-1)/m, j/m]$. Let $k_{j,m}$ denote the number of intervals $[x_{i-1,\ell}, x_{i,\ell}] \subset [(j-1)/m, j/m]$. Using convexity and $\sum_{j=1}^{m} k_{j,m} < \ell$ we get

$$\sum_{i=2}^{\ell} \varrho(\tau_{i,\ell}) \cdot (x_{i,\ell}-x_{i-1,\ell})^{2r+2}$$

$$\geq (1/m - 2\,\delta(X_\ell))^{2r+2} \cdot \sum_{j=1}^{m} \frac{c_{j,m}}{k_{j,m}^{2r+1}}$$

$$\geq (1/m - 2\,\delta(X_\ell))^{2r+2} \cdot \frac{1}{\ell^{2r+1}} \cdot \left(\sum_{j=1}^{m} c_{j,m}^{1/(2r+2)} \right)^{2r+2}.$$

We conclude that

$$\liminf_{\ell\to\infty} e_2(S_n^{K,r}, \mathrm{App}_{2,\varrho}, P)^2 \cdot \ell^{2r+1} \geq \left(d_r^H\right)^2 \cdot \left(\frac{1}{m} \cdot \sum_{j=1}^{m} c_{j,m}^{1/(2r+2)} \right)^{2r+2}.$$

Since

$$\lim_{m\to\infty} \frac{1}{m} \cdot \sum_{j=1}^{m} c_{j,m}^{1/(2r+2)} = \int_0^1 \varrho(t)^{1/(2r+2)}\, dt,$$

we have

$$\liminf_{\ell\to\infty} e_2(S_n^{K,r}, \mathrm{App}_{2,\varrho}, P)^2 \cdot \ell^{2r+1} \geq \left(d_r^H\right)^2 \cdot I_{\varrho,r}(\varrho^{1/(2r+2)})^2,$$

and the asymptotic lower bound for the minimal errors follows.

Now we consider K-spline algorithms $S_n^{\psi,K,r}$ that are based on regular sequences. Assume $\int_0^1 \psi(t)\, dt = 1$ without loss of generality. In this case

$$x_{i,\ell} - x_{i-1,\ell} = \frac{1}{\psi(\xi_{i,\ell})\,(\ell-1)}$$

with $\xi_{i,\ell} \in]x_{i-1,\ell}, x_{i,\ell}[$. If $\sup_{0<t<1} \varrho(t) < \infty$ then

$$\lim_{\ell\to\infty} A_\ell(0,1) \cdot \ell^{2r+1} = \left(d_r^H\right)^2 \cdot \lim_{\ell\to\infty} \sum_{i=2}^{\ell} \frac{\varrho(\tau_{i,\ell})}{\psi(\xi_{i,\ell})^{2r+1}} \cdot (x_{i,\ell} - x_{i-1,\ell})$$

$$= \left(d_r^H\right)^2 \cdot I_{\varrho,r}(\psi)^2.$$

For general ϱ, we take $0 < \varepsilon < 1/2$ to obtain

$$\lim_{\ell\to\infty} A_\ell(\varepsilon, 1-\varepsilon) \cdot \ell^{2r+1} = \left(d_r^H\right)^2 \cdot \int_\varepsilon^{1-\varepsilon} \varrho(t) \cdot \psi(t)^{-(2r+1)}\, dt$$

analogously. Furthermore we note that

$$\limsup_{\ell\to\infty}(A_\ell(0,\varepsilon) + A_\ell(1-\varepsilon,1)) \cdot \ell^{2r+1}$$

$$\leq \int_{[0,\varepsilon]\cup[1-\varepsilon,1]} \varrho(t)\, dt \cdot \limsup_{\ell\to\infty} \ell^{2r+1} \cdot \delta(X_\ell)^{2r+1}$$

$$\leq \int_{[0,\varepsilon]\cup[1-\varepsilon,1]} \varrho(t)\, dt \cdot \frac{1}{\inf_{0<t<1} \psi(t)}.$$

Letting ε tend to zero, we get

$$e_2(S_n^{\psi,K,r}, \mathrm{App}_{2,\varrho}, P)^2 \approx \left(d_r^H\right)^2 \cdot I_{\varrho,r}(\psi)^2 \cdot \ell^{-(2r+1)}.$$

If $\inf_{0<t<1} \varrho(t) > 0$, then the density $\psi = \varrho^{1/(2r+2)}$ yields the asymptotic upper bound for the minimal errors, see (19).

It remains to establish this upper bound for general ϱ. The restriction of $\varrho^{1/(2r+2)}$ on a subinterval $[\varepsilon, 1-\varepsilon]$ defines a regular sequence

$$\varepsilon = x_{1,m} < \cdots < x_{m,m} = 1 - \varepsilon.$$

Consider a K-spline algorithm $S_{m,\varepsilon}^{\psi,K,r}$ that uses the knots $x_{i,m}$ together with equidistant knots

$$\varepsilon \cdot \frac{j}{k} \qquad \text{and} \qquad 1 - \varepsilon \cdot \frac{j}{k},$$

for $j = 0, \ldots, k-1$. Hence $m + 2k$ is the total number of knots that is used by $S_{m,\varepsilon}^{\psi,K,r}$. Take

$$k = k(m,\varepsilon)$$

such that

$$\lim_{m\to\infty} \frac{k(m,\varepsilon)}{m} = \varepsilon^{1/2}.$$

As previously, we get

$$\limsup_{m\to\infty} e_2(S_{m,\varepsilon}^{\psi,K,r}, \mathrm{App}_{2,\varrho}, P)^2 \cdot (m + 2k(m,\varepsilon))^{2r+1}$$

$$\leq (d_r^H)^2 \cdot \int_\varepsilon^{1-\varepsilon} \varrho(t)^{1/(2r+2)}\, dt \cdot \|\varrho^{1/(2r+2)}\|_1^{2r+1} \cdot \limsup_{m\to\infty} \left(\frac{m + 2k(m,\varepsilon)}{m}\right)^{2r+1}$$

$$+ \|\varrho\|_1 \cdot \varepsilon^{2r+1} \cdot \limsup_{m\to\infty} \left(\frac{m + 2k(m,\varepsilon)}{k(m,\varepsilon)}\right)^{2r+1}$$

$$\leq (d_r^H)^2 \cdot I_{\varrho,r}(\varrho^{1/(2r+2)})^2 \cdot (1 + 2 \cdot \varepsilon^{1/2})^{2r+1} + \|\varrho\|_1 \cdot (\varepsilon \cdot (2 + \varepsilon^{-1/2}))^{2r+1}.$$

Letting ε tend to zero, the upper bound for the minimal errors with general ϱ follows. □

COROLLARY 20. *Under the assumption of Proposition 19, along with the additional assumption*

$$\inf_{0<t<1} \varrho(t) > 0,$$

the regular sequence generated by $\varrho^{1/(2r+2)}$ leads to asymptotically optimal methods.

REMARK 21. Assume that ϱ is not bounded away from zero. In this case a modification of the regular sequence generated by $\varrho^{1/(2r+2)}$ leads to good methods for L_2-approximation: close to the boundary we take equidistant points. The asymptotic constants for the resulting errors are arbitrarily close to the optimal constant $d_r^H \cdot I_{\varrho,r}(\varrho^{1/(2r+2)})$. Details are given in the proof above. See also Müller-Gronbach (1996) for results in the case $r = 0$.

3.2. L_2-Approximation with Function Values. Now we use function values for L_2-approximation. From the Hermite data case we already know that the minimal errors are of order $n^{-(r+1/2)}$, since derivatives can be approximated by divided differences. Besides stability problems, which may arise in this approach, the asymptotic constants get too large. Hence we take a different approach, as for the integration problem, to determine the asymptotic constants for the minimal errors.

Hermite data allow local considerations, which reflects, for instance, the Markov property of $f[x] = (f(x), f'(x), \dots, f^{(r)}(x))$ with respect to the r-fold integrated Wiener measure. This is no longer true if we study methods

$$S_n(f) = \sum_{i=1}^n f(x_i) \cdot a_i$$

that use function values only. Here we are only able to analyze the performance of methods that are based on regular sequences. We conjecture, however, that analoga to Corollary 20 and Remark 21 hold. As a consequence, it would be sufficient to consider regular sequences. Cf. Section 2.1.

Let ψ denote a density on $[0,1]$ that satisfies (18), and let

$$0 = x_{1,n} < \cdots < x_{n,n} = 1$$

denote the corresponding regular sequence. Put

$$X_n^{\psi} = \{x_{1,n}, \ldots, x_{n,n}\}$$

and

$$\delta(X_n^{\psi}) = \max_{i=2,\ldots,n} (x_{i,n} - x_{i-1,n}).$$

Consider the K-spline algorithm $S_n^{\psi,K}$ that uses the knots from X_n^{ψ}. Then (17) implies that

$$e_2(S_n^{\psi,K}, \mathrm{App}_{2,\varrho}, P)^2 = \int_0^1 \sup\{h(t)^2 : h \in B(K),\ h(x) = 0 \text{ for } x \in X_n^{\psi}\} \cdot \varrho(t)\, dt.$$

As previously, let B_r denote the unit ball in $W_2^{r+1}([0,1])$ with respect to the seminorm $\|h^{(r+1)}\|_2$.

First we study the contribution of a subinterval $[a,b] \subset [0,1]$ to the error of $S_n^{\psi,K}$. It turns out that the asymptotic behavior does not change if we replace the unit ball $B(K)$ in the reproducing kernel Hilbert space $H(K)$ by B_r. Furthermore, it is irrelevant whether the function is known completely or not at all outside of $[a,b]$. We put

$$Z(K, X) = \{h \in B(K) : h(x) = 0 \text{ for } x \in X\}$$

and

$$Z(W_2^{r+1}, X) = \{h \in B_r : h(x) = 0 \text{ for } x \in X\}$$

where $X \subset [0,1]$.

The following asymptotic equivalence replaces the Markov property, which holds in the Hermite data case.

LEMMA 22. *Let K satisfy Sacks-Ylvisaker regularity conditions of order r. For every regular sequence X_n^{ψ} and every $0 \le a < b \le 1$ we have*

$$\int_a^b \sup\{h(t)^2 : h \in Z(K, X_n^{\psi})\} \cdot \varrho(t)\, dt$$

$$\approx \int_a^b \sup\{h(t)^2 : h \in Z(W_2^{r+1}, X_n^{\psi} \cap [a,b])\} \cdot \varrho(t)\, dt$$

$$\approx \int_a^b \sup\{h(t)^2 : h \in Z(W_2^{r+1}, X_n^{\psi}),\ h[a] = h[b] = 0\} \cdot \varrho(t)\, dt.$$

PROOF. By c we denote different positive constants, which may only depend on K, ϱ, and ψ.

Let $[u,v] \subset [a,b]$ and assume $h \in Z(W_2^{r+1}, X_n^{\psi} \cap [a,b])$. Every subinterval of $[a,b]$ of length at least

$$L = \|\psi\|_1 \cdot \left((n-1) \cdot \inf_{0 < t < 1} \psi(t)\right)^{-1}$$

contains a zero of h since $\delta(X_n^\psi) \leq L$. Using Lemma 4 we get

$$\left(\int_a^b h'(s)^2 \, ds\right)^{1/2} \leq (r+1)! \cdot L^r \cdot \left(\int_a^b h^{(r+1)}(s)^2 \, ds\right)^{1/2} \leq c \cdot n^{-r},$$

which implies

$$|h(t)| \leq c \cdot n^{-(r+1/2)}$$

for every $t \in [a, b]$.

Put

$$A_n(u, v) = \int_u^v \sup\{h(t)^2 : h \in Z(W_2^{r+1}, X_n^\psi \cap [a, b])\} \cdot \varrho(t) \, dt$$

and

$$B_n = \int_a^b \sup\{h(t)^2 : h \in Z(W_2^{r+1}, X_n^\psi), \ h[a] = h[b] = 0\} \cdot \varrho(t) \, dt.$$

We have

(21)
$$A_n(u, v) \leq c \cdot n^{-(2r+1)} \cdot (v - u)^{1/2}.$$

Since

$$\{h \in B_r : h[x] = 0 \text{ for } x \in X_n^\psi \cup \{a, b\}\} \subset \{h \in Z(W_2^{r+1}, X_n^\psi) : h[a] = h[b] = 0\},$$

we use (20) to obtain

(22)
$$B_n \geq c \cdot n^{-(2r+1)} \cdot Q$$

with

$$Q = \min\left\{\varrho(t) : \left|t - \tfrac{1}{2}(a + b)\right| \leq \tfrac{1}{4}(b - a)\right\} \cdot (b - a)^{2r+2}.$$

Take $0 < \varepsilon < \frac{1}{2}(b - a)$. Moreover, take $g \in C^{r+1}([0, 1])$ such that $g = 0$ on $\left[0, a + \frac{1}{2}\varepsilon\right]$ and on $\left[b - \frac{1}{2}\varepsilon, 1\right]$, $g = 1$ on $[a + \varepsilon, b - \varepsilon]$, and $0 < g < 1$ otherwise. Put $M = \sup_{k=1,\ldots,r+1} \|g^{(k)}\|_\infty$, and assume that $h \in Z(W^{r+1}, X_n^\psi \cap [a, b])$. Using Lemma 4, we get

$$\|(g \cdot h)^{(r+1)}\|_2 \leq 1 + c \cdot \sup_{k=1,\ldots,r+1} \|g^{(k)}\|_\infty \cdot \left(\int_{a+\varepsilon/2}^{b-\varepsilon/2} h^{(r+1-k)}(t)^2 \, dt\right)^{1/2}$$

$$\leq 1 + c \cdot \frac{M}{n}.$$

This yields

$$A_n(a + \varepsilon, b - \varepsilon) \leq \left(1 + c \cdot \frac{M}{n}\right) \cdot B_n.$$

From (21) and (22) we obtain

$$A_n(a, a + \varepsilon) \leq c \cdot n^{-(2r+1)} \cdot \varepsilon^{1/2} \leq c \cdot \varepsilon^{1/2} \cdot \frac{B_n}{Q}.$$

The same estimate holds for $A_n(b - \varepsilon, b)$. Since $B_n \leq A_n(a, b)$ we have

$$1 \leq \limsup_{n \to \infty} \frac{A_n(a, b)}{B_n} \leq 1 + c \cdot \frac{\varepsilon^{1/2}}{Q}.$$

For ε tending to zero we obtain

$$A_n(a, b) \approx B_n.$$

If n is sufficiently large then

$$\left(\int_{a+\varepsilon/2}^{b-\varepsilon/2} h^{(r+1)}(t)^2 \, dt \right)^{1/2} \leq 1 + \frac{c}{n}$$

for every $h \in Z(K, X_n^\psi)$ due to Proposition 6. Furthermore,

$$\left(\int_a^b h^{(r+1)}(t)^2 \, dt \right)^{1/2} \leq c.$$

By the arguments used previously,

$$\limsup_{n \to \infty} \int_a^b \sup\{h(t)^2 : h \in Z(K, X_n^\psi)\} \cdot \varrho(t) \, dt \, / \, B_n \leq 1.$$

Conversely, Proposition 8 yields

$$\sup\{h(t)^2 : h \in Z(W_2^{r+1}, X_n^\psi), \, h[a] = h[b] = 0\}$$
$$\leq \sup\{h(t)^2 : h \in Z(W_2^{r+1}, X_n^\psi), \, h[0] = h[1] = 0\}$$
$$\leq (1 + c/n) \cdot \sup\{h(t)^2 : h \in Z(K, X_n^\psi)\}.$$

Hence

$$\liminf_{n \to \infty} \int_a^b \sup\{h(t)^2 : h \in Z(K, X_n^\psi)\} \cdot \varrho(t) \, dt \, / \, B_n \geq 1,$$

which completes the proof. $\square$

Because of the preceding lemma, we consider the extremal problems

$$M^r(t, X_n^\psi, a, b) = \sup\{h(t)^2 : h \in Z(W_2^{r+1}, X_n^\psi \cap [a, b])\}$$

and

$$M_0^r(t, X_n^\psi, a, b) = \sup\{h(t)^2 : h \in Z(W_2^{r+1}, X_n^\psi), \, h[a] = h[b] = 0\},$$

where $0 \leq a < b \leq 1$ and $t \in [a, b]$. Clearly

$$M_0^r(t, X_n^\psi, a, b) \leq M^r(t, X_n^\psi, a, b),$$

and both quantities equal zero if $t \in X_n^\psi$. If $t \notin X_n^\psi$ and if $X_n^\psi \cap [a, b]$ contains more than r points then the extremal functions are uniquely determined for both problems. These functions are polynomial splines of degree $2r + 1$ with natural or with Hermite boundary conditions, respectively.

In addition, we consider the extremal problem

$$M^r(\tau, \Xi)$$

$$= \sup\left\{ h(\tau)^2 : h \in W_2^{r+1}(\mathbb{R}), \int_{\mathbb{R}} h^{(r+1)}(t)^2 \, dt \leq 1, \ h(\xi_j) = 0 \text{ for } j \in \mathbb{Z} \right\}$$

on the real line. Here $\tau \in \mathbb{R}$ and $\Xi = \{\xi_j : j \in \mathbb{Z}\} \subset \mathbb{R}$ with

$$\xi_{j-1} < \xi_j, \qquad \lim_{j \to \pm\infty} \xi_j = \pm\infty.$$

If $\tau \notin \Xi$ then the extremal function is uniquely determined and is a polynomial spline of degree $2r + 1$, see DeVore and Lorentz (1993, Chapter 13.5). In the case $r = 0$ the extremal function vanishes outside of the interval $[\xi_{j-1}, \xi_j]$ that contains τ, and

$$(23) \qquad M^0(\tau, \Xi) = \frac{(\tau - \xi_{j-1}) \cdot (\xi_j - \tau)}{\xi_j - \xi_{j-1}}.$$

We relate $M^r(t, X_n^\psi, a, b)$ and $M_0^r(t, X_n^\psi, a, b)$ to $M^r(\tau, \Xi)$, where τ and Ξ are defined in the following way. Assume that

$$t \in \,]x_{i-1,n}, x_{i,n}[\, \subset [a, b]$$

and let

$$\xi_k' < \cdots < \xi_0' = x_{i-1,n} < x_{i,n} = \xi_1' < \cdots < \xi_\ell'$$

with

$$\{\xi_k', \ldots, \xi_\ell'\} = X_n^\psi \cap [a, b].$$

Define $\Xi' = \{\xi_j' : j \in \mathbb{Z}\}$ by

$$\xi_j' - \xi_{j-1}' = \frac{\|\psi\|_1}{\psi((a+b)/2) \cdot (n-1)}, \qquad j > \ell,$$

and

$$\xi_{j+1}' - \xi_j' = \frac{\|\psi\|_1}{\psi((a+b)/2) \cdot (n-1)}, \qquad j < k.$$

Finally define $\Xi = \{\xi_j : j \in \mathbb{Z}\}$ and τ by the affine transformation

$$\xi_j = \frac{\xi_j' - \xi_0'}{\xi_1' - \xi_0'}, \qquad \tau = \frac{t - \xi_0'}{\xi_1' - \xi_0'}.$$

In particular, $0 = \xi_0 < \tau < \xi_1 = 1$. Note that Ξ depends on ψ, n, t, a, and b. We get

$$(24) \qquad M^r(t, X_n^\psi, a, b) \geq M^r(t, \Xi') = (x_{i,n} - x_{i-1,n})^{2r+1} \cdot M^r(\tau, \Xi)$$

and

$$(25) \qquad M_0^r(t, X_n^\psi, a, b) \leq M^r(t, \Xi') = (x_{i,n} - x_{i-1,n})^{2r+1} \cdot M^r(\tau, \Xi).$$

If ψ is constant on $[a, b]$ then $\Xi = \mathbb{Z}$. In general we have a continuous dependence of $M^r(\tau, \Xi)$ on the global mesh ratio

$$q(\Xi) = \sup_{i,j \in \mathbb{Z}} \frac{\xi_i - \xi_{i-1}}{\xi_j - \xi_{j-1}}$$

of Ξ. By construction,

(26) $$q(\Xi) \leq \sup_{a \leq s, t \leq b} \frac{\psi(s)}{\psi(t)}.$$

LEMMA 23. *There exists a continuous nondecreasing function Q such that $Q(1) = 1$ and*

$$\frac{1}{Q(q(\Xi))} \leq \frac{M^r(\tau, \Xi)}{M^r(\tau, \mathbb{Z})} \leq Q(q(\Xi)), \qquad \tau \in \,]0, 1[,$$

for every Ξ with $\xi_0 = 0$, $\xi_1 = 1$, and $q(\Xi) < 5/4$.

PROOF. Let $\phi \in C^{r+1}([0, 1])$ with $\phi = 0$ on $\left[0, \frac{1}{5}\right]$, $\phi = 1$ on $\left[\frac{4}{5}, 1\right]$, and $0 \leq \phi' \leq 2$. As before we use c to denote different positive constants, which may only depend on ϕ and r.

Define a function Φ on $\mathbb{R}$ inductively by $\Phi(x) = x$ for $x \in [0, 1]$ and

$$\Phi(x) = \Phi(j) + x - j + \phi(x - j) \cdot (\xi_{j+1} - \xi_j - 1)$$

for $j \in \mathbb{N}$ and $x \in \,]j, j+1]$, as well as

$$\Phi(x) = \Phi(j) - j + x + \phi(j - x) \cdot (\xi_{j-1} - \xi_j + 1),$$

for $-j \in \mathbb{N}_0$ and $x \in [j-1, j[$. Then $\Phi \in C^{r+1}(\mathbb{R})$ with $\Phi(j) = \xi_j$ for every $j \in \mathbb{Z}$. Moreover,

$$\tfrac{1}{2} \leq 1 - 2 \cdot (q(\Xi) - 1) \leq \Phi'(x) \leq 1 + 2 \cdot (q(\Xi) - 1)$$

and

$$|\Phi^{(k)}(x)| \leq c \cdot (q(\Xi) - 1), \qquad k = 2, \ldots, r+1,$$

for every $x \in \mathbb{R}$.

Let $h \in W_2^{r+1}(\mathbb{R})$ with $\int_{\mathbb{R}} h^{(r+1)}(t)^2 \, dt \leq 1$ and $h(\xi_j) = 0$ for $j \in \mathbb{Z}$. Clearly $\widetilde{h} = h \circ \Phi$ satisfies $\widetilde{h}(j) = 0$ and $\widetilde{h}(\tau) = h(\tau)$ for every $\tau \in \,]0, 1[$. Furthermore,

$$|\widetilde{h}^{(r+1)}(x)| \leq |h^{(r+1)}(\Phi(x))| \cdot \Phi'(x)^{r+1} + c \cdot \sum_{k=1}^{r} |h^{(k)}(\Phi(x))| \cdot (q(\Xi) - 1),$$

and since $\int_{\mathbb{R}} h^{(k)}(t)^2 \, dt \leq c$ we obtain

$$\int_{\mathbb{R}} \widetilde{h}^{(r+1)}(t)^2 \, dt \leq Q(q(\Xi))$$

with

$$Q(u) = \left((1 + 2(u-1))^{r+1/2} + c \cdot (u-1)\right)^2.$$

Thus

$$M^r(\tau, \Xi) \leq M^r(\tau, \mathbb{Z}) \cdot Q(q(\Xi)).$$

A reverse estimate is obtained in the same manner.   □

Let

$$d_r = \left(\int_0^1 M^r(\tau, \mathbb{Z}) \, d\tau \right)^{1/2}$$

where

(27) $M^r(\tau, \mathbb{Z})$

$$= \sup\left\{ h(\tau)^2 : h \in W_2^{r+1}(\mathbb{R}), \int_{\mathbb{R}} h^{(r+1)}(t)^2 \, dt \leq 1, \; h(j) = 0 \text{ for } j \in \mathbb{Z} \right\}.$$

Note that (23) yields

$$d_0 = \frac{1}{\sqrt{6}} = d_0^H.$$

Now we determine the asymptotic behavior of the errors of K-spline algorithms $S_n^{\psi, K}$ that are based on the regular sequence generated by a density ψ. As previously, the asymptotic constant consists of two parts. One part captures the dependence on the density ψ and the weight function ϱ. The other part only depends on the smoothness r; here, its value is determined by the values of the extremal problems (27).

PROPOSITION 24. *Let P be a zero mean measure with covariance kernel K that satisfies Sacks-Ylvisaker conditions of order $r \in \mathbb{N}_0$. Then*

$$e_2(S_n^{\psi, K}, \mathrm{App}_{2,\varrho}, P) \approx d_r \cdot I_{\varrho, r}(\psi) \cdot n^{-(r+1/2)}$$

for every density ψ that satisfies (18).

PROOF. Let $n \in \mathbb{N}$ and choose $0 < a < b < 1$ such that

$$u(a, b) = \sup_{a \leq s, t \leq b} \frac{\psi(s)}{\psi(t)}$$

satisfies $u(a, b) < \frac{5}{4}$. By $\sum_i$ we denote summation over all indices i such that $[x_{i-1,n}, x_{i,n}] \subset [a, b]$.

From (24), (26), and Lemma 23 we obtain

$$\int_a^b M^r(t, X_n^\psi, a, b) \cdot \varrho(t) \, dt \geq \sum_i \varrho(s_{i,n}) \cdot \int_{x_{i-1,n}}^{x_{i,n}} M^r(t, X_n^\psi, a, b) \, dt$$

$$\geq \sum_i \varrho(s_{i,n}) \cdot (x_{i,n} - x_{i-1,n})^{2r+2} \int_0^1 M^r(\tau, \Xi) \, d\tau$$

$$\geq \frac{d_r^2}{Q(u(a,b))} \cdot \sum_i \varrho(s_{i,n}) \cdot (x_{i,n} - x_{i-1,n})^{2r+2}$$

for sufficiently large n with $s_{i,n} \in \,]x_{i-1,n}, x_{i,n}[$. As in the proof of Proposition 19, this yields

$$
\liminf_{n \to \infty} n^{2r+1} \cdot \int_a^b M^r(t, X_n^\psi, a, b) \cdot \varrho(t)\, dt
$$
$$
\geq \frac{d_r^2 \cdot \|\psi\|_1^{2r+1}}{Q(u(a,b))} \cdot \int_a^b \varrho(t) \cdot \psi(t)^{-(2r+1)}\, dt.
$$

On the other hand, let a_n and b_n denote the smallest and largest element in $X_n^\psi \cap [a,b]$, respectively. By (25), (26), and Lemma 23,

$$
\int_{a_n}^{b_n} M_0^r(t, X_n^\psi, a, b) \cdot \varrho(t)\, dt = \sum_i \varrho(s_{i,n}) \cdot \int_{x_{i-1,n}}^{x_{i,n}} M_0^r(t, X_n^\psi, a, b)\, dt
$$
$$
\leq \sum_i \varrho(s_{i,n}) \cdot (x_{i,n} - x_{i-1,n})^{2r+2} \int_0^1 M^r(\tau, \Xi)\, d\tau
$$
$$
\leq d_r^2 \cdot Q(u(a,b)) \cdot \sum_i \varrho(s_{i,n}) \cdot (x_{i,n} - x_{i-1,n})^{2r+2}
$$

for sufficiently large n with $s_{i,n} \in \,]x_{i-1,n}, x_{i,n}[$. Furthermore,

$$
\int_a^{a_n} M_0^r(t, X_n^\psi, a, b) \cdot \varrho(t)\, dt \leq c \cdot n^{-(2r+1)} \cdot (a_n - a)^{1/2},
$$

see (21), and the same estimate holds on the interval $[b_n, b]$. We conclude that

$$
\limsup_{n \to \infty} n^{2r+1} \cdot \int_a^b M_0^r(t, X_n^\psi, a, b) \cdot \varrho(t)\, dt
$$
$$
\leq d_r^2 \cdot \|\psi\|_1^{2r+1} \cdot Q(u(a,b)) \cdot \int_a^b \varrho(t) \cdot \psi(t)^{-(2r+1)}\, dt.
$$

Let $0 < \varepsilon < \frac{1}{4}$ and consider a finite covering of $[\varepsilon, 1 - \varepsilon]$ by non-overlapping subintervals $[a_\ell, b_\ell]$ with

$$
u(a_\ell, b_\ell) < 1 + \varepsilon.
$$

Estimating the contribution of $[0, \varepsilon] \cup [1 - \varepsilon, 1]$ by (21) as before, we obtain

$$
\limsup_{n \to \infty} n^{2r+1} \cdot \int_0^1 M_0^r(t, X_n^\psi, a, b) \cdot \varrho(t)\, dt
$$
$$
\leq d_r^2 \cdot \|\psi\|_1^{2r+1} \cdot Q(1 + \varepsilon) \cdot \int_\varepsilon^{1-\varepsilon} \varrho(t) \cdot \psi(t)^{-(2r+1)}\, dt + c \cdot \varepsilon^{1/2}.
$$

The asymptotic upper bound for $e_2(S_n^{\psi, K}, \mathrm{App}_{2,\varrho}, P)$ follows from Lemma 22, since Q is continuous with $Q(1) = 1$.

Using M^r instead of M_0^r, the asymptotic lower bound follows analogously. $\qquad\square$

REMARK 25. By the previous proposition, we can determine regular sequences that yield small asymptotic constants for the errors of the corresponding K-spline algorithms. We have

$$\inf_{\psi} \lim_{n \to \infty} n^{r+1/2} \cdot e_2(S_n^{\psi,K}, \text{App}_{2,\varrho}, P) = d_r \cdot I_{\varrho,r}(\varrho^{1/(2r+2)})$$

and the infimum is attained for the density $\psi = \varrho^{1/(2r+2)}$ if $\inf_{0<t<1} \varrho(t) > 0$, cf. Corollary 20.

In general the modification that is discussed in Remark 21 yields constants that are arbitrarily close to the infimum above. Hence we get the upper bound

$$\limsup_{n \to \infty} n^{r+1/2} \cdot e_2(n, \Lambda^{\text{std}}, \text{App}_{2,\varrho}, P) \leq d_r \cdot I_{\varrho,r}(\varrho^{1/(2r+2)})$$

for the minimal errors of linear methods that are based on function values only. We conjecture that this bound is sharp.

We add that Proposition 19 yields the lower bound

$$\liminf_{n \to \infty} n^{r+1/2} \cdot e_2(n, \Lambda^{\text{std}}, \text{App}_{2,\varrho}, P) \geq d_r^H \cdot I_{\varrho,r}(\varrho^{1/(2r+2)}).$$

For $r > 1$, however, $d_r^H < d_r$.

EXAMPLE 26. Even for linear univariate problems, optimal methods and closed expressions for nth minimal errors are usually unknown. Therefore only asymptotic results are established. Let us check whether such results are relevant for small values of n for a specific example.

We take the twice integrated Wiener measure and study L_2-approximation with $\varrho = 1$. Let S_n denote the Q_2-spline algorithm that uses the knots k/n for $k = 1, \ldots, n$. Recall that $S_n(f)$ is an interpolating polynomial spline of degree 3 in this case. From Proposition 24 we know that

$$\kappa_n = \frac{e_2(S_n, \text{App}_2, w_2)}{d_2 \cdot n^{-5/2}}$$

tends to one as n tends to ∞. The following numbers already reflect this fact:

n	5	10	15	20	25	30
κ_n	1.1501	1.0784	1.0529	1.0399	1.0321	1.0268

Here (III.19) was used in the numerical computation of $e_2(S_n, \text{App}_2, w_2)$. Moreover, an inclusion method for computing d_2 shows that

$$d_2 = 0.0184434 \ldots .$$

3.3. L_2-Approximation with Continuous Linear Functionals.

Finally we consider the class Λ^{all} of permissible functionals. This class contains not only function and derivative evaluations. but also weighted integrals. For instance, wavelet approximation of random functions is studied under various regularity assumptions. See, e.g., Istas (1992), Benassi, Jaffard, and Roux (1993), and Cambanis and Masry (1994).

PROPOSITION 27. *Under the assumption of Proposition 24,*

$$e_2(n, \Lambda^{\mathrm{all}}, \mathrm{App}_2, P) \approx ((2r+1)\pi)^{-1/2} \cdot (n\pi)^{-(r+1/2)}$$

for the constant weight function $\varrho = 1$ and

$$e_2(n, \Lambda^{\mathrm{all}}, \mathrm{App}_{2,\varrho}, P) \asymp n^{-(r+1/2)}$$

for general ϱ.

PROOF. The first asymptotic result is an immediate consequence of Propositions III.24 and 10. The upper bound for general ϱ follows from Proposition 24, and by restriction to $\left[\frac{1}{3}, \frac{2}{3}\right]$ we get the lower bound, since ϱ is bounded away from zero on this subinterval. □

Proposition 27 yields a sharp bound for minimal errors on the class Λ^{all}. It remains, however, to determine (asymptotically) optimal methods that are based on continuous linear functionals. In fact, the solution of the corresponding eigenvalue problem is known only in exceptional cases. See Section III.2.3.

REMARK 28. We compare the asymptotic constants for L_2-approximation based on Hermite data, function values, or continuous linear functionals. For simplicity, we consider the constant weight function $\varrho = 1$. We take $\psi = 1$, which yields an asymptotically optimal method in the Hermite data case and generates at least the best regular sequence in the case of function values. Let $d_r^* = (2r+1)^{-1/2}\pi^{-(r+1)}$ denote the asymptotic constant for the class Λ^{all}, see Proposition 27. Since $n \cdot (r+1)$ bounded linear functionals give Hermite data of order r at n knots, it is reasonable to compare $d_r^H \cdot (r+1)^{r+1/2}$ with d_r and d_r^*, see Propositions 19 and 24. Letting

$$s_r = d_r/d_r^* \qquad \text{and} \qquad h_r = d_r^H \cdot (r+1)^{r+1/2}/d_r^*,$$

we obtain

r	s_r	h_r
0	1.2825	1.2825
1	1.2886	2.3592
2	1.2787	4.5901
3	1.2724	9.1588
4	1.2685	18.5403

We see a clear disadvantage for Hermite data, which gets worse with increasing smoothness r. Cf. Remark 17. The numbers d_r are computed by means of an inclusion method. Therefore we know that s_r does depend on r.

3.4. L_p-Approximation, $p < \infty$. Now we turn to L_p-approximation with $p \notin \{2, \infty\}$. These problems do not belong to the second-order theory of random functions. However, methods of proof from L_2-approximation are applicable if we assume that P is a zero mean Gaussian measure.

First we consider linear methods that use function values and that are based on regular sequences. We sketch how to obtain a generalization of Proposition 24 for finite $p \neq 2$ if P is Gaussian. Thereafter we report on results for arbitrary linear methods.

Consider q-average errors with $q = p$, and let $D \subset \mathbb{R}^d$ for the moment. Fubini's Theorem yields

$$e_p(S_n, \mathrm{App}_{p,\varrho}, P)^p = \int_D \int_F |f(t) - S_n(f)(t)|^p \, dP(f) \cdot \varrho(t) \, dt.$$

Due to our assumption on P, the difference $f(t) - S_n(f)(t)$ is normally distributed with zero mean for every t and every linear method S_n. Therefore

$$(28) \quad e_p(S_n, \mathrm{App}_{p,\varrho}, P)^p = \nu_p^p \cdot \int_D \left(\int_F |f(t) - S_n(f)(t)|^2 \, dP(f) \right)^{p/2} \cdot \varrho(t) \, dt,$$

where ν_p^p is the pth absolute moment of the standard normal distribution, cf. Remark 12.

Given fixed knots x_i, the inner integral in (28) is minimized by the K-spline algorithm for approximation of δ_t, see Section III.3.4. Hence the K-spline algorithm S_n^K for L_2-approximation is also the best method for L_p-approximation. The corresponding error is

$$(29) \quad e_p(S_n^K, \mathrm{App}_{p,\varrho}, P)^p$$
$$= \nu_p^p \int_0^1 \sup\{|h(t)|^p : h \in B(K), \ h(x_i) = 0 \text{ for } i = 1, \ldots, n\} \cdot \varrho(t) \, dt,$$

which generalizes (17). We add that the optimality of K-spline algorithms is also a consequence of Proposition III.34.

The following result is obtained in the same way as Proposition 24; hence we skip the proof. We put

$$d_{r,p} = \left(\int_0^1 M^r(\tau, \mathbb{Z})^{p/2} \, d\tau \right)^{1/p},$$

where $M^r(\tau, \mathbb{Z})$ is defined by (27), and

$$I_{\varrho,r,p}(\psi) = \left(\int_0^1 \varrho(t) \cdot \psi(t)^{-p(r+1/2)} \, dt \right)^{1/p} \cdot \left(\int_0^1 \psi(t) \, dt \right)^{r+1/2}.$$

As previously, we use $S_n^{\psi,K}$ to denote the K-spline algorithms that use the knots from the regular sequence generated by ψ.

PROPOSITION 29. *Let P be a zero mean Gaussian measure with covariance kernel K that satisfies Sacks-Ylvisaker conditions of order $r \in \mathbb{N}_0$. Then*

$$e_p(S_n^{\psi,K}, \mathrm{App}_{p,\varrho}, P) \approx \nu_p \cdot d_{r,p} \cdot I_{\varrho,r,p}(\psi) \cdot n^{-(r+1/2)}$$

for every $1 \leq p < \infty$ and every density ψ that satisfies (18).

Good regular sequences are obtained by minimizing $I_{\varrho,r,p}(\psi)$, as discussed in Remark 25 for $p = 2$. We get the upper bound

$$\limsup_{n \to \infty} n^{r+1/2} \cdot e_p(n, \Lambda^{\mathrm{std}}, \mathrm{App}_{p,\varrho}, P) \leq \nu_p \cdot d_{r,p} \cdot I_{\varrho,r,p}(\varrho^{1/(p(r+1/2)+1)})$$

by taking

(30) $\psi = \varrho^{1/(p(r+1/2)+1)}.$

This bound is sharp for regularity $r = 0$, and we conjecture that the same is true for general r. The order $n^{-(r+1/2)}$ of minimal errors for the class Λ^{std} is easily shown.

Speckman (1979) studies L_p-approximation with $1 \le p = q < \infty$ for autoregressive Gaussian processes, see Remark 18. In the Hermite data case he proves the asymptotic optimality of K-spline algorithms based on the regular sequence corresponding to (30).

Observe that the error $e_q(S_n, \mathrm{App}_{p,\varrho}, P)$ depends monotonically on p and q. Therefore we conclude

$$e_q(n, \Lambda^{\mathrm{std}}, \mathrm{App}_{p,\varrho}, P) \asymp n^{-(r+1/2)}$$

for all $1 \le p, q < \infty$ under Sacks-Ylvisaker conditions.

Now we turn to the class Λ^{all} of permissible functionals. The lower bound $n^{-(r+1/2)}$ from Proposition 27 extends to $p, q \ge 2$. The same bound is obtained for $p = q = 1$ and the r-fold integrated Wiener measure w_r in Maiorov (1992) and Sun and Wang (1994). The proof of the following result additionally relies on Proposition 8.

PROPOSITION 30. *Under the assumption of Proposition 29,*

$$e_q(n, \Lambda^{\mathrm{all}}, \mathrm{App}_{p,\varrho}, P) \asymp n^{-(r+1/2)}$$

for all $1 \le p, q < \infty$.

3.5. L_∞-Approximation. A different approach is taken in the analysis of L_∞-approximation. As previously we assume that P is a zero mean Gaussian measure that satisfies Sacks-Ylvisaker conditions of order r.

Consider a sequence of knots $x_{i,n}$, where $i = 1, \ldots, n$, that becomes dense in $[0, 1]$. Let S_n^K and $S_n^{Q_r}$ denote the corresponding spline algorithms for the kernel K of P and for the kernel Q_r of the r-fold integrated Wiener measure w_r, respectively. If the knots form a regular sequence generated by ψ, then we use $S_n^{\psi, K}$ to denote the corresponding K-spline algorithm.

By means of Propositions 6, 8 and Anderson's inequality one can show that

(31) $e_q(S_n^K, \mathrm{App}_{\infty,\varrho}, P) \approx e_q(S_n^{Q_r}, \mathrm{App}_{\infty,\varrho}, w_r).$

Anderson's inequality states that

$$Q_{\Sigma_1}(A) \ge Q_{\Sigma_2}(A)$$

for every convex and symmetric set $A \subset \mathbb{R}^k$ and centered normal distributions Q_{Σ_i} on $\mathbb{R}^k$ with covariance matrices Σ_i such that $\Sigma_2 - \Sigma_1$ is nonnegative definite, see Tong (1980, p. 55). Estimates for the distribution of $\|f - S_n^{Q_0}(f)\|_\infty$ with respect to the Wiener measure yield the following result.

PROPOSITION 31. *Let P be a zero mean Gaussian measure with covariance kernel K that satisfies Sacks-Ylvisaker conditions of order $r = 0$. Assume that ϱ is positive and continuous on $[0,1]$. Then the regular sequence generated by $\psi = \varrho^2$ leads to asymptotically optimal methods with*

$$e_q(S_n^{\varrho^2,K}, \mathrm{App}_{\infty,\varrho}, P) \approx \left(1/2 \cdot \int_0^1 \varrho^2(t)\, dt\right)^{1/2} \cdot n^{-1/2} \cdot (\ln n)^{1/2}$$

$$\approx e_q(n, \Lambda^{\mathrm{std}}. \mathrm{App}_{\infty,\varrho}, P)$$

for every $1 \leq q < \infty$. Equidistant knots, i.e., $\psi = 1$, yield

$$e_q(S_n^{1,K}, \mathrm{App}_{\infty,\varrho}, P) \approx 2^{-1/2} \cdot \sup_{t \in [0,1]} \varrho(t) \cdot n^{-1/2} \cdot (\ln n)^{1/2}.$$

For regularity $r > 0$ only the order of minimal errors and order optimal methods are known. The following results are originally obtained for w_r, their generalization is due to (31) and Proposition 8. In the proofs, additional knots with Hermite data are used in order to allow local considerations.

PROPOSITION 32. *Let P be a zero mean Gaussian measure that satisfies Sacks-Ylvisaker conditions of order $r \in \mathbb{N}_0$. Assume that ϱ is positive and continuous on $[0,1]$. Then*

$$e_q(n, \Lambda^{\mathrm{std}}, \mathrm{App}_{\infty}, P) \asymp e_q(n, \Lambda^{\mathrm{all}}, \mathrm{App}_{\infty}, P) \asymp n^{-(r+1/2)} \cdot (\ln n)^{1/2}$$

for every $1 \leq q < \infty$.

REMARK 33. Due to the previous results, sampling is order optimal for L_p-approximation for every $1 \leq p \leq \infty$, see Remark III.27. The classes Λ^{std} and Λ^{all} yield the same order of minimal errors. For every choice of p, q, and ϱ interpolation with natural splines of degree $2r+1$ at equidistant knots $x_{i,n} = (i-1)/(n-1)$ is order optimal. However, the orders of minimal errors differ slightly in the cases $p < \infty$ and $p = \infty$. Linear polynomial operators S_n that yield the optimal order in Propositions 30 and 32 are studied in Sun and Wang (1995).

These conclusions remain correct if we drop the boundary condition (D) in our assumptions on the covariance kernel K. The proof relies on Proposition 6, which says, in particular, that $H(K)$ is is a closed and finite-codimensional subspace in $H(R_r)$.

3.6. Notes and References. 1. Proposition 19, with generalization to finite $p = q$, is obtained in Speckman (1979) for autoregressive random functions. For regularity $r = 0$ and $p = q = 2$ the result is also obtained by Su and Cambanis (1993) and Müller-Gronbach (1996) under Sacks-Ylvisaker conditions.

2. Propositions 24 and 29 are due to Ritter (1996c).

3. Papageorgiou and Wasilkowski (1990) establish Proposition 27 in the case of the r-fold integrated Wiener measure w_r and $\varrho = 1$.

4. Proposition 30 is due to Maiorov (1992) and Sun and Wang (1994) for w_r and $\varrho = 1$. It is generalized to a class of measures containing w_r in Wang (1994).

5. Proposition 31 is due to Müller-Gronbach and Ritter (1996). The particular case w_0 and $\varrho = 1$ is studied in Ritter (1990), and kernels of the form $K(s,t) = u(\min(s,t)) \cdot v(\max(s,t))$ are considered in Müller-Gronbach (1994).

6. The results for $\Lambda = \Lambda^{\text{std}}$ and $\Lambda = \Lambda^{\text{all}}$ in Proposition 32 are due to Wasilkowski (1992) and Maiorov and Wasilkowski (1996), respectively, for w_r.

4. Unknown Mean and Covariance Kernel

In the previous two sections, we have determined asymptotically optimal knots or best regular sequences of knots. These knots depend only on the weight function ϱ and the smoothness r of a zero mean random function. Therefore we need not precisely know the covariance kernel K of the measure P in order to find good knots.

Once these knots $x_i = x_{i,n}$ are found, however, we have used K-spline algorithms. The corresponding coefficients depend on $K(x_i, x_j)$ as well as on $\int_0^1 K(s, x_i) \cdot \varrho(s)\, ds$ for integration and on $K(\cdot, x_i)$ for approximation. See Section III.3. Therefore K has to be known to determine these coefficients. Moreover, a system of linear equations with the covariance matrix $(K(x_i, x_j))_{1 \le i,j \le n}$ has to be solved, if $K(\cdot, x_1), \ldots, K(\cdot, x_n)$ is used as a basis for the spline space. These systems usually are ill-conditioned even for moderate values of n.

Furthermore, we have studied average errors with respect to zero mean measures, which is basically the same as assuming that a possibly nonzero mean m is known. If $m \ne 0$ then $f \mapsto f - m$ yields a zero mean measure without changing the covariance kernel. Knowing m we may take a good linear method for this measure, apply it to the centered data $f(x_i) - m(x_i)$, and simply add $S(m)$ for any linear problem with operator S. This leads to a good affine linear method in the non-centered case. Cf. Section III.4.1.

In Section 4.1 we show how good coefficients can be found in a stable and effective way, without precisely knowing the covariance kernel K. In Section 4.2 we discuss parametric and nonparametric models for an unknown mean m. The results in Sections 4.1 and 4.2 specifically deal with integration and approximation of univariate functions. A general result for prediction problems with unknown mean or covariance kernel is presented in Section 4.3.

Partial knowledge about K and m may lead to uncertainty regarding the local smoothness of the random function. This problem will be addressed in Section VII.3, where adaptive methods are studied.

4.1. Unknown Covariance Kernel. Suppose that P is a zero mean measure that satisfies Sacks-Ylvisaker conditions of order $r \in \mathbb{N}_0$. At first we assume that r is known.

For a low degree of smoothness, namely for $r = 0$, the use of K-splines can be avoided by taking piecewise linear interpolation. In particular, the asymptotically best regular sequence and piecewise linear interpolation yield asymptotically optimal methods for L_p-approximation. Analogously, trapezoidal rules based on the best regular sequences are asymptotically optimal for integration.

For approximation with $p = q = 2$ this is shown in Su and Cambanis (1993), and their result can be extended to $p = q < \infty$ for Gaussian measures as indicated in Section 3.4. A different proof is needed in the case $p = \infty$ for Gaussian measures, see Müller-Gronbach and Ritter (1996). The result for integration is due to Sacks and Ylvisaker (1970b).

Because of the results of Section 1 on reproducing kernel Hilbert spaces, interpolation with natural splines of degree $2r + 1$ is a reasonable choice for integration and approximation, also if $r > 0$. We do not know, however, whether the asymptotic behavior remains the same, if we switch from K-splines to natural splines.

Now we consider the integration problem with arbitrary $r \in \mathbb{N}_0$ and a weight function $\varrho \in C^{r+2}([0, 1])$. Let a regular sequence of knots be given. Benhenni and Cambanis (1992a) show that the asymptotic behavior does not change, if we switch from K-spline algorithms for Int_ϱ to weighted Gregory formulas of order r. These formulas depend only on ϱ, r, and on the knots x_i. Taking the asymptotically best regular sequence of knots, we end up with an asymptotically optimal sequence of quadrature formulas that only depends on the weight function ϱ and the smoothness r of the random function. For quadrature based on Hermite data, a weighted Rodriguez formula enjoys the same properties, see Benhenni and Cambanis (1992b). Numerical experiments are presented in Benhenni and Cambanis (1992a, 1992b). Further results are discussed in Cambanis (1985).

The constructions above of asymptotically optimal methods require the knowledge of the smoothness r in the sense of Sacks and Ylvisaker. Istas and Laredo (1994, 1997) present quadrature formulas that work well if only an upper bound for the smoothness is known.

Their construction is based on a regular sequence of knots and a kernel function $e \in L_\infty(\mathbb{R})$ with compact support such that, for some $\ell \in \mathbb{N}$,

$$(32) \qquad \sum_{k=-\infty}^{\infty} e(t - k) \cdot (k - t)^p = \begin{cases} 1 & \text{if } p = 0, \\ 0 & \text{if } p = 1, \dots, \ell \end{cases}$$

for all $t \in \mathbb{R}$. Examples of such kernel functions include linear combinations of B-splines and suitable scale functions of a multiresolution analysis, e.g., coiflets. A simple example is the B-spline e of degree one with breakpoints -1, 0, and 1 and $e(0) = 1$. In this case (32) holds for $\ell = 1$.

Scaled translates of e are used to approximate the integrand; however, a modification is needed near the boundary of $[0, 1]$. Let $\mathrm{supp}\, e \subset [-A, A]$ for some $A \in \mathbb{N}$, and take sequences $(g_m(k))_{k \in \mathbb{N}_0}$ for $m = 1, \dots, A$ such that $g_m(k) \neq 0$ for only finitely many k and

$$\sum_{k=0}^{\infty} g_m(k) \cdot k^p = (-m)^p, \qquad p = 0, \dots \ell.$$

Let $\psi \in C^{\ell+2}([0,1])$ denote the positive density that generates the regular sequence, and put $\Psi(t) = \int_0^t \psi(s)\,ds / \int_0^1 \psi(s)\,ds$. To simplify the notation, we consider knots $x_{i,n}$ for $i = 0,\ldots,n$ that are determined by $\Psi(x_{i,n}) = i/n$. Furthermore there are auxiliary knots $x_{i,n} = i/n$, say, for $i = -A,\ldots,-1$ and $i = n+1,\ldots,n+A$. An integrand f is evaluated at $x_{0,n},\ldots,x_{n,n}$, and auxiliary function values are defined as

$$f(x_{-m,n}) = \sum_{i=0}^{n} f(x_{i,n}) \cdot g_m(i)$$

and

$$f(x_{n+m,n}) = \sum_{i=0}^{n} f(x_{i,n}) \cdot g_m(n-i)$$

for $m = 1,\ldots,A$. Instead of an interpolating spline in $H(K)$ we use

$$f_n^{\psi,e}(t) = \sum_{i=-A}^{n+A} f(x_{i,n}) \cdot e(n\,\Psi(t) - i)$$

as an approximation of the integrand. A quadrature formula $S_n^{\psi,e}$ is given by

(33) $$S_n^{\psi,e}(f) = \mathrm{Int}_\varrho(f_n^{\psi,e}).$$

Note that $S_n^{\psi,e}$ depends on ϱ and $g_1,\ldots,g_A$, in addition to ψ, n, and e.

Assume that P satisfies (A) and (B) from the Sacks-Ylvisaker conditions with $r \in \mathbb{N}_0$. If $\ell \geq r+1$ and if ϱ is piecewise continuous, then

(34) $$e_2(S_n^{\psi,e}, \mathrm{Int}_\varrho, P) \approx c_{e,r} \cdot J_{\varrho,r}(\psi) \cdot n^{-(r+1)}$$

with a constant $c_{e,r}$, which depends only on the kernel e and the smoothness r of the random function. See Istas and Laredo (1994, 1997). We conclude that $S_n^{\psi,e}$ is order optimal under Sacks-Ylvisaker conditions for every $r = 0,\ldots,\ell-1$. The asymptotic constant for the error of $S_n^{\psi,e}$ is not optimal in general, even in the case $\varrho = \psi = 1$.

Note, however, that piecewise linear interpolation corresponds to the simplest version of the method $S_n^{\psi,e}$, where e is the B-spline of degree one with breakpoints -1, 0, and 1 such that $e(0)=1$. Furthermore, $A = 1$ and $g_1 = (1,0,0,\ldots)$.

4.2. Unknown Mean.

In a nonparametric approach one imposes smoothness conditions on the unknown mean m. The available results are instances of the following fact. If m is not known precisely but sufficiently smooth, then m is irrelevant asymptotically. One takes a good method from the case $m = 0$ and ignores the presence of a nonzero mean. We mention two examples.

Consider piecewise linear interpolation based on knots that get dense in $[0,1]$, and assume Sacks-Ylvisaker conditions with $r = 0$. The average errors for approximation are not affected asymptotically by any mean m that is Hölder continuous with exponent $\beta > \frac{1}{2}$. See Su and Cambanis (1993) for the case

$p = 2$ with stronger assumptions for m. For L_∞-approximation it suffices to have $\beta = 1/2$, see Müller-Gronbach and Ritter (1996).

Consider the quadrature formulas $S_n^{\psi,e}$, see (33). Istas and Laredo (1994, 1997) show that (34) remains true for every mean $m \in C^{r+1}([0,1])$.

In a linear parametric approach we assume that m is of the form

$$m = \sum_{k=1}^{\ell} \beta_k \cdot f_k$$

with known and linearly independent functions $f_k : [0,1] \to \mathbb{R}$ and unknown coefficients $\beta_k \in \mathbb{R}$. We have already studied this model in Section III.4.2. Suppose that K is known and consider the approximation problem. We may use the best linear unbiased predictor for $f(t)$ at every point $t \in [0,1]$, see Proposition III.36. Hereby we get a linear method S_n^1 for approximation. Alternatively, one can simply ignore the unknown mean and use the K-spline algorithm S_n^0. This method is biased in general.

Both methods have been compared asymptotically under Sacks-Ylvisaker conditions of order $r = 0$, if

$$f_k(t) = \int_0^1 K(s,t) \cdot \varrho_k(s)\, ds$$

with known functions $\varrho_k \in C([0,1])$. The set of all functions of this form is a dense subset of the Hilbert space $H(K)$, and any such function is Lipschitz continuous. Consider a regular sequence of points and the corresponding methods S_n^0 and S_n^1. Then

$$e_q(S_n^1, \mathrm{App}_p, P) \approx e_q(S_n^0, \mathrm{App}_p, P).$$

The same asymptotical behavior is also achieved by piecewise linear interpolation under weaker assumptions for f_k, namely, Hölder continuity with exponent greater than $\frac{1}{2}$ or at least $\frac{1}{2}$ for L_2-approximation or L_∞-approximation, respectively. See Su and Cambanis (1993) for L_2-approximation with $q = 2$ and Müller-Gronbach and Ritter (1996) for L_∞-approximation and Gaussian measures P.

4.3. Compatible Second-Order Structures. Sections 4.1 and 4.2 deal with a specific problem, integration or approximation of univariate functions under particular smoothness assumptions. In a series of papers, Stein studies the effect of a misspecified mean m and covariance function K for general prediction problems, see Stein (1990a, 1990b). Consider an arbitrary compact set $D \subset \mathbb{R}^d$, instead of $D = [0,1]$, and an arbitrary linear functional $S \in \Lambda^{\mathrm{all}}$, instead of $S = \mathrm{Int}_\varrho$. Moreover, consider a fixed sequence of knots $x_{1,n}, \ldots, x_{n,n} \in D$ that gets dense in D. Two second-order structures (m_0, K_0) and (m_1, K_1) on D are called *compatible*, if the corresponding Gaussian measures on $\mathbb{R}^D$ are mutually absolutely continuous. Stein shows that a wrong but compatible second-order structure (m_0, K_0) instead of the correct second-order structure (m_1, K_1) still yields asymptotically efficient methods.

More precisely, let $T_{n,j}(f) - m_j$ denote the spline in $H(K_j)$ that interpolates the data $f(x_{i,n}) - m_j(x_{i,n})$ at the knots $x_{i,n}$ for $i = 1, \ldots, n$. Furthermore, let P_j denote a measure on a class of real-valued functions on D with mean m_j and covariance kernel K_j. Clearly $S(T_{n,j}(f))$ defines an affine linear method. It has minimal average error $e_2(S \circ T_{n,j}, S, P_j)$ among all affine linear methods that use the knots $x_{1,n}, \ldots, x_{n,n}$ for every such measure P_j. The correct second-order structure (m_1, K_1) is assumed to be compatible to the structure (m_0, K_0) that is actually used for prediction. Stein shows that the methods $S \circ T_{n,0}$ are still uniformly asymptotically efficient, i.e.,

$$\lim_{n \to \infty} \sup_{S \in \Lambda^{\mathrm{all}}} \frac{e_2(S \circ T_{n,0}, S, P_1)}{e_2(S \circ T_{n,1}, S, P_1)} = 1.$$

Compatibility of (m_1, K_1) and (m_0, K_0), or, more generally (m_0, cK_0) for some constant $c > 0$, is rather restrictive. For instance, if the kernels K_j satisfy the Sacks-Ylvisaker conditions of order r_j with $r_0 \neq r_1$, then $(0, K_0)$ and $(0, K_1)$ are not compatible.

We add that Wahba (1974) gives a detailed analysis of the role of compatibility for integration of univariate functions using Hermite data, see Remark 18.

5. Further Smoothness Conditions and Results

A variety of smoothness conditions is used for the analysis of integration or approximation of univariate random functions. So far, we have considered the Sacks-Ylvisaker conditions. These conditions uniquely determine the order of minimal errors for integration and, at least if $r = 0$, for approximation.

In this section we survey other smoothness conditions and present a selection of results. At first we discuss Hölder conditions and locally stationary random functions. In both cases one can only derive upper bounds for the order of minimal errors. Then we discuss stationary random functions. It is unknown whether the respective smoothness conditions for stationary random functions determine the asymptotical behavior of minimal errors. See Notes and References 5.4.6–8 for problems with multifractional or analytic random functions or smoothness in the L_q-sense with $q \neq 2$.

Throughout this section, we let P denote a zero mean measure on a class F of real-valued functions on $[0, 1]$. We assume $K \in C^{r,r}([0, 1]^2)$ with $r \in \mathbb{N}_0$ for the covariance kernel K of P. This property holds iff the corresponding random function has continuous derivatives in quadratic mean up to order r, see Lemma III.14.

5.1. Hölder Conditions. We say that P or K satisfies a *Hölder condition* if

$$(35) \qquad \int_F (f^{(r)}(s) - f^{(r)}(t))^2 \, dP(f) \leq c \cdot |s - t|^{2\beta}$$

for all $s, t \in [0, 1]$ with constants $c > 0$ and $0 \leq \beta \leq 1$. Recall that the left-hand side can be expressed in terms of $K^{(r,r)}$, see (1).

A Hölder condition yields upper bounds for the minimal errors for integration and approximation. See Section VI.1.1, where the multivariate case $s, t \in [0, 1]^d$ is studied. In the univariate case these upper bounds are as follows.

PROPOSITION 34. *Assume that* (35) *holds with* $r \in \mathbb{N}$ *and* $0 \leq \beta \leq 1$ *for a Gaussian measure* P. *Let* $1 \leq q < \infty$. *Then*

$$e_q(n, \Lambda^{\mathrm{std}}, \mathrm{Int}, P) = O(n^{-(r+\beta+1/2)})$$

for integration,

$$e_q(n, \Lambda^{\mathrm{std}}, \mathrm{App}_p, P) = O(n^{-(r+\beta)})$$

for L_p-*approximation with* $p < \infty$, *and*

$$e_q(n, \Lambda^{\mathrm{std}}, \mathrm{App}_\infty, P) = O(n^{-(r+\beta)} \cdot (\ln n)^{1/2})$$

for L_∞-*approximation.*

For integration or L_2-approximation with $p = 2$ we may clearly drop the assumption that P is Gaussian.

In the univariate case one often uses a more precise specification of the local smoothness in order to determine the strong asymptotic behavior of certain linear methods. We discuss two approaches in the sequel.

5.2. Locally Stationary Random Functions. *Local stationarity* means that $f^{(r)}$ is precisely Hölder continuous with exponent β and *local Hölder constant* $\gamma(t)$ at every point t. Formally,

$$(36) \qquad \lim_{h \to 0} \ \sup_{0 < |s - t| \leq h} \left(\int_F (f^{(r)}(s) - f^{(r)}(t))^2 \, dP(f)/|s - t|^{2\beta} - \gamma(t)^2 \right) = 0,$$

where γ is positive and continuous on $[0, 1]$. For $r = 0$ this notion was introduced by Berman (1974), who also required $K(t, t) = 1$. The smoothness of the random function is thus defined by r, β, and the function γ.

EXAMPLE 35. If P satisfies the Sacks-Ylvisaker conditions (A) and (B), then (36) holds with $\beta = \frac{1}{2}$ and $\gamma = 1$. Cf. Remarks 1 and 2.

Let $0 < \beta < 1$ and consider the covariance kernel

$$K(s, t) = \tfrac{1}{2} \left(|s|^{2\beta} + |t|^{2\beta} - |s - t|^{2\beta} \right).$$

Since

$$K(s, s) - 2K(s, t) + K(t, t) = |s - t|^{2\beta},$$

we have (36) with $r = 0$ and $\gamma = 1$ in this case. The zero mean Gaussian measure with the covariance kernel K corresponds to a *fractional Brownian motion*, see Lifshits (1995, p. 36). For $\beta = \frac{1}{2}$ we get $K(s, t) = \min(s, t)$, which is the covariance kernel of the Wiener measure.

Examples with $r > 0$ are obtained by r-fold integration from examples with $r = 0$. Multiplication with sufficiently smooth deterministic functions yields examples with non-constant γ.

Suppose that Hermite data $f(x_{i,n}), \ldots, f^{(r)}(x_{i,n})$ of order r are available at the knots $x_{1,n}, \ldots, x_{n,n}$ from a regular sequence generated by ψ. Piecewise interpolation by polynomials of degree at most $2r+1$ defines a simple method $S_n^{\psi,r}$ for L_p-approximation. For finite p and the constant weight function $\varrho = 1$ the errors of $S_n^{\psi,r}$ are known asymptotically.

PROPOSITION 36. *Let $1 \leq p < \infty$, and assume that (36) holds for a Gaussian measure P with $0 < \beta < 1$. Then there exists a constant $c_{r,\beta,p} > 0$ such that*

$$e_p(S_n^{\psi,r}, \mathrm{App}_p, P) \approx c_{r,\beta,p} \cdot \left(\int_0^1 \gamma(t)^p \cdot \psi(t)^{-p(r+\beta)} \, dt \right)^{1/p} \cdot n^{-(r+\beta)},$$

if ψ is positive and continuous with $\int_0^1 \psi(t) \, dt = 1$.

For $p = 2$ we may clearly drop the assumption that P is Gaussian. For fixed r, β, p, and γ the asymptotic constant is minimized if ψ is proportional to γ^θ with

$$\theta = \frac{1}{r + \beta + 1/p}.$$

In this case the constant is

$$c_{r,\beta,p} \cdot \left(\int_0^1 \gamma(t)^{1/(r+\beta+1/p)} \, dt \right)^{r+\beta+1/p}.$$

REMARK 37. Neither (35) nor (36) imply nontrivial lower bounds for the nth minimal errors. In fact, $e_q(n, \Lambda^{\mathrm{std}}, \mathrm{App}_p, P) = 0$ might hold for every n. A corresponding Gaussian measure P is constructed as follows. Take a deterministic function $f : [0,1] \to \mathbb{R}$ with local Hölder constant $\gamma(t)$ for Hölder exponent β at t, and let the random variable X have a standard normal distribution. Then a single evaluation of $X \cdot f$ at a point t with $f(t) \neq 0$ suffices to recover $X \cdot f$ on the interval $[0,1]$.

On the other hand the upper bound $e_p(n, \Lambda^{\mathrm{std}}, \mathrm{App}_p, P) = O(n^{-(r+\beta)})$ from Proposition 36 cannot be improved in general. See Section 3.4 for the case $\beta = \frac{1}{2}$ and Section VI.1.2 for the case $0 < \beta < 1$.

5.3. Stationary Random Functions.

Stationarity (in the wide sense) means invariance of the covariance kernel K with respect to shifts, i.e.,

$$K(s,t) = K(0, |t - s|)$$

for all $s, t \in [0,1]$. The corresponding random function is also called (wide sense) stationary. In general, the behavior at all points (t, t) on the diagonal is crucial for the smoothness of covariance kernels. In case of stationarity it suffices to look at the behavior around the origin $(0,0)$.

Assuming stationarity, we put

$$k(t) = K(0, |t|), \qquad t \in [-1, 1].$$

If

$$k \in C^{2r}([-1, 1])$$

for some $r \in \mathbb{N}_0$, then $K \in C^{r,r}([0,1]^2)$ and

$$K^{(i,j)}(s,t) = (-1)^j \cdot k^{(i+j)}(s - t)$$

for $i, j \in \{0, \ldots, r\}$. Furthermore, assume that

(37) $$k^{(2r)}(t) = k^{(2r)}(0) - b \cdot (-1)^r \cdot |t|^{2\beta} + v(t)$$

for $t \in [-1, 1]$, where $b > 0$,

$$0 < \beta < 1,$$

and

$$v(t) = o(|t|^{2\beta}).$$

In addition, v has to be sufficiently smooth; here we require that $v|_{[0,1]}$ is three times continuously differentiable. The smoothness of the random function is thus defined by the parameters r, β, and b.

From (37) we get

$$\int_F (f^{(r)}(s) - f^{(r)}(t))^2 \, dP(f) = 2 \cdot (b \cdot |s - t|^{2\beta} - (-1)^r \cdot v(s - t)),$$

so that (36) holds with $\gamma^2 = 2b$. For convenience we use the normalization

$$b = \tfrac{1}{2}.$$

It is easy to see that the Sacks-Ylvisaker conditions (A) and (B) are satisfied iff $\beta = \tfrac{1}{2}$.

EXAMPLE 38. The covariance kernel $K(s,t) = \tfrac{1}{2}\exp(-|s-t|^{2\beta})$ has a singularity of the form (37) with $r = 0$. The particular case $\beta = \tfrac{1}{2}$, which corresponds to an Ornstein-Uhlenbeck process, was already considered in Example 3. Suitable r-fold integration that preserves the stationarity yields examples for every $r \geq 1$. Cf. Example 3 and Remark 9.

Consider the integration problem with a weight function ϱ, and assume that a density ψ is given. Let $S_n^{\psi,r}$ denote the weighted Gregory formula of order r that uses the knots $x_{1,n}, \ldots, x_{n,n}$ from the regular sequence generated by ψ, see Section 4.1. Note that this quadrature formula also depends on ϱ, in addition to n, r, and ψ.

PROPOSITION 39. Let $\varrho, \psi \in C^{r+3}([0,1])$ with $\psi > 0$ and $\int_0^1 \psi(t) \, dt = 1$, and assume that (37) holds. Then

$$e_2(S_n^{\psi,r}, \mathrm{Int}_\varrho, P) \approx c_{r,\beta} \cdot \left(\int_0^1 \varrho(t)^2 \cdot \psi(t)^{-(2(r+\beta)+1)} \, dt \right)^{1/2} \cdot n^{-(r+\beta+1/2)},$$

where

$$c_{r,\beta}^2 = \frac{|\zeta(-2(r+\beta))|}{(2r+2\beta) \cdot (2r-1+2\beta) \ldots (1+2\beta)}$$

and ζ denotes the Riemann zeta function.

Suppose additionally that $\varrho > 0$. Then the density ψ that is proportional to ϱ^θ with

$$\theta = \frac{1}{r+\beta+1}$$

leads to the smallest asymptotic constant for given r, β, and $\varrho > 0$, namely

$$c_{r,\beta} \cdot \left(\int_0^1 \varrho(t)^{1/(r+\beta+1)} \, dt \right)^{r+\beta+1}.$$

It is unknown whether the corresponding weighted Gregory formulas $S_n^{\psi,r}$ are asymptotically optimal in general. We conjecture that this is indeed the case.

The conjecture holds at least for $\beta = \frac{1}{2}$ and v satisfying the following property. Let $L = K^{(r,r)}$. For every $s \in [0,1]$ the function $v''(s - \cdot)$ belongs to the reproducing kernel Hilbert space $H(L)$ and $\sup_{0 < s < 1} \|v''(s - \cdot)\|_L < \infty$. In this case the Sacks-Ylvisaker conditions of order r are satisfied. Moreover,

$$|\zeta(-2r - 1)| = \frac{|b_{2r+2}|}{2r+2}$$

with b_k denoting the kth Bernoulli number. Hence

$$c_{r,1/2}^2 = \frac{|b_{2r+2}|}{(2r+2)!},$$

and Proposition 15 yields

$$e_2(n, \Lambda^{\mathrm{std}}, \mathrm{Int}_\varrho, P) \approx e_2(S_n^{\psi,r}, \mathrm{Int}_\varrho, P).$$

In Section VI.1.2 we study stationary random functions in the multivariate case. In particular, we will see that the upper bound $e_p(n, \Lambda^{\mathrm{std}}, \mathrm{Int}_\varrho, P) = O(n^{-(r+\beta+1/2)})$ from Proposition 39 cannot be improved.

5.4. Notes and References. 1. For L_∞-approximation Proposition 34 is due to Buslaev and Seleznjev (1999). See Section VI.1.1 for a multivariate version of Proposition 34 in the other cases. Hölder conditions are used in Cambanis and Masry (1994) for L_2-approximation of stationary random functions.

2. Proposition 36 is due to Seleznjev (2000) in a more general form.

3. Proposition 39 is due to Stein (1995c) for $r + \beta < \frac{3}{2}$ and due to Benhenni (1997, 1998) for arbitrary $r \in \mathbb{N}_0$ and $0 < \beta < 1$. According to Stein (1995c) the same result holds for the fractional Brownian motion, where $r = 0$ and $0 < \beta < 1$. For $r + \beta < \frac{3}{2}$, Pitt, Robeva, and Wang (1995) obtained essentially the same result. In their work the integral is a line integral of a stationary two-dimensional random field. We add that Stein (1995c) analyzes so-called median sampling, where

$$\int_0^{x_{i,n}} \psi(t)\, dt = \frac{i - 1/2}{n} \cdot \int_0^1 \psi(t)\, dt$$

for $i = 1, \ldots, n$.

4. Istas and Laredo (1994, 1997) analyze quadrature formulas that work well if only an upper bound for $r + \beta$ in the smoothness condition (37) for stationary random functions is known. See Section 4.1 for the respective quadrature formulas and an asymptotic result in the particular case $\beta = \frac{1}{2}$.

5. Seleznjev (1991) considers L_∞-approximation for periodic zero mean Gaussian random functions that are stationary and satisfy (37). He determines the asymptotic behavior of the distribution function

$$F_n(u) = P(\{f : \|f - S_n(f)\|_\infty \le u\}),$$

where $S_n(f)$ are trigonometric Jackson polynomials. The method S_n uses Fourier coefficients, i.e., functionals from $\Lambda^{\mathrm{all}} \setminus \Lambda^{\mathrm{std}}$. See Piterbarg and Seleznjev (1994), Seleznjev (1996), and Hüsler (1999) for similar results on L_∞-approximation by piecewise linear interpolation.

6. Istas (1997) studies integration of multifractional random functions by means of quadrature formulas from Stein (1995c). Multifractional random functions satisfy a Hölder condition where the exponent β and the constant γ depend on the location t.

7. See Benhenni and Istas (1998) for results on integration of analytic random functions.

8. So far we have always studied smoothness in quadratic mean. Weba (1991a, 1991b, 1992) considers random functions with continuous derivatives up to some order in L_q-sense with $q \neq 2$. He obtains upper bounds for the q-average errors of specific methods for integration or L_∞-approximation.

6. Local Smoothness and Best Regular Sequences

The relation between the local smoothness of random functions and the optimal selection of knots becomes more transparent if we slightly change our point of view on integration and L_p-approximation. Suppose that a zero mean random function g, whose kernel satisfies Sacks-Ylvisaker conditions of order r, and a positive and sufficiently smooth function γ is given. The random function $f = \gamma \cdot g$ is locally stationary, as (36) holds with $\beta = \frac{1}{2}$.

Suppose that a positive and sufficiently smooth weight function ϱ is given, and consider regular sequences of knots and corresponding spline algorithms. For integration and, at least if $r = 0$, for approximation the best choice of a density ψ leads to asymptotically optimal methods.

Clearly the problem Int_ϱ for the random function g is equivalent to the integration problem with weight function one for the random function $\gamma \cdot g$, where

$$\gamma = \varrho.$$

According to Proposition 15 the best density is

$$\psi = \gamma^{1/(r+3/2)}.$$

Analogously, the problem $\mathrm{App}_{p,\varrho}$ with $p < \infty$ for g is equivalent to L_p-approximation with weight function one for the random function $\gamma \cdot g$, where

$$\gamma = \varrho^{1/p}.$$

The best density is

$$\psi = \gamma^{1/(r+1/2+1/p)},$$

see (30).

Finally, $\mathrm{App}_{\infty,\varrho}$ for the random function g is equivalent to $\mathrm{App}_{\infty,1}$ for the random function $\gamma \cdot g$, where

$$\gamma = \varrho.$$

We conjecture that

$$\psi = \gamma^{1/(r+1/2)}$$

is the best density and leads to asymptotically optimal methods for L_∞-approximation. According to Proposition 31 the conjecture holds for $r = 0$.

The results and conjectures suggest to choose

$$(38) \qquad\qquad\qquad \psi = \gamma^\theta,$$

where

$$\theta = \begin{cases} (r + \beta + 1/p)^{-1} & \text{for } L_p\text{-approximation,} \\ (r + \beta + 1)^{-1} & \text{for integration.} \end{cases}$$

Hence there is continuous dependence of the (best) density ψ on the parameter p, if ψ is expressed as a power of the local Hölder constant γ. Moreover, the same exponent θ works for integration and L_1-approximation. We stress, however, that the orders of minimal errors are different for integration, L_p-approximation with $1 \leq p < \infty$, and L_∞-approximation.

Now we turn to the case $0 < \beta < 1$. The previous choice for θ is also best possible for integration of stationary random functions, at least if weighted Gregory formulas are used, see Proposition 39. The same conclusion holds for approximation of locally stationary random functions if we have Hermite data available and use piecewise polynomial interpolation, see Proposition 36.

Equidistant knots lead to order optimal methods for the integration and approximation problems under Sacks-Ylvisaker conditions. However, they are asymptotically optimal only if the weight function ϱ is constant.

In fact, they are arbitrarily bad with respect to the asymptotic constant for some weight functions ϱ. We compare the asymptotic constants for the errors of K-spline algorithms that are based on the best regular sequence and on equidistant knots, respectively. Take ψ according to (38), and put

$$\Theta(S) = \lim_{n\to\infty} e_q(S_n^{\psi,K}, S, P)/e_q(S_n^{1,K}, S, P)$$

for $S = \text{Int}_\varrho$ or $S = \text{App}_{p,\varrho}$. Then

$$\Theta(\text{Int}_\varrho) = \left(\int_0^1 \gamma^\theta(t)\, dt\right)^{1/\theta} / \left(\int_0^1 \gamma^2(t)\, dt\right)^{1/2}$$

for integration,

$$\Theta(\text{App}_{p,\varrho}) = \left(\int_0^1 \gamma^\theta(t)\, dt\right)^{1/\theta} / \left(\int_0^1 \gamma^p(t)\, dt\right)^{1/p}$$

for L_p-approximation with $p = q < \infty$, and

$$\Theta(\text{App}_{\infty,\varrho}) = \left(\int_0^1 \gamma^\theta(t)\, dt\right)^{1/\theta} / \sup_{t\in[0,1]} \gamma(t)$$

for L_∞-approximation.

CHAPTER V

Linear Problems for Univariate Functions with Noisy Data

So far we have studied errors that arise because the data $\lambda_i(f)$ do not determine the solution $S(f)$ uniquely. Now we take into account a second source of error by studying problems with inaccurate or noisy data $\lambda_i(f) + \varepsilon_i$. As previously, we consider the average case with respect to the functions f, and analogously we make stochastic assumptions on the noise $\varepsilon = (\varepsilon_1, \ldots, \varepsilon_n)$. We restrict our attention to linear problems with values in a Hilbert space.

In Section 1 we define q-average errors for $q = 2$ and we show that smoothing splines yield optimal algorithms. In Section 2 we report on results for integration and L_2-approximation of univariate functions. In Section 3 we study differentiation of univariate functions.

1. Optimality of Smoothing Spline Algorithms

Consider a linear problem with operator $S : C^k(D) \to G$ and a zero mean measure P on $F \subset C^k(D)$. Let K denote the covariance kernel of P and assume, for simplicity, that $K \in C^{r,r}(D^2)$ for some $r \geq k$. Furthermore, let Q denote a zero mean measure on $\mathbb{R}^n$.

A linear method that is based on noisy values of the functionals $\lambda_i \in \Lambda^{\text{all}}$ is of the form

$$S_n(f, \varepsilon) = \sum_{i=1}^n (\lambda_i(f) + \varepsilon_i) \cdot a_i.$$

We study *average errors*

$$e_2(S_n, S, P, Q) = \left(\int_{F \times \mathbb{R}^n} \|S(f) - S_n(f, \varepsilon)\|_G^2 \, d(P \otimes Q)(f, \varepsilon) \right)^{1/2}$$

$$= \left(\int_F \int_{\mathbb{R}^n} \|S(f) - S_n(f, \varepsilon)\|_G^2 \, dQ(\varepsilon) \, dP(f) \right)^{1/2}.$$

Hence we consider a random function f being distributed according to P and a noise ε independent of f and distributed according to Q.

We may formally view the linear problem as a problem on $F \times \mathbb{R}^n$, where exact data $\lambda_i(f) + \varepsilon_i$ are available. From Section III.2 we know that the error $e_2(S_n, S, P, Q)$ depends on P and Q only through the covariance kernel K of P and the covariance matrix of Q. As usual, we assume that the covariance matrix of Q is a multiple $\sigma^2 E_n$ of the identity matrix E_n and $\sigma^2 > 0$. Thus $\varepsilon_1, \ldots, \varepsilon_n$ are pairwise uncorrelated with zero mean and common variance σ^2. We put

$$e_2(S_n, S, P, \sigma^2) = e_2(S_n, S, P, Q).$$

Suppose that we are given fixed functionals λ_i and we wish to minimize the average error with respect to the coefficients $a_i \in G$. For noisy data, smoothing splines, instead of interpolating splines, lead to optimal coefficients. Let

$$y_i = \lambda_i(f) + \varepsilon_i.$$

The corresponding *smoothing spline* $h \in H(K)$ is the unique solution of the minimization problem

(1) $$\|h\|_K^2 + \frac{1}{\sigma^2} \cdot \sum_{i=1}^n (\lambda_i(h) - y_i)^2 \quad \to \quad \min$$

in the reproducing kernel Hilbert space $H(K)$. Recall that $H(K) \subset C^k(D)$ by Proposition III.11. The following proof shows, in particular, that the smoothing spline depends linearly on the data y_i. The variance σ^2 determines how much we should trust the data. For a large variance, $\lambda_i(h)$ and y_i may differ significantly and small norms $\|h\|_K$ are preferred. Conversely, if σ^2 tends to zero then the smoothing spline tends to the interpolating spline, as defined in Section III.3.1. The smoothing spline algorithm is defined by

$$S_n(f) = S(h),$$

i.e., we apply the operator S to the smoothing spline. The smoothing spline algorithm clearly depends on the covariance kernel K and the variance σ^2, in addition to $S, \lambda_1, \ldots, \lambda_n$. The next result is the analogue of Proposition III.31, which deals with exact data.

PROPOSITION 1. *Consider a linear problem with a Hilbert space G, and assume that functionals $\lambda_1, \ldots, \lambda_n \in \Lambda^{\text{all}}$ are given. Then the smoothing spline algorithm has minimal error $e_2(S_n, S, P, \sigma^2)$ among all linear methods S_n that use the functionals λ_i.*

PROOF. First we assume that $S : C^k(D) \to \mathbb{R}$ is a bounded linear functional. Put $\xi = \Im S$ and $\xi_i = \Im \lambda_i$, where $\Im$ denotes the fundamental isomorphism with respect to K, see Section III.1.2. Consider the Hilbert space $H(K) \times \mathbb{R}^n$, equipped with the scalar product

$$\langle (h_1, p_1), (h_2, p_2) \rangle = \langle h_1, h_2 \rangle_K + \sigma^2 \cdot \sum_{i=1}^n p_{1,i} \cdot p_{2,i}$$

and the induced norm $\| \cdot \|$. Let e_i denote the ith unit vector in $\mathbb{R}^n$, and let B denote the unit ball in $H(K) \times \mathbb{R}^n$.

The error of every linear method S_n satisfies

$$
\begin{aligned}
e(S_n, S, P, \sigma^2)^2 &= \int_F \left(S(f) - \sum_{i=1}^n \lambda_i(f) \cdot a_i \right)^2 dP(f) + \sigma^2 \sum_{i=1}^n a_i^2 \\
&= \left\| \xi - \sum_{i=1}^n a_i \cdot \xi_i \right\|_K^2 + \sigma^2 \sum_{i=1}^n a_i^2 \\
&= \left\| (\xi, 0) - \sum_{i=1}^n a_i \cdot (\xi_i, e_i) \right\|^2,
\end{aligned}
$$

see Corollary III.7. We conclude that

$$
(2) \qquad e(S_n, S, P, \sigma^2) = \sup_{(h,p) \in B} \left| S(h) - \sum_{i=1}^n (\lambda_i(h) + \sigma^2 p_i) \cdot a_i \right|.
$$

The right-hand side corresponds to a linear problem on the unit ball in a Hilbert space, where, formally, exact data are available. In this situation, the optimality of a spline algorithm is known, see Traub, Wasilkowski, and Woźniakowski (1988, Section 4.5.7). In our particular case the interpolating spline $(h, p) \in H(K) \times \mathbb{R}^n$ is given as the unique solution of the minimization problem

$$
\|(h, p)\| \to \min \qquad \text{on} \qquad \{ (h, p) \in H(K) \times \mathbb{R}^n : \lambda_i(h) + \sigma^2 p_i = y_i \}.
$$

The spline algorithms is defined as the application of $(S, 0)$ to the interpolating spline (h, p). The algorithm is linear, since (h, p) depends linearly on $y_1, \ldots, y_n$. It remains to observe that h is the unique solution of the minimization problem (1).

We proceed as in Section III.2.2, if S takes values in an arbitrary Hilbert spaces G. □

1.1. Notes and References.

1. Numerical problems with noisy data are analyzed in several settings. The worst case or average case with respect to the functions and the worst case or average case with respect to the noise is studied. We refer in particular to Plaskota (1996).

The worst case for f and the average case for ε is frequently analyzed in the statistical literature. The approximation problem, where S is the embedding into an L_p-space, is then called nonparametric regression.

2. Proposition 1 is due to Plaskota (1996) and, for the particular case $S(f) = f(t)$, due to Kimeldorf and Wahba (1970a, 1970b). Optimality properties of smoothing spline algorithms are also known in other settings. See Eubank (1988), Traub et al. (1988), Wahba (1990), and Plaskota (1996).

In this section the smoothing parameter σ^2 in (1) is determined by the variance of the noise. Other choices, in particular in the case of an unknown variance σ^2, are studied in the literature. See Eubank (1988) and Wahba (1990).

2. Integration and Approximation of Univariate Functions

Now we turn to the particular problems $S = \text{Int}$ of integration and $S = \text{App}_2$ of L_2-approximation and we report on results for univariate functions under Sacks-Ylvisaker conditions. The *nth minimal error* is defined by

$$e_2(n, \Lambda, S, P, \sigma^2) = \inf_{S_n} e_2(S_n, S, P, \sigma^2),$$

where S_n varies over all linear methods that use n functionals from a class Λ. It makes no sense to study the whole class Λ^{all}, since a noise with given variance becomes irrelevant for functionals $c\lambda$ if c is large. Hence we consider the ball

$$\Lambda_K^{\text{all}} = \left\{ \lambda \in \Lambda^{\text{all}} : \int_F \lambda(f)^2 \, dP(f) \leq \sup_{0 \leq t \leq 1} K(t, t) \right\},$$

noting that

$$\Lambda^{\text{std}} \subset \Lambda_K^{\text{all}}.$$

For simplicity we fix $\sigma^2 > 0$ and study the order of minimal errors as the number n of functionals tends to ∞. Stronger results concerning the asymptotic dependence on σ and n simultaneously and estimates for asymptotic constants are sometimes known. However, asymptotically optimal methods or asymptotic constants for minimal errors seem to be unknown in the average case setting with noisy data.

PROPOSITION 2. *Let P be a zero mean measure that satisfies Sacks-Ylvisaker conditions of order $r \in \mathbb{N}_0$. Then the minimal errors for integration are of order*

$$e_2(n, \Lambda^{\text{std}}, \text{Int}, P, \sigma^2) \asymp e_2(n, \Lambda_K^{\text{all}}, \text{Int}, P, \sigma^2) \asymp n^{-1/2}$$

for all r.

We see that sampling is order optimal for integration with noisy data, in contrast to the exact data case where integration is trivial for Λ_K^{all}. A lower bound of order $n^{-1/2}$ holds in general for linear problems with noisy data and $\Lambda \subset \Lambda_K^{\text{all}}$, see Plaskota (1996, p. 177). For integration, an error of order $n^{-1/2}$ is clearly achieved by averaging n noisy observations of $\lambda_i = \text{Int}$.

The minimal errors depend on the regularity r for L_2-approximation.

PROPOSITION 3. *Let P be a zero mean measure that satisfies Sacks-Ylvisaker conditions of order $r \in \mathbb{N}_0$. Then the minimal errors for L_2-approximation are of order*

$$e_2(n, \Lambda^{\text{std}}, \text{App}_2, P, \sigma^2) \asymp n^{-\frac{1}{2} + \frac{1}{4r+4}}$$

for the class $\Lambda = \Lambda^{\text{std}}$ and

$$e_2(n, \Lambda_K^{\text{all}}, \text{App}_2, P, \sigma^2) \asymp \begin{cases} n^{-1/2} \cdot \ln n & \text{if } r = 0, \\ n^{-1/2} & \text{if } r > 0 \end{cases}$$

for the class $\Lambda = \Lambda_K^{\text{all}}$.

While sampling is order optimal for approximation in the exact data case, see Remark IV.33, this is no longer true in the presence of noise. Optimal functionals $\lambda_1, \ldots, \lambda_n \in \Lambda_K^{\text{all}}$ are determined for general linear problems in Plaskota (1996, Section 3.8.1). The construction involves the eigenvalues μ_j and eigenfunctions ξ_j of the integral operator with kernel K, see Proposition III.24. However, specific linear combinations of eigenfunctions are optimal, while

$$\text{span}\{\lambda_1, \ldots, \lambda_n\} = \text{span}\{\xi_1, \ldots, \xi_n\}$$

is sufficient for optimality in the exact data case.

REMARK 4. Equidistant knots $x_i = (i-1)/(n-1)$ lead to order optimal methods for the class Λ^{std} and integration or approximation under Sacks-Ylvisaker conditions. Furthermore, we may replace the smoothing spline in $H(K)$ by a polynomial smoothing spline. The latter is defined by the minimization problem

$$(3) \qquad \|h^{(r+1)}\|_2^2 + \frac{1}{\sigma^2} \cdot \sum_{i=1}^{n} (h(x_i) - y_i))^2 \quad \to \quad \min$$

in $W_2^{r+1}([0,1])$ with noisy data $y_i = f(x_i) + \varepsilon_i$. If $n \geq r + 1$, then a natural polynomial spline of degree $2r + 1$ is the unique solution of the minimization problem. These splines or their integrals define order optimal methods.

2.1. Notes and References. 1. Proposition 2 is due to Plaskota (1992) for the Wiener measure. Using Proposition IV.8, we get an upper bound of order $n^{-1/2}$ under Sacks-Ylvisaker conditions for every r. See Plaskota (1996, p. 177) for the general lower bound of order $n^{-1/2}$.

2. Proposition 3 is due to Plaskota for the class $\Lambda = \Lambda_K^{\text{all}}$ and the r-fold integrated Wiener measure, see Plaskota (1996, p. 195). By Proposition IV.10, the same bounds hold under Sacks-Ylvisaker conditions. See Plaskota (1992) and Ritter (1996b) for the case $\Lambda = \Lambda^{\text{std}}$ with $r = 0$ and $r > 0$, respectively. We add that for $\Lambda = \Lambda^{\text{std}}$ the proof of Proposition 5 applies with $k = 0$.

3. L_∞-approximation from noisy function values is studied in Plaskota (1998). The order of minimal errors with respect to the r-fold integrated Wiener measure is given by $n^{-1/2+1/(4r+4)} \cdot (\ln n)^{1/2}$. We observe that the additional term $(\ln n)^{1/2}$ appears both in the noisy and exact data cases, when we switch from the L_2-norm to the L_∞-norm. Cf. Sections IV.3.2 and IV.3.5.

4. We discuss L_p-approximation with finite $p \neq 2$ in Remark 7.

5. Deterministic or stochastic noise may be used to model round-off errors for the input data. If we can fully control the round-off, then quantization is a more appropriate model. See Benhenni and Cambanis (1996) for the analysis of integration of univariate random functions on the basis of quantized observations.

3. Differentiation of Univariate Functions

It makes sense to study numerical differentiation in the presence of noise. Without noise, derivatives $f^{(k)}(t)$ can be approximated (at least theoretically) by divided differences with arbitrary precision under suitable regularity assumptions, e.g., Sacks-Ylvisaker conditions in an average case analysis. Therefore differentiation and a certain approximation problem on a class of functions of lower regularity are basically equivalent for $\Lambda = \Lambda^{\text{std}}$. This is no longer true if function values are corrupted by noise.

We wish to approximate a derivative of a univariate function globally on $[0, 1]$, and we consider the distance between $f^{(k)}$ and its approximation in the L_2-norm. Hence the differentiation problem is formally defined by the linear operator

$$\text{Diff}_k : C^r([0, 1]) \to L_2([0, 1])$$

with $0 < k \leq r$ and

$$\text{Diff}_k(f) = f^{(k)}.$$

PROPOSITION 5. *Let P be a zero mean measure that satisfies Sacks-Ylvisaker conditions of order $r \in \mathbb{N}_0$. Then the minimal errors for differentiation are of order*

$$e_2(n, \Lambda^{\text{std}}, \text{Diff}_k, P, \sigma^2) \asymp n^{-\frac{1}{2} + \frac{k+1/2}{2r+2}}$$

for $k \leq r$.

PROOF. First we establish the lower bound. For every linear method S_n,

$$(4) \quad e_2(S_n, \text{Diff}_k, P, \sigma^2) = \int_0^1 \int_F \int_{\mathbb{R}^n} (f^{(k)}(t) - S_n(f, \varepsilon)(t))^2 \, dQ(f) \, dP(\varepsilon) \, dt.$$

In particular, for the smoothing spline algorithm with knots $x_1, \ldots, x_n$, we have

$$(5) \quad \int_F \int_{\mathbb{R}^n} (f^{(k)}(t) - S_n(f, \varepsilon)(t))^2 \, dQ(f) \, dP(\varepsilon)$$

$$= \sup\{h^{(k)}(t)^2 : (h, p) \in B, \ h(x_i) = -\sigma^2 \, p_i \text{ for } i = 1, \ldots, n\}$$

$$= \sup\left\{h^{(k)}(t)^2 : h \in H(K), \ \|h\|_K + 1/\sigma^2 \cdot \sum_{i=1}^n h(x_i)^2 \leq 1\right\},$$

see (2) and (III.14).

Put $q = \sigma \cdot n^{-1/2}$. Let $\psi \in C^{r+1}(\mathbb{R})$ with

$$\int_{\mathbb{R}} \psi^{(r+1)}(s)^2 \, ds \leq 1, \qquad \sup_{s \in \mathbb{R}} |\psi(s)| \leq 1, \qquad \psi^{(k)}(0) > 0,$$

and

$$(6) \quad \psi(s) = 0 \quad \text{if} \quad |s| \geq \tfrac{1}{2}.$$

By c we denote different positive constants, which only depend on K and ψ. For every $t \in [0, 1]$ we define

$$h_t(s) = q^{\frac{2r+1}{2r+2}} \cdot \psi\left((s - t)/q^{\frac{1}{r+1}}\right).$$

Clearly

$$\|h_t\|_\infty \le q^{\frac{2r+1}{2r+2}}, \qquad h_t^{(k)}(t) = q^{\frac{2(r-k)+1}{2r+2}} \cdot \psi^{(k)}(0),$$

and

$$h_t(s) = 0 \quad \text{if} \quad |s - t| \ge \tfrac{1}{2} q^{\frac{1}{r+1}}.$$

In particular, h_t vanishes with all derivatives at $s = 0$ or $s = 1$. From Proposition IV.8 we conclude that $h_t \in H(K)$ and

$$\|h_t\|_K \le c \cdot \|h_t^{(r+1)}\|_2 \le c.$$

Put $X = \{x_1, \ldots, x_n\} \subset [0,1]$. Without loss of generality we may assume that the points x_i are pairwise different. Put $v = \min(q^{1/(r+1)}, 1/4)$ and define $I_j =](j-1)2v, j2v[$ for $j = 1, \ldots, \lfloor 1/(2v) \rfloor$. Observe that

$$\#\{j : \#(I_j \cap X) \le 4nv\} \ge \lfloor 1/(4v) \rfloor,$$

where $\#$ is used to denote the number of elements in a set. Taking into account only points $t \in](j-1)2v + v/2, j2v - v/2[$, we conclude that the Lebesgue measure of the set

$$U = \{t \in [0,1] : \#(]t - v/2, t + v/2[\cap X) \le 4nv\}$$

is at least $v \cdot \lfloor 1/(4v) \rfloor \ge 1/8$. For every $t \in U$ the function h_t satisfies

$$\frac{1}{\sigma^2} \cdot \sum_{i=1}^n h_t(x_i)^2 \le \frac{1}{\sigma^2} \cdot 4nv \cdot \|h_t\|_\infty^2 \le 4.$$

From (4) and (5) we get the lower bound

$e(S_n, \mathrm{Diff}_k, P, \sigma^2)^2$

$$\ge \int_U \sup\Big\{ h^{(k)}(t)^2 : h \in H(K),\ \|h\|_K^2 + \frac{1}{\sigma^2} \cdot \sum_{i=1}^n h(x_i)^2 \le 1 \Big\} \, dt$$

$$\ge c \cdot q^{\frac{2(r-k)+1}{r+1}}.$$

Now we establish the upper bound. Let $x_i = (i-1)/(n-1)$. From (2), (4) and Proposition IV.8 we obtain

$$e(S_n, \mathrm{Diff}_k, P, \sigma^2)^2 = \int_0^1 \sup_{(h,p) \in B} |h^{(k)}(t) - S_n(h, \sigma^2 p)(t)|^2 \, dt$$

$$\le c \int_0^1 \sup \Big\{ |h^{(k)}(t) - S_n(h, \sigma^2 p)(t)|^2 : (h, p) \in \tilde{B} \Big\} \, dt$$

for every linear method S_n, where

$$\tilde{B} = \Big\{ (h, p) \in W_2^{r+1}([0,1]) \times \mathbb{R}^n : \|h^{(r+1)}\|_2^2 + \sigma^2 \sum_{i=1}^n p_i^2 \le 1 \Big\}.$$

Assume that $n \geq \max(r+1, \sigma^2)$ and let h be the solution of (3). Note that h also solves the minimization problem

$$\|h^{(r+1)}\|_2^2 + \sigma^2 \sum_{i=1}^{n} p_i^2 \quad \rightarrow \quad \min$$

on

$$\{(h, p) \in W_2^{r+1}([0,1]) \times \mathbb{R}^n : h(x_i) + \sigma^2 p_i = y_i\}.$$

Consider the method $S_n(f, \varepsilon) = h^{(k)}$. A general theorem on optimality of spline algorithms, see Traub *et al.* (1988, Section 4.5.7), implies the following. For fixed $t \in [0,1]$, the method $(h, p) \mapsto (S_n(h, \sigma^2 p))^{(k)}(t)$ is worst case optimal on $\widetilde{B}$ to recover $h^{(k)}(t)$ from the data $h(x_i) + \sigma^2 p_i$. Furthermore,

$$\sup\{|h^{(k)}(t) - S_n(h, \sigma^2 p)(t)| : (h, p) \in \widetilde{B}\}$$
$$= \sup\{|h^{(k)}(t)| : (h, p) \in \widetilde{B}, \ h(x_i) = -\sigma^2 p_i\}$$
$$\leq \sup\{\|h^{(k)}\|_\infty : (h, p) \in \widetilde{B}, \ h(x_i) = -\sigma^2 p_i\}.$$

We conclude that

$$e(S_n, \mathrm{Diff}_k, P, \sigma^2) \leq c \cdot \sup\Big\{\|h^{(k)}\|_\infty : \|h^{(r+1)}\|_2 \leq 1, \ \sum_{i=1}^{n} h(x_i)^2 \leq \sigma^2\Big\}.$$

According to Ragozin (1983, Theorem 3.2),

$$\|h\|_2^2 \leq c \cdot \left(\frac{1}{n} \cdot \sum_{i=1}^{n} h(x_i)^2 + n^{-(2r+2)} \cdot \|h^{(r+1)}\|_2^2\right)$$

for every $h \in W_2^{r+1}([0,1])$. Therefore

$$e(S_n, \mathrm{Diff}_k, P, \sigma^2) \leq c \cdot \sup\{\|h^{(k)}\|_\infty : \|h^{(r+1)}\|_2 \leq 1, \ \|h\|_2 \leq q + n^{-(r+1)}\}.$$

Take $\psi \in C^{r+1}(\mathbb{R})$ with

$$\sup_{s \in \mathbb{R}} |\psi(s)| \leq 1, \qquad \psi(0) = 1, \qquad \psi^{(j)}(0) = 0 \quad \text{for } j = 1, \ldots, r,$$

and (6). For $h \in W_2^{r+1}([0,1])$ with $|h^{(k)}(t^*)| = \|h^{(k)}\|_\infty$ we define

$$h_0(s) = \psi(s - t^*) \cdot h(s).$$

Note that h_0 vanishes with all derivatives at $s = 0$ or 1. If $\|h^{(r+1)}\|_2 \leq 1$ then $\|h_0^{(r+1)}\|_2 \leq c$. Moreover, $\|h_0\|_2 \leq \|h\|_2$ and $\|h_0^{(k)}\|_\infty \geq |h_0^{(k)}(t^*)| = \|h^{(k)}\|_\infty$. Hence

$$e(S_n, \mathrm{Diff}_k, P, \sigma^2) \leq c \cdot \sup\{\|h^{(k)}\|_\infty : h(s) = \cdots = h^{(r)}(s) = 0 \text{ for } s = 0 \text{ or } 1,$$
$$\|h^{(r+1)}\|_2 \leq 1, \ \|h\|_2 \leq q + n^{-(r+1)}\}.$$

Due to Gabushin (1967) and Kwong and Zettl (1992, p. 21), a Landau inequality

$$\|h^{(k)}\|_\infty \leq c \cdot \|h\|_2^{\frac{2(r-k)+1}{2r+2}} \cdot \|h^{(r+1)}\|_2^{\frac{2k+1}{2r+2}}$$

holds for all $h \in W_2^{r+1}([0,1])$ such that $h, \ldots, h^{(r)}$ have zeros in $[0,1]$. This implies

$$e(S_n, \text{Diff}_k, P, \sigma^2) \leq c \cdot (q + n^{-(r+1)})^{\frac{2(r-k)+1}{2r+2}},$$

and the upper bound follows. $\square$

REMARK 6. In the previous proof we have shown, in particular, that the following methods are order optimal to recover $f^{(k)}$ if $n \geq \max(r+1, \sigma^2)$. Take equidistant knots $x_i = (i-1)/(n-1)$ and define $S_n(f, \varepsilon) = h^{(k)}$ where h is the natural smoothing spline of degree $2r+1$ from (3).

As for L_2-approximation, sampling is not order optimal for differentiation in the presence of noise. Errors of order $n^{-1/2} \cdot \ln n$ for $k = r$ and $n^{-1/2}$ for $k < r$ are achievable using functionals from Λ_K^{all}, see Proposition 3.

We add that error bounds are also known for recovering derivatives $f^{(k)}$ at a single point $t \in [0,1]$. The minimal errors are of order

$$n^{-\frac{1}{2} + \frac{k}{2r+1}}$$

under Sacks-Ylvisaker conditions of order $r \in \mathbb{N}_0$. The order is achieved by polynomial interpolation of sample means at points that are suitably concentrated around t, see Ritter (1996b).

REMARK 7. Consider differentiation with errors in the L_p-norm with $p \neq 2$. Formally this problem is defined by the operator Diff_k acting between $C^r([0,1])$ and $G = L_p([0,1])$. If $p < \infty$ and if P and the noise Q are Gaussian, then the error bounds from Proposition 5 hold as well for q-average errors. For a proof an analogue to (IV.28) can be used. Proposition 3 can be generalized in the same way.

3.1. Notes and References. 1. Proposition 5 is due to Ritter (1996b).

2. Relations between Landau problems and numerical differentiation are used by several authors, see Arestov (1975), Micchelli (1976), Kwong and Zettl (1992), and Donoho (1994) for results and further references.

CHAPTER VI

Integration and Approximation
of Multivariate Functions

In this chapter we study integration and L_2-approximation of functions that depend on d variables. As in the univariate case, we are mainly interested in methods that use function values only, and we derive bounds for average errors from smoothness properties of the covariance kernel. We treat isotropic smoothness conditions and tensor product problems in Sections 1 and 2, respectively. In the latter case, the smoothness is characterized by the tensor product structure of the covariance kernel on $\left([0,1]^d\right)^2$ together with smoothness properties of the d underlying covariance kernels on $[0,1]^2$. Moreover, general techniques are available that relate error bounds for the tensor product problem to error bounds for the underlying univariate problems, see Sections 2.1–2.3. It turns out that tensor product methods are not well suited for tensor product problems. Instead, Smolyak formulas are much better and are often order optimal. On the other hand, tensor product methods are often order optimal under isotropic smoothness conditions.

We use Hölder conditions to derive upper bounds for the minimal errors, see Section 1.1 for the isotropic case and Section 2.4 for the tensor product case. Since Hölder conditions are too weak to determine the asymptotic behavior of minimal errors, we introduce more specific assumptions that determine the order of the minimal errors. First of all, these are isotropic as well as tensor product decay conditions for spectral densities of stationary kernels, see Sections 1.2 and 2.5. Furthermore, we use Sacks-Ylvisaker conditions in the tensor product case in Section 2.6. In Section 1.3 we briefly address the integration problem for stationary random fields in dimension $d = 2$. Section 1.4 deals with specific isotropic random functions, namely, fractional Brownian fields. In Section 2.7 we discuss L_2-discrepancy and its relation to the average case analysis of the integration problem.

Smoothness properties and the corresponding orders of minimal errors are summarized at the beginning of Section 3. We then explain that the order of minimal errors only partially reveals the dependence of the minimal errors on d, especially when d is large and only a moderate accuracy is required. We indicate how the dimension and accuracy can be analyzed simultaneously, and we present a selection of results in Section 3.

For both problems, integration and approximation using only function values, asymptotically optimal methods seem to be unknown in dimensions $d \geq 2$. For the analysis of L_p-approximation with $p \neq 2$ we need to strengthen our assumptions on the measure P, since these problems do not belong to the second-order theory of random functions. If P is Gaussian, then the error bounds from this chapter are valid also for q-average errors and L_p-approximation with $1 \leq p, q < \infty$. It seems interesting to obtain results for L_∞-approximation in the multivariate case.

1. Isotropic Smoothness

Let D denote a centered ball in $\mathbb{R}^d$ or $D = \mathbb{R}^d$. A zero mean random field on D with covariance kernel K is called *(wide sense) isotropic*, if K is invariant with respect to orthogonal transformations, i.e.,

$$K(s, t) = K(Qs, Qt)$$

for all $s, t \in D$ and every orthogonal matrix $Q \in \mathbb{R}^{d \times d}$. The kernel is called isotropic (in the wide sense), too, if this invariance holds.

We study different kinds of quadratic mean smoothness conditions that are isotropic in the following sense. The smoothness condition holds for the kernel

$$K_Q(s, t) = K(Qs, Qt),$$

where Q is any orthogonal matrix, if the condition holds for the kernel K. We do not require isotropy of K itself, and therefore $K_Q \neq K$ in general.

In Section 1.1 we use Hölder conditions to establish upper bounds for the minimal errors. As in the univariate case, these conditions do not determine the asymptotic behavior of minimal errors. Therefore we analyze more specific smoothness conditions in Sections 1.2–1.4. For stationary random fields, which are studied in Section 1.2, we determine the order of minimal errors and order optimal methods in terms of the decay of the spectral density. Integration of stationary isotropic random fields in dimension $d = 2$ is considered in Section 1.3. For this problem, cubature formulas based on local lattice designs are used. These designs can be regarded as a generalization of regular sequences to the multivariate case. The order of minimal errors and order optimal methods for fractional Brownian fields are determined in Section 1.4. These random fields are isotropic but non-stationary. In particular, we cover Lévy's Brownian motion. We also present average case results for solving Poisson equations in Remark 20.

For simplicity we study error bounds on

$$D = [0, 1]^d,$$

although results hold for more general domains. In many cases the random fields are naturally defined on the whole domain $\mathbb{R}^d$, in which case we then formally consider the restriction of the random field onto the unit cube D.

1.1. Hölder Conditions. Henceforth the Euclidean norm and scalar product in $\mathbb{R}^d$ are denoted by $\|\cdot\|$ and $\langle\cdot,\cdot\rangle$, respectively. Let $r \in \mathbb{N}_0$ and let P denote a zero mean measure on a class $F \subset C^r(D)$ with covariance kernel K. We consider the smoothness properties

$$(1) \qquad\qquad K \in C^{r,r}(D^2)$$

and

$$(2) \qquad L(s,s) - 2\,L(s,t) + L(t,t) \le c \cdot \|s - t\|^{2\beta}$$

for all $s, t \in D$ and each kernel

$$L = K^{(\alpha,\alpha)}, \qquad |\alpha| = r.$$

Here c is a positive constant,

$$0 \le \beta \le 1,$$

and $\alpha \in \mathbb{N}_0^d$ with $|\alpha| = \sum_{\nu=1}^d \alpha_\nu$. We then say that K and P satisfy *Hölder conditions* of order (r, β).

Recall that (1) was previously introduced in Section III.1.3, and (1) is equivalent to quadratic mean continuity of all partial derivatives $f^{(\alpha)}$ up to order r with respect to P, see Lemma III.14. The covariance kernel of the random field $f^{(\alpha)}$ is given by $K^{(\alpha,\alpha)}$, and

$$\int_F (f^{(\alpha)}(s) - f^{(\alpha)}(t))^2 \, dP(f) = K^{(\alpha,\alpha)}(s,s) - 2\,K^{(\alpha,\alpha)}(s,t) + K^{(\alpha,\alpha)}(t,t)$$

follows immediately. Therefore (2) states that all partial derivatives $f^{(\alpha)}$ of order $|\alpha| = r$ are Hölder continuous with exponent β in quadratic mean. In the univariate case, $d = 1$, we have already studied Hölder conditions in Section IV.5.1.

EXAMPLE 1. For every d and $0 < \beta < 1$, the covariance kernel

$$K(s,t) = \tfrac{1}{2} \left(\|s\|^{2\beta} + \|t\|^{2\beta} - \|s - t\|^{2\beta} \right)$$

satisfies Hölder conditions of order $(0, \beta)$, since

$$K(s,s) - 2\,K(s,t) + K(t,t) = \|s - t\|^{2\beta}.$$

We add that K is in fact nonnegative definite, and we refer to Ossiander and Waymire (1989) for a simple proof.

For every nonnegative definite function K on D^2 that satisfies (1) and (2) with $\beta > 0$, a zero mean Gaussian measure on $C^r(D)$ with covariance kernel K exists, see Adler (1981, Theorem 3.4.1). Conversely, (2) holds if the partial derivatives $f^{(\alpha)}$ of order r form random fields of index β, see Adler (1981) and Kahane (1985).

The zero mean Gaussian random field corresponding to the kernel K is called a *fractional Brownian field* or fractional Brownian motion with multidimensional time. For $d = 1$ and $\beta = \tfrac{1}{2}$ we get the classical Brownian motion kernel $K(s,t) = \min(s,t)$. Fractional Brownian fields and related random fields are further investigated in Section 1.4.

Hölder conditions yield the following upper bounds for minimal errors.

PROPOSITION 2. *Let P be a zero mean measure that satisfies Hölder conditions of order (r, β). Then*

$$e_2(n, \Lambda^{\text{std}}, \text{App}_{2,\varrho}, P) = O\left(n^{-\frac{r+\beta}{d}}\right)$$

for the minimal error of L_2-approximation. Furthermore, if additionally $\varrho \in L_2(D)$, then

$$e_2(n, \Lambda^{\text{std}}, \text{Int}_\varrho, P) = O\left(n^{-\left(\frac{r+\beta}{d}+\frac{1}{2}\right)}\right)$$

for the minimal error of integration.

PROOF. Let $C^{r,\beta}(D)$ denote the space of all functions from $C^r(D)$ whose partial derivatives of order r are Hölder continuous with exponent β.

First we show that Hölder conditions of order (r, β) imply

$$H(K) \subset C^{r,\beta}(D)$$

for the Hilbert space with reproducing kernel K. In fact, $H(K)$ consists of functions having continuous partial derivatives of order up to r and

$$h^{(\alpha)}(t) = \langle h, K^{(0,\alpha)}(\cdot, t)\rangle_K$$

for $h \in H(K)$ and $t \in D$ if $|\alpha| \le r$. This follows from (1), see Proposition III.11. Furthermore,

$$|h^{(\alpha)}(s) - h^{(\alpha)}(t)| \le \|h\|_K \cdot \|K^{(0,\alpha)}(\cdot, s) - K^{(0,\alpha)}(\cdot, t)\|_K$$
$$= \|h\|_K \cdot \left(K^{(\alpha,\alpha)}(s, s) - 2K^{(\alpha,\alpha)}(s, t) + K^{(\alpha,\alpha)}(t, t)\right)^{1/2},$$

so that the partial derivatives of h of order r are Hölder-continuous with exponent β because of (2).

Using Proposition III.19 and the continuity of the embedding $H(K) \hookrightarrow C^{r,\beta}(D)$ we get

$$e_2(n, \Lambda^{\text{std}}, \text{App}_{2,\varrho}, P) \le \|\varrho\|_1^{1/2} \cdot e_{\max}(n, \Lambda^{\text{std}}, \text{App}_\infty, B(K))$$
$$= O\left(e_{\max}(n, \Lambda^{\text{std}}, \text{App}_\infty, B(C^{r,\beta}))\right).$$

Here

$$B(C^{r,\beta}) = \{g \in C^{r,\beta}(D) : |g^{(\alpha)}(s) - g^{(\alpha)}(t)| \le \|s - t\|^\beta \text{ for } s, t \in D \text{ if } |\alpha| = r\}$$

denotes the unit ball in the Hölder space. The minimal errors for the latter worst case problem are of order $n^{-(r+\beta)/d}$, see, e.g., Novak (1988, p. 34).

For integration we apply Proposition II.4, which yields an additional factor $n^{-1/2}$ in the order of the minimal errors. □

Now we describe a simple linear method that yields the upper bound for approximation.

REMARK 3. It is known that piecewise polynomial interpolation leads to errors of optimal order for L_∞-approximation on $B(C^{r,\beta})$, see Novak (1988, p. 34). We briefly describe this construction, which will also be used for integration later.

Let $\mathbb{P}_{\ell,d}$ denote the space of polynomials of degree at most ℓ in d variables. Put

$$v = \dim \mathbb{P}_{\ell,d} = \binom{d+\ell}{\ell}$$

and choose

$$x_1, \ldots, x_v \in D$$

such that interpolation in these knots is uniquely possible in $\mathbb{P}_{\ell,d}$. Furthermore, let $p_1, \ldots, p_v \in \mathbb{P}_{\ell,d}$ denote the corresponding Lagrange polynomials. Define $p_i(x) = 0$ if $x \notin D$.

Consider a partition

$$D = \bigcup_{j=1}^{M} Q_j$$

with non-overlapping cubes Q_j, whose edges are parallel to the coordinate axes, and let $T_j(t) = v_j t + s_j$, where v_j and s_j denote the sidelength and the lower left corner of Q_j, respectively. Thus $T_j(D) = Q_j$.

For approximation we define the linear methods

$$S_v^j(f) = \sum_{i=1}^{v} f(T_j(x_i)) \cdot p_i \circ T_j^{-1}$$

and

$$S_n(f) = \sum_{j=1}^{M} S_v^j(f).$$

Clearly S_n uses at most

$$n = M \cdot v$$

knots. On the cube Q_j, the function $S_v^j(f)$ coincides with the polynomial from $\mathbb{P}_{\ell,d}$ that interpolates f in the knots $T_j(x_i)$. Outside of Q_j the function $S_v^j(f)$ vanishes.

For integration the construction yields composite cubature formulas, namely

$$S_n(f) = \sum_{j=1}^{M} S_v^j(f)$$

with

$$S_v^j(f) = \sum_{i=1}^{v} f(T_j(x_i)) \cdot \int_{Q_j} p_i(T_j^{-1}(t)) \cdot \varrho(t)\, dt.$$

Clearly

$$S_v^j(p) = \int_{Q_j} p(t) \cdot \varrho(t) \, dt, \qquad p \in \mathbb{P}_{\ell,d}.$$

Specifically, for approximation under Hölder conditions of order (r, β), take polynomials of degree

$$\ell = \begin{cases} r - 1 & \text{if } \beta = 0, \\ r & \text{if } 0 < \beta \leq 1 \end{cases}$$

and take a uniform partition with $M = k^d$ cubes of sidelength $\nu_j = 1/k$. It is easy to verify that

$$e_{\max}(S_n, \mathrm{App}_\infty, B(C^{r,\beta})) \leq O(k^{-(r+\beta)}).$$

By Proposition III.19,

$$e_2(S_n, \mathrm{App}_{2,\varrho}, P) \leq \|\varrho\|_1^{1/2} \cdot O\left(n^{-\frac{r+\beta}{d}}\right).$$

We mention two other classes of methods S_n that also satisfy this estimate. Interpolation by piecewise polynomials of degree ℓ in each variable allows to use function values on regular grids

$$\left\{ \frac{i}{k\ell} : i = 0, \dots, k\ell \right\}^d.$$

Approximation by smooth functions is provided by K-spline algorithms that are based, for instance, on these regular grids.

For integration the upper bound in Proposition 2 is proven non-constructively, as it relies on Proposition II.4. We obtain a simple Monte Carlo method, integrating the piecewise polynomial approximation $S_n(f)$ and applying the classical Monte Carlo method to $f - S_n(f)$, which has an average Monte Carlo error of the proper order. We give constructive proofs under more specific assumptions in the sequel.

REMARK 4. Hölder conditions measure the smoothness of random fields by $r + \beta$ with r and β chosen maximally to satisfy (1) and (2). However, for every value of $r+\beta$ there exists a kernel of the proper smoothness such that the minimal errors are zero, even for $n = 1$. Consider the random field $f = Z g$, where Z is a random variable with standard normal distribution and g is a deterministic function of the proper smoothness. In this case $K(s,t) = g(s) g(t)$, and $f(x_1)$ allows to recover f exactly if $g(x_1) \neq 0$. Therefore conditions (1) and (2) alone are not sufficient to derive lower bounds for minimal errors.

On the other hand, we already know that for $d = 1$ and $\beta = \frac{1}{2}$ the upper bounds from Proposition 2 cannot be improved in general; the bounds are sharp for every measure satisfying Sacks-Ylvisaker conditions, see Remark IV.1, Proposition IV.15 and Remark IV.25.

Lemma 5 and Proposition 8 from the following section show that the bounds cannot be improved in general for any $d \in \mathbb{N}$, $r \in \mathbb{N}_0$, and $0 < \beta < 1$. If $\beta = 0$ or 1, then at least no general improvement is possible in terms of powers of n.

1.2. Stationary Random Fields. A second-order zero mean random field on $\mathbb{R}^d$ is called *(wide sense) stationary*, if its covariance kernel $\mathfrak{K}$ is invariant under translations, i.e.,

$$\mathfrak{K}(s, t) = \mathfrak{K}(0, t - s),$$

for all $s, t \in \mathbb{R}^d$. In this case the kernel is called stationary, too, and we put

$$\mathfrak{k}(t) = \mathfrak{K}(0, t).$$

If $\mathfrak{k}$ is continuous then Bochner's Theorem implies that

$$\mathfrak{k}(t) = \int_{\mathbb{R}^d} \exp(\imath \langle t, u \rangle) \, d\mu(u)$$

for some finite measure μ on $\mathbb{R}^d$. If μ is absolutely continuous with respect to the Lebesgue measure then

$$\mathfrak{k}(t) = \int_{\mathbb{R}^d} \exp(\imath \langle t, u \rangle) \cdot \mathfrak{f}(u) \, du,$$

where $\mathfrak{f} \in L_1(\mathbb{R}^d)$ is nonnegative with $\mathfrak{f}(u) = \mathfrak{f}(-u)$ since $\mathfrak{k}$ is real-valued. The function $\mathfrak{f}$ is called the *spectral density* and μ is called the spectral measure of the random field or the covariance kernel. Conversely, a kernel $\mathfrak{K}$ that is constructed from a symmetric density $\mathfrak{f}$ in this way is the covariance kernel of a stationary Gaussian random field.

We are interested in functions on the cube D, and hence we consider the restriction of stationary random fields to D. Hereby we get a probability measure P on a class F of real-valued functions f on D. By K and k we denote the restrictions of $\mathfrak{K}$ and $\mathfrak{k}$ to D^2 and D, respectively.

Suppose, for instance, that we wish to approximate $f(t)$ by $S_1(f) = af(x)$ where $t, x \in D$. In case of stationarity, the coefficient $a = K(t, x)/K(x, x)$ of the K-spline algorithm depends on t and x only through $t - x$.

Recall that $B(K)$ denotes the unit ball in the reproducing kernel Hilbert space $H(K)$, and analogously we use $B(\mathfrak{K})$ to denote the unit ball in $H(\mathfrak{K})$, which consists of real-valued function on $\mathbb{R}^d$. From the characterization of $H(K)$ in terms of $H(\mathfrak{K})$ we get

(3) $$e_2(S_n, \mathrm{Int}_\varrho, P) = e_{\max}(S_n, \mathrm{Int}_\varrho, B(K)) = e_{\max}(S_n, \mathrm{Int}_\varrho, B(\mathfrak{K}))$$

for every cubature formula S_n, see Lemma III.4 and Proposition III.17. Analogously,

(4) $$e_2(S_n, \mathrm{App}_{2,\varrho}, P)^2 = \int_D \sup_{h \in B(K)} |h(t) - S_n(h)(t)|^2 \cdot \varrho(t) \, dt$$

$$= \int_D \sup_{h \in B(\mathfrak{K})} |h(t) - S_n(h)(t)|^2 \cdot \varrho(t) \, dt$$

for every linear method S_n for approximation, see Section III.2.3. For convenience, we sometimes prefer to work with the space $H(\mathfrak{K})$.

Since continuity or differentiability properties of covariance kernels follow from the respective properties along the diagonal, we have to study the smoothness of $\mathfrak{k}$ at the origin. In particular, K satisfies (1) if

$$(5) \qquad\qquad \mathfrak{k}^{(2\alpha)} \in C(\mathbb{R}^d), \qquad |\alpha| \leq r.$$

The covariance kernel corresponding to the partial derivative $f^{(\alpha)}$ is given by

$$\mathfrak{L}(s,t) = \mathfrak{K}^{(\alpha,\alpha)}(s,t) = (-1)^{|\alpha|} \cdot \mathfrak{k}^{(2\alpha)}(t-s).$$

Thus L, the restriction of $\mathfrak{L}$ to D^2, satisfies (2) if

$$(6) \qquad\qquad \mathfrak{l}(0) - \mathfrak{l}(t) \leq \tfrac{1}{2} c \cdot \|t\|^{2\beta}$$

for all $t \in \mathbb{R}^d$, where

$$\mathfrak{l}(t) = \mathfrak{L}(0,t) = (-1)^{|\alpha|} \cdot \mathfrak{k}^{(2\alpha)}(t).$$

Suppose that (5) holds and (6) is satisfied with $\beta > 0$ for all $\alpha \in \mathbb{N}_0^d$ with $|\alpha| = r$. Then a zero mean Gaussian measure P on $C^r(D)$ with covariance kernel K exists, see Adler (1981, p. 62).

It is well known that the smoothness of $\mathfrak{k}$ at the origin and the decay of the spectral density $\mathfrak{f}$ at infinity are related. We consider the conditions

$$(7) \qquad\qquad \mathfrak{f}(u) \leq c_1 \cdot \|u\|^{-2\gamma}, \qquad \|u\| \geq c_2,$$

and

$$(8) \qquad\qquad c_3 \cdot (1 + \|u\|^2)^{-\gamma} \leq \mathfrak{f}(u), \qquad u \in \mathbb{R}^d.$$

Here c_i are positive constants and we require integrability of the bounds on $\mathfrak{f}$ on their domains, i.e.,

$$\gamma > \tfrac{1}{2} d.$$

Note that the smoothness conditions (7) and (8) are isotropic.

Put

$$\lfloor v \rfloor = \max\{j \in \mathbb{N}_0 : j \leq v\}$$

for $v \geq 0$.

LEMMA 5. *Suppose that $\gamma - \tfrac{1}{2} d \notin \mathbb{N}$ and put*

$$r = \lfloor \gamma - \tfrac{1}{2} d \rfloor, \qquad\qquad \beta = \gamma - \tfrac{1}{2} d - r.$$

Then $0 < \beta < 1$ and (7) imply Hölder conditions of order (r, β) for the covariance kernel corresponding to $\mathfrak{f}$.

PROOF. We have

$$\int_{\mathbb{R}^d} u_1^{2\alpha_1} \ldots u_d^{2\alpha_d} \cdot \mathfrak{f}(u)\, du \leq \int_{\mathbb{R}^d} \|u\|^{2|\alpha|} \cdot \mathfrak{f}(u)\, du < \infty$$

if $|\alpha| < \gamma - \tfrac{1}{2} d$, which implies the existence of $\mathfrak{k}^{(2\alpha)}$ if $|\alpha| \leq r$.

Assume that $|\alpha| = r$ and $\|t\| \leq 1/c_2$. Then

$$\mathfrak{l}(0) - \mathfrak{l}(t)$$
$$\leq \int_{\|u\| \leq 1/\|t\|} (1 - \cos(\langle t, u \rangle)) \cdot \|u\|^{2r} \cdot \mathfrak{f}(u)\, du + 2c_1 \int_{\|u\| > 1/\|t\|} \|u\|^{2(r-\gamma)}\, du.$$

The second integral is bounded, up to a multiplicative constant, by $\|t\|^{2(\gamma-r)-d} = \|t\|^{2\beta}$. For the first integral,

$$\int_{\|u\| \leq 1/\|t\|} (1 - \cos(\langle t, u \rangle)) \cdot \|u\|^{2r} \cdot \mathfrak{f}(u)\, du$$
$$\leq \int_{\|u\| \leq 1/\|t\|} \langle t, u \rangle^2 \cdot \|u\|^{2r} \cdot \mathfrak{f}(u)\, du$$
$$\leq \|t\|^2 \int_{\|u\| < c_2} \|u\|^{2r+2} \cdot \mathfrak{f}(u)\, du + \|t\|^2 \int_{c_2 \leq \|u\| \leq 1/\|t\|} \|u\|^{2(r-\gamma)+2}\, du$$
$$\leq c_2^{2r+2} \|t\|^2 + \|t\|^{2(\gamma-r)-d} \int_{\|u\| \leq 1} \|u\|^{2(r-\gamma)+2}\, du$$
$$\leq c \cdot \|t\|^{2\beta}$$

with a constant c, since $r > \gamma - \frac{1}{2}d - 1$. $\square$

The upper bound (7) for the spectral density, together with Proposition 2 and Lemma 5, already yields upper bounds for the minimal errors if $\gamma - \frac{1}{2}d \notin \mathbb{N}$. In order to avoid the latter restriction and to have a constructive proof for integration we proceed differently.

EXAMPLE 6. The spectral densities

$$\mathfrak{f}_\gamma(u) = (1 + \|u\|^2)^{-\gamma}$$

with $\gamma > \frac{1}{2}d$ play a crucial role in this section. We use $\mathfrak{K}_\gamma$ to denote the corresponding covariance kernel and put

$$\mathfrak{k}_\gamma(t) = \int_{\mathbb{R}^d} \exp(\imath \langle t, u \rangle) \cdot \mathfrak{f}_\gamma(u)\, du.$$

Observe that $\mathfrak{K}_\gamma$ corresponds to a stationary and isotropic random field, since $\mathfrak{f}_\gamma(Qu) = \mathfrak{f}_\gamma(u)$ and hereby $\mathfrak{K}_\gamma(Qs, Qt) = \mathfrak{K}_\gamma(s, t)$ for every orthogonal matrix Q.

We summarize some facts on function spaces related to $\mathfrak{k}_\gamma$, see Nikolskij (1975), Triebel (1983), and Wloka (1987) for details. The function $\mathfrak{k}_\gamma$ is called a *Bessel-MacDonald kernel*, and $\mathfrak{k}_\gamma(t)$ is positive and tends to zero exponentially if $\|t\|$ tends to infinity. If $d = \gamma = 1$ then $\mathfrak{k}_1(t) = \pi \exp(-|t|)$, and therefore $\mathfrak{f}_1$ is the spectral density of an Ornstein-Uhlenbeck process, see Example IV.3.

Let $h \in L_2(\mathbb{R}^d)$ and put

$$|h|_\gamma^2 = (2\pi)^{-2d} \int_{\mathbb{R}^d} \mathfrak{f}_\gamma(u)^{-1} \cdot |\widehat{h}(u)|^2\, du.$$

Here $\widehat{h}$ denotes the Fourier transform of h, induced by

$$\widehat{h}(t) = \int_{\mathbb{R}^d} \exp(-\imath \langle t, u \rangle) \cdot h(u) \, du$$

if $h \in L_1(\mathbb{R}^d)$. It is known that $H(\mathfrak{K}_\gamma)$ is the *Bessel potential space*

$$H^\gamma = \{ h \in L_2(\mathbb{R}^d) : |h|_\gamma < \infty \}$$

equipped with the norm $|\cdot|_\gamma$. For a proof, note that

$$\mathfrak{k}_\gamma = \widehat{\mathfrak{f}}_\gamma, \qquad \widehat{\mathfrak{k}}_\gamma = (2\pi)^d \cdot \mathfrak{f}_\gamma$$

since $\mathfrak{f}_\gamma$ and $\mathfrak{k}_\gamma$ are symmetric. Hence $\mathfrak{k}_\gamma \in H^\gamma$, which implies $\mathfrak{K}_\gamma(\cdot, t) = \mathfrak{k}_\gamma(\cdot - t) \in H^\gamma$ for all $t \in \mathbb{R}^d$. Moreover, every function

$$h \in C_0^\infty(\mathbb{R}^d) = \{ h \in C^\infty(\mathbb{R}^d) : \operatorname{supp} h \text{ compact} \}$$

satisfies

$$h(t) = (2\pi)^{-d} \int_{\mathbb{R}^d} \exp(\imath \langle t, u \rangle) \cdot \widehat{h}(u) \, du$$

$$= (2\pi)^{-2d} \int_{\mathbb{R}^d} \mathfrak{f}_\gamma(u)^{-1} \cdot \widehat{h}(u) \cdot \exp(\imath \langle t, u \rangle) \widehat{\mathfrak{k}}_\gamma(u) \, du.$$

Therefore $h(t)$ is given as the scalar product of h and $\mathfrak{K}_\gamma(\cdot, t)$ in H^γ, as claimed. Since $C_0^\infty(\mathbb{R}^d)$ is dense in H^γ we conclude that $H^\gamma = H(\mathfrak{K}_\gamma)$. We add that Lemma 5, applied to $\mathfrak{f} = \mathfrak{f}_\gamma$, reflects the Sobolev Embedding Theorem.

Let $\gamma_1 < \gamma_2$ and $0 < \lambda < 1$. Clearly $H^{\gamma_2} \subset H^{\gamma_1}$. From the Hölder inequality we obtain the *interpolation inequality*

$$|h|_{\lambda \gamma_1 + (1-\lambda)\gamma_2} \leq |h|_{\gamma_1}^\lambda \cdot |h|_{\gamma_2}^{1-\lambda}$$

if $h \in H^{\gamma_2}$. Let Z be a Banach space and let $T : H^{\gamma_1} \to Z$ be a bounded linear operator. The interpolation inequality implies

$$\|T\|_{\lambda \gamma_1 + (1-\lambda)\gamma_2} \leq \|T\|_{\gamma_1}^\lambda \cdot \|T\|_{\gamma_2}^{1-\lambda}$$

where $\|T\|_\gamma$ denotes the norm of the restriction of T onto H^γ.

For arbitrary domains $A \subset \mathbb{R}^d$ we define seminorms $I_\gamma(\cdot, A)$ by

$$I_\gamma(h, A)^2 = \sum_{|\alpha| = \gamma} \int_A h^{(\alpha)}(u)^2 \, du, \qquad \gamma \in \mathbb{N},$$

and

$$I_\gamma(h, A)^2 = \sum_{|\alpha| = \lfloor \gamma \rfloor} \int_{A \times A} \frac{(h^{(\alpha)}(u) - h^{(\alpha)}(v))^2}{\|u - v\|^{d + 2(\gamma - \lfloor \gamma \rfloor)}} \, d(u, v), \qquad \gamma \notin \mathbb{N}.$$

The *Sobolev-Slobodetskiĭ spaces* are given by

$$W^\gamma(A) = \{ h \in L_2(A) : h^{(\alpha)} \text{ exists if } |\alpha| \leq \lfloor \gamma \rfloor, \ I_\gamma(h, A) < \infty \}$$

and are equipped with the norms

$$|h|_{\gamma, A}^2 = \int_A h(t)^2 \, dt + I_\gamma(h, A)^2.$$

We have

$$H^\gamma = W^\gamma(\mathbb{R}^d)$$

and the norms $|\cdot|_\gamma$ and $|\cdot|_{\gamma,\mathbb{R}^d}$ are equivalent. Consider the cube $A = D$, and let K_γ denote the restriction of $\mathfrak{K}_\gamma$ onto D^2. Then

$$H(K_\gamma) = \{h|_D : h \in H^\gamma\} \subset W^\gamma(D)$$

with equality if $\gamma \in \mathbb{N}$.

LEMMA 7. *Under condition* (7),

$$H(K) \subset H(K_\gamma).$$

Under condition (8),

$$H(K_\gamma) \subset H(K).$$

PROOF. Let $\mathfrak{K}$ be defined by an arbitrary spectral density $\mathfrak{f}$. For all $b_i \in \mathbb{R}$ and $t_i \in \mathbb{R}^d$ we have

$$\sum_{i,j=1}^k b_i b_j \cdot \mathfrak{K}(t_i, t_j) = \sum_{i,j=1}^k b_i b_j \int_{\mathbb{R}^d} \exp(\imath\langle t_i - t_j, u\rangle \cdot \mathfrak{f}(u)) \, du$$

$$= \int_{\mathbb{R}^d} \left| \sum_{i=1}^k b_i \cdot \exp(\imath\langle t_i, u\rangle) \right|^2 \cdot \mathfrak{f}(u) \, du.$$

Consider two spectral densities $\mathfrak{f}$ and $\tilde{\mathfrak{f}}$ such that $\mathfrak{f} \leq \tilde{\mathfrak{f}}$. We conclude that $\tilde{\mathfrak{K}} - \mathfrak{K}$ is nonnegative definite and $H(\mathfrak{K}) \subset H(\tilde{\mathfrak{K}})$ by Lemma III.3. Since (8) states that $c_3 \mathfrak{f}_\gamma \leq \mathfrak{f}$, we get $H(\mathfrak{K}_\gamma) \subset H(\mathfrak{K})$, which yields $H(K_\gamma) \subset H(K)$.

Assume that $\mathfrak{f}$ satisfies (7) and put

$$\mathfrak{f}_\infty(u) = \begin{cases} \mathfrak{f}(u) & \text{if } \|u\| < c_2, \\ 0 & \text{otherwise.} \end{cases}$$

Let $H(\mathfrak{K}_\infty)$ and $H(K_\infty)$ denote the reproducing kernel Hilbert spaces of functions on $\mathbb{R}^d$ and D, respectively. Since $\mathfrak{f}_\infty$ satisfies (7) for all $\gamma > 0$, we get $H(K_\infty) \subset C^r(D)$ for all $r \in \mathbb{N}$, see Lemma 5. Thus $H(K_\infty) \subset H(K_\gamma)$. On the other hand, $\mathfrak{f} - \mathfrak{f}_\infty \leq c \mathfrak{f}_\gamma$ for some constant c, and therefore $H(\mathfrak{K} - \mathfrak{K}_\infty) \subset H(\mathfrak{K}_\gamma)$, which yields $H(K - K_\infty) \subset H(K_\gamma)$. Hence $H(K) \subset H(K_\gamma)$ follows from $K = K_\infty + (K - K_\infty)$ by Lemma III.2. □

The decay conditions for the spectral density determine the order of the minimal errors as follows.

PROPOSITION 8. *Let P be a zero mean measure whose spectral density satisfies* (7) *and* (8). *Then*

$$e_2(n, \Lambda^{\text{std}}, \text{Int}_\varrho, P) \asymp n^{-\frac{\gamma}{d}}$$

for the minimal error of integration and

$$e_2(n, \Lambda^{\text{std}}, \text{App}_{2,\varrho}, P) \asymp n^{-\left(\frac{\gamma}{d} - \frac{1}{2}\right)}$$

for the minimal error of L_2-approximation.

PROOF. By Lemma 7 along with (3) and (4), it is sufficient to prove the bounds for the spectral densities f_γ of Example 6. Furthermore, Proposition II.4 yields the lower bound for the approximation problem, once we have the lower bound for the integration problem. We let c denote different positive constants, which only depend on γ and d.

Consider the integration problem in the particular case $\varrho = 1$. Fix a function $g \in C^\infty(\mathbb{R}^d)$ such that

$$\int_{\mathbb{R}^d} g(u)\, du = 2$$

and

$$g(u) = 0 \qquad \text{if} \quad \|u\| > \tfrac{1}{4}.$$

Let S_n be an arbitrary cubature formula using knots $x_1, \ldots, x_n$. Assume, without loss of generality, that $n = \tfrac{1}{2} m^d$ for some integer m. Consider a partition of D into m^d non-overlapping cubes Q_i with sidelength m^{-1}. Furthermore assume, without loss of generality, that the cubes $Q_1, \ldots, Q_n$ do not contain any knot x_i. Let q_i denote the center of Q_i and define

$$h(u) = \sum_{i=1}^{n} g(m\,(u - q_i)).$$

Then $\text{Int}(h) = 1$ and h vanishes at all knots x_i as well as on a neighborhood of the origin. Therefore

$$e_2(S_n, \text{Int}, P) \geq \sup\{|\text{Int}(h)| : h \in B(\mathfrak{K}_\gamma),\ h(x_i) = 0 \text{ for } i = 1, \ldots, n\}$$

$$\geq c \cdot \frac{\text{Int}(h)}{|h|_\gamma} \geq c \cdot \frac{1}{|h|_{\gamma, \mathbb{R}^d}},$$

see (3). The functions $g(m\,(\cdot - q_i))$ have pairwise disjoint supports. For $\gamma \in \mathbb{N}_0$ this implies

$$|h|^2_{\gamma, \mathbb{R}^d} = \sum_{i=1}^{n} |g(m\,(\cdot - q_i))|^2_{\gamma, \mathbb{R}^d} = n \cdot |g(m \cdot)|^2_{\gamma, \mathbb{R}^d}.$$

Using the scaling property

$$|g(m \cdot)|^2_{\gamma, \mathbb{R}^d} \leq m^{2\gamma - d} \cdot |g|^2_{\gamma, \mathbb{R}^d},$$

we get

$$|h|^2_{\gamma, \mathbb{R}^d} \leq n \cdot m^{2\gamma - d} \cdot |g|^2_{\gamma, \mathbb{R}^d} = c \cdot n^{2\gamma/d} \cdot |g|^2_{\gamma, \mathbb{R}^d}.$$

By the interpolation inequality, an analogous estimate holds for each $\gamma \geq 0$, and the lower bound on $e_2(S_n, \text{Int}, P)$ follows.

In the general case we assume $\varrho \in L_1(D)$ and $\varrho \geq c > 0$ on a nonempty open subset of D. The previous arguments work if we replace D by a small subcube, where ϱ is bounded away from zero. This completes the proof of the lower bounds.

The upper bounds for integration are obtained by composite formulas that are defined in Remark 3. Here we take polynomials of degree $\ell = \gamma - 1$ if $\gamma \in \mathbb{N}$ and $\ell = \lfloor \gamma \rfloor$ otherwise, and we fix admissible knots $x_i \in D$ and Lagrange polynomials p_i. Put

$$\mu = \max_{i=1,\dots,v} \sup_{t \in D} |p_i(t)|.$$

Consider a cube $Q_j \subset D$ and put $v_j = \mathrm{vol}(Q_j)$. For every $h \in W^\gamma(D)$ there exists a polynomial $q_j \in \mathbb{P}_{\ell,d}$ such that

$$\sup_{t \in Q_j} |h(t) - q_j(t)| \leq c \cdot v_j^{\gamma/d - 1/2} \cdot I_\gamma(h, Q_j),$$

see Birman and Solomjak (1967, Lemma 3.1). The corresponding cubature formula S_v^j satisfies

$$\left| \int_{Q_j} h(t) \cdot \varrho(t)\, dt - S_v^j(h) \right| = \left| \int_{Q_j} (h - q_j)(t) \cdot \varrho(t)\, dt - S_v^j(h - q_j) \right|$$

$$\leq c \cdot v_j^{\gamma/d - 1/2} \cdot I_\gamma(h, Q_j) \cdot \left(\int_{Q_j} |\varrho(t)|\, dt + \sum_{i=1}^v \int_{Q_j} |p_i(T_j^{-1}(t))| \cdot |\varrho(t)|\, dt \right)$$

$$\leq c \cdot v_j^{\gamma/d - 1/2} \int_{Q_j} |\varrho(t)|\, dt \cdot \mu \cdot I_\gamma(h, Q_j).$$

For every $m \in \mathbb{N}$ and every $\varrho \in L_1(D)$ there exists a partition $Q_1, \dots, Q_M$ of D with $M \leq m$ and

$$(9) \qquad \max_{j=1,\dots,M} v_j^{\gamma/d - 1/2} \int_{Q_j} |\varrho(t)|\, dt \leq c \cdot m^{-(\gamma/d + 1/2)} \cdot \|\varrho\|_1,$$

see Birman and Solomjak (1967, Theorem 2.1). The cubature formula S_n based on this partition satisfies

$$|\mathrm{Int}_\varrho(h) - S_n(h)| \leq c\mu \|\varrho\|_1 \cdot m^{-(\gamma/d + 1/2)} \cdot \sum_{j=1}^M I_\gamma(h, Q_j).$$

Since

$$\sum_{j=1}^M I_\gamma(h, Q_j) \leq M^{1/2} \left(\sum_{j=1}^M I_\gamma(h, Q_j)^2 \right)^{1/2} \leq m^{1/2} \cdot I_\gamma(h, D)$$

we obtain

$$|\mathrm{Int}_\varrho(h) - S_n(h)| \leq c\mu \|\varrho\|_1 \cdot m^{-\gamma/d} \cdot I_\gamma(h, D).$$

This completes the proof for integration, because $I_\gamma(h, D) \leq c$ for all $h \in B(K_\gamma)$.

The upper bounds for the minimal errors of L_2-approximation in the case $\gamma - \frac{1}{2}d \notin \mathbb{N}$ are a consequence of Proposition 2 and Lemma 5. The bounds extend to the case $\gamma - \frac{1}{2}d \in \mathbb{N}$ by means of (4) and the interpolation inequality.   □

REMARK 9. Under the assumptions of Proposition 8, the following methods are order optimal. For approximation we use a uniform partition of D and polynomials of degree

$$\ell = \lfloor \gamma - \tfrac{1}{2}d \rfloor,$$

see Remark 3. Polynomials of degree ℓ in each variable or K-splines that are based on regular grids work as well.

For integration we use composite formulas of degree

$$\ell = \begin{cases} \gamma - 1 & \text{if } \gamma \in \mathbb{N}, \\ \lfloor \gamma \rfloor & \text{otherwise.} \end{cases}$$

The partition of D is chosen according to the construction of Birman and Solom-jak (1967), see (9). Clearly we can take a uniform partition if ϱ is continuous on D.

REMARK 10. Essentially conditions (7) and (8) say that the spectral density $\mathfrak{f}$ tends to zero like $\|u\|^{-2\gamma}$ as $\|u\|$ tends to infinity. Stein (1993, 1995a) uses the following more general conditions on $\mathfrak{f}$ to study the integration problem.

A function $\mathfrak{g} :]0,\infty[\to]0,\infty[$ is called *regularly varying* with exponent λ if

$$\lim_{u \to \infty} \mathfrak{g}(cu)/\mathfrak{g}(u) = c^\lambda$$

for every $c > 0$. Assume that $\mathfrak{f}(u)/\mathfrak{g}(\|u\|)$ is bounded away from zero and infinity for $\|u\|$ sufficiently large, where $\mathfrak{g}$ is regularly varying with exponent -2γ. Furthermore, assume that $\mathfrak{f}$ is bounded and positive. The particular choice

(10) $$\mathfrak{g}(u) = u^{-2\gamma}$$

corresponds to (7) and (8).

Stein studies cubature formulas that are based on grids

$$G_k = \left\{ \frac{2i-1}{2k} : i = 1, \ldots, k \right\}^d,$$

assuming that the weight function ϱ has bounded partial derivatives of order $d + 2 + \lfloor \gamma \rfloor$. He constructs formulas S_n with $n = k^d$ such that

$$e_2(S_n, \mathrm{Int}_\varrho, P) \approx \inf\{e_2(S_{k^d}, \mathrm{Int}_\varrho, P) : S_{k^d} \text{ is based on } G_k\}$$

$$\approx \left(\int_{[-k\pi, k\pi]^d} \sum_{j \in \mathbb{Z}^d \setminus \{0\}} \mathfrak{f}(u + 2\pi k j) \cdot |\widehat{\varrho}(u)|^2 \, du \right)^{1/2}.$$

Hence these formulas are asymptotically optimal in the class of all methods that are based on the grids G_k.

In the case (10), the order of $e_2(S_n, \mathrm{Int}_\varrho, P)^2$ is given by

$$\int_{[-k\pi, k\pi]^d} \sum_{j \in \mathbb{Z}^d \setminus \{0\}} \mathfrak{f}(u + 2\pi\,kj) \cdot |\widehat{\varrho}(u)|^2 \, du$$

$$= O\left(\int_{[-k\pi, k\pi]^d} \sum_{j \in \mathbb{Z}^d \setminus \{0\}} \|kj\|^{-2\gamma} \cdot |\widehat{\varrho}(u)|^2 \, du \right)$$

$$= O(k^{-2\gamma}) = O(n^{-2\gamma/d}).$$

The cubature formulas S_n are therefore order optimal in the class of all methods, at least in the particular case (10), see Proposition 8.

For $d = 2$ these results are extended in Section 1.3.

We now show that sampling is order optimal for the approximation problem.

PROPOSITION 11. *Under the assumptions of Proposition 8,*

$$e_2(n, \Lambda^{\mathrm{std}}, \mathrm{App}_{2,\varrho}, P) \asymp e_2(n, \Lambda^{\mathrm{all}}, \mathrm{App}_{2,\varrho}, P)$$

if ϱ is continuous.

PROOF. It is sufficient to consider the case $\varrho = 1$ and, due to Lemma 7 and (4), the spectral density $\mathfrak{f} = \mathfrak{f}_\gamma$. Let B denote the unit ball in $W^\gamma(D)$ with respect to the seminorm $I_\gamma(\cdot, D)$, and let μ_n denote the nth eigenvalue of the integral operator with kernel K_γ.

The order of the minimal errors for L_2-approximation on B is known to be

$$e_{\max}(n, (W^\gamma(D))^*, \mathrm{App}_2, B) \asymp n^{-\gamma/d}$$

for $\Lambda = (W^\gamma(D))^*$ consisting of all bounded linear functionals on $W^\gamma(D)$, see Pinkus (1985, Chapter 7). If $\gamma \in \mathbb{N}$ then $W^\gamma(D) = H(K_\gamma)$ and we get

$$\mu_{n+1}^{1/2} = e_{\max}(n, (H(K_\gamma))^*, \mathrm{App}_2, B(K_\gamma)) \asymp n^{-\gamma/d},$$

see Remark III.26. The interpolation inequality for the spaces $H(\mathfrak{K}_\gamma)$ yields $\mu_n \asymp n^{-2\gamma/d}$ for all $\gamma > \frac{1}{2}d$, and Proposition III.24 implies

$$e_2(n, \Lambda^{\mathrm{all}}, \mathrm{App}_2, P) = \left(\sum_{j=n+1}^\infty \mu_j \right)^{1/2} \asymp n^{-\left(\frac{\gamma}{d} - \frac{1}{2}\right)}.$$

By Proposition 8 this order is also achieved for the class Λ^{std}. $\square$

REMARK 12. A stationary and isotropic kernel $\mathfrak{K}$ is also called a radial basis function. For irregularly spaced knots x_i the corresponding $\mathfrak{K}$-spline algorithm yields a scattered data interpolation. Instead of nonnegative definiteness, the more general conditional positive definiteness is studied, and worst case error bounds for the approximation problem on the corresponding Hilbert spaces are derived. See Powell (1992) and Schaback (1997).

1.3. Integration for Stationary Isotropic Random Fields, $d = 2$. A good sampling design for integration or approximation should have a greater density of knots where the weight function ϱ is large in absolute value. In the univariate case we have used regular sequences to this end, and Chapter IV shows how to relate the density and the weight function in an asymptotically optimal way.

Unfortunately, we do not control asymptotic constants for minimal errors in the multivariate case. With respect to order optimality, however, the weight function ϱ has an impact on the design only for integration with ϱ having singularities, see, e.g., Proposition 8 and Remark 9.

Now we consider integration with smooth weight functions for stationary isotropic random fields in dimension $d = 2$. We study cubature formulas that use *local lattice designs*. The latter are defined by a sequence of real numbers $\omega_n > 0$ with

$$(11) \qquad \lim_{n \to \infty} w_n = 0, \qquad \lim_{n \to \infty} n^{1/2} \cdot w_n = \infty,$$

a 2×2 matrix B with $\det B = 1$, and a continuous positive function ψ on D with

$$(12) \qquad \int_D \psi(x)\, dx = 1.$$

For $j = (j_1, j_2)' \in \mathbb{Z}^2$ we put

$$P_j^n = \{B(\omega_n \cdot x) : x \in [j_1, j_1 + 1] \times [j_2, j_2 + 1]\}$$

to obtain a covering of the plane $\mathbb{R}^2$ by non-overlapping parallelograms P_j^n with area $\mathrm{vol}(P_j^n) = \omega_n^2$. Let $R_j^n = P_j^n \cap D$ and $r_j^n = \mathrm{vol}(R_j^n)$. In the sequel we only consider those indices j with $r_j^n > 0$. Put

$$(13) \qquad m_j^n = \left\lfloor w_n \cdot \left(n/r_j^n \cdot \int_{R_j^n} \psi(x)\, dx \right)^{1/2} \right\rfloor,$$

and observe that $\inf_j m_j^n$ tends to infinity as n tends to infinity. We subdivide P_j^n into $(m_j^n)^2$ parallelograms as follows. For

$$k = (k_1, k_2)' \in \{1, \dots, m_j^n\}^2$$

let

$$P_{j,k}^n = \{B(\omega_n \cdot (j + 1/m_j^n \cdot x)) : x \in [k_1 - 1, k_1] \times [k_2 - 1, k_2]\}.$$

We put $R_{j,k}^n = P_{j,k}^n \cap D$, and in the sequel we only consider those indices (j, k) with $r_{j,k}^n = \mathrm{vol}(R_{j,k}^n) > 0$. Let $x_{j,k}^n$ denote the center of mass of $R_{j,k}^n$ and consider the cubature formulas

$$(14) \qquad S_n(f) = \sum_{j,k} \varrho(x_{j,k}^n) \cdot r_{j,k}^n \cdot f(x_{j,k}^n).$$

Note that S_n depends on n as well as on B, ω_n, ψ, and ϱ. Locally, on each set $P_j^n \subset D$, the knots $x_{j,k}^n$ form a lattice, whose shape is controlled by B and whose density is controlled by ψ via (13). The total number of knots that are used by S_n is different from n in general. It is, however, strongly equivalent to n.

In the simplest case, B is the identity matrix and $1/\omega_n \in \mathbb{N}$. Then we obtain squares $P_j^n = R_j^n$ and $P_{j,k}^n = R_{j,k}^n$ with edges parallel to the coordinate axes. The knots

$$x_{j,k}^n = \omega_n \cdot \left(j + 1/m_j^n \cdot \left(k - (\tfrac{1}{2}, \tfrac{1}{2})'\right)\right)$$

form grids on the squares P_j^n, and for every j the mesh-size of the grid is roughly the square-root of $r_j^n/(n \cdot \int_{P_j^n} \psi(x)\,dx)$.

The spectral density $\mathfrak{f}$ of a stationary and isotropic random field satisfies

(15) $$\mathfrak{f}(u) = \mathfrak{g}(\|u\|)$$

where $\mathfrak{g} : [0, \infty[\to [0, \infty[$. Recall that $\|\cdot\|$ denotes the Euclidean norm. Moreover, recall the notion of regularly varying functions, which was introduced in Remark 10. The following result is proven in Stein (1995b).

PROPOSITION 13. *Let P denote a zero mean measure whose spectral density $\mathfrak{f}$ satisfies (15) with a regularly varying function $\mathfrak{g}$ with exponent $-\lambda$ for*

$$\lambda \in \,]2, 4[.$$

Moreover, assume that ϱ has continuous derivatives up to order 3 and ψ is continuous and positive on D and satisfies (12). Then

$$e_2(S_n, \mathrm{Int}_\varrho, P) \approx (2\pi)^{1-\lambda/2} \cdot M_{\varrho,\lambda}(\psi) \cdot L_\lambda(B) \cdot g(n^{1/2})^{1/2},$$

where

$$M_{\varrho,\lambda}(\psi) = \left(\int_D \varrho(x)^2 \cdot \psi(x)^{-\lambda/2}\,dx\right)^{1/2}$$

and

$$L_\lambda(B) = \left(\sum_{j \in \mathbb{Z}^2 \setminus \{0\}} \|(B')^{-1}j\|^{-\lambda}\right)^{1/2}.$$

The asymptotic quality of local lattice designs is therefore determined by the choice of ψ and B, while the scaling factors ω_n are irrelevant as long as (11) holds. If ϱ is positive then $M_{\varrho,\lambda}(\psi)$ is minimal for

(16) $$\psi(x) = \varrho(x)^{4/(\lambda+2)} / \int_D \varrho(x)^{4/(\lambda+2)}\,dx,$$

which can be verified using the Hölder inequality, cf. Lemma IV.13. Concerning B, we observe that

(17) $$L_\lambda^2(B) = \sum_{j \in \mathbb{Z}^2 \setminus \{0\}} |j' \cdot A \cdot j|^{-\lambda/2},$$

where $A = (B'B)^{-1}$. The right-hand side in (17) defines an Epstein zeta-function, which is well defined for all $\lambda > 2$. It is known that for fixed λ the minimum of this function with respect to all symmetric 2×2 matrices A with $\det A = 1$ is attained for

$$A = \frac{2}{3^{1/2}} \cdot \begin{pmatrix} 1 & \frac{1}{2} \\ \frac{1}{2} & 1 \end{pmatrix}.$$

Therefore an optimal choice of B is

(18) $$B = \frac{2^{1/2}}{3^{1/4}} \cdot \begin{pmatrix} 1 & -\frac{1}{2} \\ 0 & 3^{1/2}/2 \end{pmatrix},$$

which locally yields equilateral triangular lattices. We are led to the following conclusion.

COROLLARY 14. *If ϱ is positive, in addition to the assumptions of Proposition 13, then (16) and (18) yield the asymptotically best cubature formulas of the form (14).*

Stein (1995b) conjectures that the cubature formulas from Corollary 14 are in fact asymptotically optimal.

The previous results are applicable in particular for the Bessel-MacDonald kernel $\mathfrak{K}_\gamma$ with $\gamma \in \,]1, 2[$, see Example 6. The corresponding function $\mathfrak{g}(u) = (1 + u^2)^{-\gamma}$ is regularly varying with exponent -2γ.

1.4. Fractional Brownian Fields. In this section we use the isotropic Wiener measure or closely related measures to analyze integration and approximation. These measures are naturally constructed on classes of functions defined on the whole space $\mathbb{R}^d$ or on a centered ball in $\mathbb{R}^d$. The corresponding random fields are non-stationary but isotropic.

First we look at functions of low regularity. The *fractional Brownian field* with parameter

$$0 < \beta < 1$$

is the zero mean Gaussian random field on $\mathbb{R}^d$ with covariance kernel

$$\mathfrak{K}^\beta(s, t) = \tfrac{1}{2} \left(\|s\|^{2\beta} + \|t\|^{2\beta} - \|s - t\|^{2\beta} \right), \qquad s, t \in \mathbb{R}^d,$$

see Example 1. For $\beta = \frac{1}{2}$ this kernel belongs to the so-called *Lévy's Brownian motion*, which is a generalization of the univariate Brownian motion to multidimensional time, since

$$\mathfrak{K}^{1/2}(s, t) = \begin{cases} \min(|s|, |t|) & \text{if } s \cdot t > 0, \\ 0 & \text{otherwise} \end{cases}$$

for $d = 1$.

By restriction onto the cube D we get a Gaussian measure on the space $F = C(D)$. For $\beta = \frac{1}{2}$ this measure is also called *isotropic Wiener measure*. In Example 1 we have noticed that the corresponding covariance kernels satisfy Hölder conditions of order $(0, \beta)$. On each line through the origin the random functions form a univariate fractional Brownian motion. Let $d = 1$ and $0 \leq s < t < u \leq 1$. The corresponding increments satisfy

$$\int_F (f(u) - f(t)) \cdot (f(t) - f(s)) \, dP(f) = \tfrac{1}{2} \left((u - s)^{2\beta} - (u - t)^{2\beta} - (t - s)^{2\beta} \right).$$

Hence the increments are uncorrelated if $\beta = \tfrac{1}{2}$, positively correlated if $\beta > \tfrac{1}{2}$ and negatively correlated if $\beta < \tfrac{1}{2}$.

Measures on spaces of smooth functions are often constructed from measures of spaces on less regular functions by some kind of smoothing. For instance, in the univariate case we have used r-fold integration. For fractional Brownian fields we wish to preserve isotropy, and therefore we use a power of the inverse Laplacian for smoothing.

Let $B \subset \mathbb{R}^d$ denote a centered ball containing the unit cube D in its interior, and let G denote the Green's function of the domain B. Thus

$$(19) \qquad (Tf)(s) = \int_B G(s, t) \cdot f(t) \, dt$$

fulfills

$$\Delta(Tf) = f \qquad \text{and} \qquad (Tf)|_{\partial B} = 0$$

for every Hölder continuous function f. Consider an isotropic kernel $\mathfrak{K}_\theta$ on B^2 that satisfies Hölder conditions of order (r, β) for some $\beta > 0$ and $r \in \mathbb{N}_0$. Define

$$\mathfrak{K}_{\theta+1}(s, t) = \int_{B^2} \mathfrak{K}_\theta(u, v) \cdot G(s, u) \, G(t, v) \, d(u, v).$$

Since G is isotropic, $\mathfrak{K}_{\theta+1}$ is isotropic, too. We add that $\mathfrak{K}_\theta$ is obtained from $\mathfrak{K}_{\theta+1}$ by applying Δ in both arguments. Therefore $\mathfrak{K}_{\theta+1}$ satisfies Hölder conditions of order $(r + 2, \beta)$.

Beginning with

$$\mathfrak{K}_0(s, t) = \mathfrak{K}_0^\beta(s, t) = \mathfrak{K}^\beta(s, t), \qquad s, t \in B,$$

we obtain a system of isotropic kernels $\mathfrak{K}_\theta^\beta$ on B^2. The smoothness increases with θ and β, and $\mathfrak{K}_\theta^\beta$ satisfies Hölder conditions of order $(2\theta, \beta)$.

Consider the zero mean Gaussian measure on $C(B)$ that is induced by a fractional Brownian field. Its image under the mapping T^θ is the zero mean Gaussian measure on $C^{2\theta}(B)$ with covariance kernel $\mathfrak{K}_\theta^\beta$. By restriction we get Gaussian measures on $C^{2\theta}(D)$.

Note that

$$\mathfrak{K}_\theta^\beta(t, t) = \cdots = \mathfrak{K}_1^\beta(t, t) = 0, \qquad t \in \partial B.$$

Smoothing with suitable stochastic boundary conditions yields isotropic kernels that do not vanish at the boundary, see Ritter and Wasilkowski (1996a) for details. Proposition 18 extends to this case.

First we assume that $\theta = 0$, and we put

$$\gamma = \beta + \tfrac{1}{2} d.$$

The Hilbert space $H(\mathfrak{K}^\beta)$ consists of real-valued functions on $\mathbb{R}^d$, and it is characterized as follows, see Singer (1994).

LEMMA 15. *Let*

$$Z = \{g \in C_0^\infty(\mathbb{R}^d) : 0 \notin \operatorname{supp} g\}.$$

Then $Z \subset H(\mathfrak{K}^\beta)$ *as a dense subset, and there exists a constant* $c > 0$ *such that*

$$\|h\|_{\mathfrak{K}^\beta}^2 = c \cdot \int_{\mathbb{R}^d} \|u\|^{2\gamma} \cdot |\widehat{h}(u)|^2 \, du$$

for every $h \in Z$.

We add that the constant c is known explicitly. Moreover, the norm $\|h\|_{\mathfrak{K}^\beta}$ for $h \in Z$ coincides with the L_2-norm of $\Delta^{\gamma/2}h$. Here $\Delta^{\gamma/2}$ denotes a fractional power of the Laplacian.

Let K^β denote the restriction of $\mathfrak{K}^\beta$ onto D^2. See Example 6 for the definition and a brief discussion of the Sobolev-Slobodetskiĭ spaces $W^\gamma(A)$.

LEMMA 16. *The reproducing kernel Hilbert space of a fractional Brownian field on the unit cube D satisfies*

$$H(K^\beta) = \{h \in W^\gamma(D) : h(0) = 0\}$$

with equivalence of norms.

PROOF. We let c denote different positive constants, which depend only on β and d. A norm equivalent to $\| \cdot \|_{\mathfrak{K}^\beta}$ is given by $I_\gamma(\cdot, \mathbb{R}^d)$ on Z, see Lemma 15. This follows from Mazja (1985, Corollary 7.1.2.1, Theorem 7.1.2.4). Moreover, if $h \in H(\mathfrak{K}^\beta)$ then

$$\sup_{t \in D} |h(t)| \leq \sup_{t \in D} \mathfrak{K}^\beta(t, t)^{1/2} \cdot \|h\|_{\mathfrak{K}^\beta} \leq c \cdot \|h\|_{\mathfrak{K}^\beta}.$$

Using Lemma 15 we conclude that $h \mapsto h|_D$ defines a continuous mapping between $H(\mathfrak{K}^\beta)$ and $W^\gamma(D)$. Hence $H(K^\beta) \subset W^\gamma(D)$.

From $\gamma < \frac{1}{2}d + 1$ it follows that the closure of Z in $W^\gamma(\mathbb{R}^d)$ is given by

$$Z_1 = \{h \in W^\gamma(\mathbb{R}^d) : h(0) = 0\}.$$

Since

$$\|h\|_{\mathfrak{K}^\beta}^2 \leq c \cdot \int_{\mathbb{R}^d} (1 + \|u\|^2)^\gamma \cdot |\widehat{h}(u)|^2 \, du$$

for $h \in Z$, we get $Z_1 \subset H(\mathfrak{K}^\beta)$, which implies $h \in H(K^\beta)$ for every $h \in W^\gamma(D)$ with $h(0) = 0$. $\qquad\square$

Note that $H(\mathfrak{K}^\beta) \subset W^\gamma(\mathbb{R}^d)$ does not hold, since $\mathfrak{K}^\beta(\cdot, t) \notin L_2(\mathbb{R}^d)$ if $t \neq 0$. In the general case $\theta \in \mathbb{N}_0$ we put

$$\gamma = 2\theta + \beta + \tfrac{1}{2}d.$$

The Hilbert space $H(\mathfrak{K}_\theta^\beta)$ consists of real-valued functions on B, and its elements are given as follows.

LEMMA 17.

$$H(\mathfrak{K}^\beta_\theta) = \{h \in W^\gamma(B) : \Delta^\theta h(0) = 0, \ h|_{\partial B} = \cdots = (\Delta^{\theta-1}h)|_{\partial B} = 0\}$$

with equivalence of norms.

PROOF. If $\theta = 0$ then the proof of Lemma 16 may be used, with D being replaced by B.

For $\theta > 0$ the operator Δ^θ defines a bounded linear bijection between the spaces

$$Y = \{h \in W^\gamma(B) : \Delta^\theta h(0) = 0, \ h|_{\partial B} = \cdots = (\Delta^{\theta-1}h)|_{\partial B} = 0\}$$

and $H(\mathfrak{K}^\beta_0)$. We complete the proof by verifying that $\mathfrak{K}^\beta_\theta$ is the reproducing kernel in the space Y, equipped with the scalar product

$$\langle h_1, h_2 \rangle_Y = \langle \Delta^\theta h_1, \Delta^\theta h_2 \rangle_{\mathfrak{K}^\beta_0}.$$

Using Mercer's Theorem, see Section III.2.3, we obtain

$$\mathfrak{K}^\beta_0(s, t) = \sum_{j=1}^\infty h_j(s) \cdot h_j(t)$$

where the series is uniformly convergent on B^2 and the functions h_j form an orthonormal basis in $H(\mathfrak{K}^\beta_0)$. This implies

$$\mathfrak{K}^\beta_\theta(s, t) = \sum_{j=1}^\infty T^\theta h_j(s) \cdot T^\theta h_j(t),$$

where T is given by (19). Clearly $T^\theta h_j$ is an orthonormal basis of Y with respect to the scalar product $\langle \cdot, \cdot \rangle_Y$, and therefore $\mathfrak{K}^\beta_\theta(\cdot, t) \in Y$ for every $t \in B$. Since function evaluation is continuous with respect to $|\cdot|_{\gamma,B}$, it is continuous also with respect to $\|\cdot\|_Y$. Hence we get

$$h(t) = \sum_{j=1}^\infty \langle h, T^\theta h_j \rangle_Y \cdot T^\theta h_j(t) = \langle h, \mathfrak{K}^\beta_\theta(\cdot, t) \rangle_Y$$

for all $t \in B$ and $h \in Y$. $\qquad\square$

Let K^β_θ denote the restriction of $\mathfrak{K}^\beta_\theta$ onto D^2. We show that the upper bounds from Proposition 2 with $r = 2\theta$ are sharp for these kernels and we give a constructive proof for the upper bound for integration.

PROPOSITION 18. Let P be a zero mean measure with covariance kernel K^β_θ. Then

$$e_2(n, \Lambda^{\mathrm{std}}, \mathrm{Int}_\varrho, P) \asymp n^{-\left(\frac{2\theta+\beta}{d} + \frac{1}{2}\right)}$$

for the minimal error of integration and

$$e_2(n, \Lambda^{\mathrm{std}}, \mathrm{App}_{2,\varrho}, P) \asymp e_2(n, \Lambda^{\mathrm{all}}, \mathrm{App}_{2,\varrho}, P) \asymp n^{-\frac{2\theta+\beta}{d}}$$

for the minimal error of L_2-approximation.

PROOF. Consider the function spaces from Example 6 with $\gamma = 2\theta + \beta + \frac{1}{2}d$. Lemma 17 and Lemma III.4 imply

$$\{h \in H(K_\gamma) : \Delta^\theta h(0) = 0\} \subset H(K_\theta^\beta) \subset \{h \in W^\gamma(D) : \Delta^\theta h(0) = 0\}.$$

Thus the proofs of Propositions 8 and 11 are also applicable for the kernels K_θ^β.

□

REMARK 19. The methods discussed in Remark 9 are order optimal also under the assumptions of Proposition 18. Thus piecewise polynomials of degree 2θ for approximation and composite cubature formulas of degree

$$\ell = \begin{cases} 2\theta + \frac{1}{2}d & \text{for even } d, \\ 2\theta + \frac{1}{2}(d-1) & \text{for odd } d,\ 0 < \beta \leq \frac{1}{2}, \\ 2\theta + \frac{1}{2}(d+1) & \text{for odd } d,\ \frac{1}{2} < \beta < 1 \end{cases}$$

yield the optimal order.

REMARK 20. Ritter and Wasilkowski (1996b) study the problem of solving the *Poisson equation*

$$\Delta g = f \quad \text{and} \quad g|_{\partial\Omega} = 0$$

on a smooth domain $\Omega \subset \mathbb{R}^d$. Starting with Lévy's Brownian field they obtain a zero mean Gaussian measure P_θ on $F = C^{2\theta}(\Omega)$, analogously to the construction presented in this section. Two variants of the problem, approximating the global solution u on Ω and approximating a local solution $\xi(u)$, are studied. For instance, the value $\xi(u) = u(t)$ at a fixed point $t \in \Omega$ or a weighted integral of u is a local solution. Both variants form a linear problem, see Section II.2.3.

The authors consider linear methods

$$S_n(f) = \sum_{i=1}^{n} f(x_i) \cdot a_i$$

that use function values at knots $x_i \in \Omega$ and where $a_i \in \mathbb{R}$ for approximating a local solution and $a_i \in L_2(\Omega)$ for approximating the global solution. The average errors are defined by

$$e_2(S_n, \Delta^{-1}, P_\theta) = \left(\int_F \|\Delta^{-1}f - S_n(f)\|_2^2 \, dP_\theta(f) \right)^{1/2}$$

and

$$e_2(S_n, \xi \circ \Delta^{-1}, P_\theta) = \left(\int_F |\xi(\Delta^{-1}f) - S_n(f)|^2 \, dP_\theta(f) \right)^{1/2}.$$

It turns out the minimal errors are of order

$$e_2(n, \Lambda^{\text{std}}, \Delta^{-1}, P_\theta) \asymp n^{-\frac{2\theta + (1+\min(d,4))/2}{d}}, \qquad d \neq 4,$$

for approximating the global solution and

$$e_2(n, \Lambda^{\text{std}}, \xi \circ \Delta^{-1}, P_\theta) \asymp n^{-\frac{2\theta + (1+d)/2}{d}}, \qquad d \in \mathbb{N},$$

for approximating a local solution. For the global solution with $d = 4$ the estimate

$$c_1 \cdot n^{-\frac{2\theta+5/2}{4}} \leq e_2(n, \Lambda^{\text{std}}, \Delta^{-1}, P_\theta) \leq c_2 \cdot n^{-\frac{2\theta+5/2}{4}} \cdot (\ln n)^{1/2}$$

holds with positive constants c_1 and c_2.

The orders of minimal errors for both problems coincide if $d \leq 3$. For large d and fixed smoothness θ the orders differ significantly.

See Werschulz (1991) for further average case results for elliptic equations and for Fredholm integral equations.

1.5. Notes and References. 1. Proposition 2 is due to Ritter, Wasilkowski, and Woźniakowski (1993).

2. Weba (1995) studies integration of L_q-random fields with partial derivatives up to order $r = 4$ in L_q-sense. He derives an upper bound for the q-average error of product Simpson rules in dimension $d = 3$.

3. We refer to Adler (1981), Yadrenko (1983), and Ivanov and Leonenko (1989) for general facts on stationary random fields.

4. Propositions 8 and 11 are due to Ritter (1996c). For smooth weight functions ϱ the upper bound for integration in Proposition 8 follows from Stein (1993, 1995a), where more general conditions on spectral densities are used. Furthermore, Stein (1990a, 1990b) uses conditions (7) and (8) for linear problems with values in $\mathbb{R}$. See also Stein (1999).

5. Local lattice designs, as well as Proposition 13 and Corollary 14, are due to Stein (1995b) in the more general case of a compact and Jordan-measurable set $D \subset \mathbb{R}^2$.

6. Lemma 15 is due to Singer (1994). In particular for $\beta = \frac{1}{2}$ the result is due to Ciesielski (1975), and for $\beta = \frac{1}{2}$ and odd dimension d it is due to Molchan (1967).

7. Proposition 18 is due to Wasilkowski (1993) for $\theta = 0$ and $\beta = \frac{1}{2}$, and due to Ritter and Wasilkowski (1996a) for $\theta > 0$ and $\beta = \frac{1}{2}$. The upper bounds for integration were proven non-constructively in these papers. The general case, as well as a constructive proof for integration, is due to Ritter (1996c).

2. Tensor Product Problems

Now we analyze integration and L_2-approximation for random fields under the following assumptions.

(A) The set D is a compact rectangular subset of $\mathbb{R}^d$. For simplicity we take

$$D = [0, 1]^d.$$

(B) The weight function ϱ is of tensor product form,

$$\varrho = \varrho_1 \otimes \cdots \otimes \varrho_d.$$

(C) The covariance kernel of the zero mean measure P on $F \subset C^r(D)$ is of tensor product form, $K = \bigotimes_{\nu=1}^{d} K_\nu$.

In (B) the operator $\otimes$ denotes the tensor product of real-valued functions on $[0,1]$,

$$(\varrho_1 \otimes \cdots \otimes \varrho_d)(\tau_1, \ldots, \tau_d) = \varrho_1(\tau_1) \cdots \varrho_d(\tau_d).$$

Tensor products of covariance kernels are defined in Section 2.1. In contrast to the analysis of isotropic smoothness the rectangular shape of the domain D is crucial now.

In Section 2.1 we present some basic facts concerning tensor product kernels and the associated random fields and Hilbert spaces. We discuss tensor product methods in Section 2.2. Actually, we have already used a particular tensor product method in Section 1, namely, interpolation by (piecewise) polynomials on a grid. The error analysis reveals that tensor product methods are not well suited for tensor product problems. Instead, Smolyak formulas are much better and are often order optimal. These formulas are linear combinations of tensor product methods, and their error can be expressed in terms of the errors of the underlying univariate methods, see Section 2.3.

Then we study specific problems that are defined by smoothness properties of the covariance kernels K_ν in (C). Hölder conditions, decay conditions for univariate spectral densities, and Sacks-Ylvisaker conditions are used in Sections 2.4–2.6. The Wiener sheet is an important example for Sacks-Ylvisaker conditions. We obtain upper bounds for minimal errors under Hölder conditions, and we determine the order of minimal errors and provide order optimal methods in the other cases. In Section 2.7 we discuss L_2-discrepancy and its relation to the average case analysis of the integration problem.

In the average case setting a *tensor product problem* is a linear problem for multivariate functions with the following properties. The domain D is the unit cube or, more general, a rectangular subset of $\mathbb{R}^d$, and (C) holds. The bounded linear operator $S : C^r(D) \to G$ takes values in a Hilbert space G that is the tensor product of Hilbert spaces $G_1, \ldots, G_d$. Moreover, there exist bounded linear operators $S^\nu : C^r([0,1]) \to G_\nu$ such that

(20) $$S(f_1 \otimes \cdots \otimes f_d) = S^1(f_1) \otimes \cdots \otimes S^d(f_d)$$

for $f_1, \ldots, f_d \in C^r([0,1])$.

Consider the integration problem, where formally $G = \mathbb{R}$ is the d-fold tensor product of the Hilbert space $G_\nu = \mathbb{R}$. Clearly (B) implies Fubini's Theorem (20) with

$$S = \mathrm{Int}_\varrho, \qquad S^\nu = \mathrm{Int}_{\varrho_\nu}.$$

Weight functions of tensor product form also turn L_2-approximation into a tensor product problem. Now G_ν is the L_2-space on $[0,1]$ with weight function ϱ_ν, and G is the L_2-space on D with weight function ϱ. Property (20) trivially holds, since the mappings

$$S = \mathrm{App}_{2,\varrho}, \qquad S^\nu = \mathrm{App}_{2,\varrho_\nu}$$

are embeddings.

We do not study general tensor product problems. Instead we focus on integration and L_2-approximation. We add that the results from Sections 2.2 and 2.3 hold for general tensor product problems.

2.1. Tensor Product Kernels. In the sequel let

$$s = (\sigma_1, \ldots, \sigma_d) \in D, \qquad t = (\tau_1, \ldots, \tau_d) \in D.$$

A covariance kernel K on D^2 is of *tensor product* form, if

$$K(s,t) = \prod_{\nu=1}^{d} K_\nu(\sigma_\nu, \tau_\nu)$$

for all $s, t \in D$, where $K_1, \ldots, K_d$ are covariance kernels on $[0,1]^2$. Thus K corresponds to a random field on D, while the kernels K_ν correspond to univariate random functions on $[0,1]$. We use the notation

$$K = \bigotimes_{\nu=1}^{d} K_\nu$$

for tensor product kernels. Constructively, one may take arbitrary kernels K_ν and define K in the above way. Since every kernel K_ν is nonnegative definite, the same property holds for K, see Vakhania, Tarieladze, and Chobanyan (1987, p. 187).

EXAMPLE 21. The zero mean Gaussian random function with covariance kernel

$$K(s,t) = \prod_{\nu=1}^{d} \min(\sigma_\nu, \tau_\nu)$$

is called the *Wiener sheet* or Wiener-Chentsov random field. The corresponding measure on $C(D)$ is called *Wiener sheet measure*. Both the Wiener sheet and Lévy's Brownian motion are generalizations of the Brownian motion to multivariate time. However, they enjoy strongly different properties for $d > 1$, see Adler (1981) and Lifshits (1995). This is reflected in the error bounds from Proposition 18 with $\theta = 0$ and $\beta = \frac{1}{2}$ and Proposition 37 with $r = 0$.

Now we briefly discuss the related notions of tensor product random fields and tensor product Hilbert spaces. Let P_ν denote a zero mean measure on a class F_ν of univariate functions with covariance kernel K_ν. Consider the product measure $P_1 \otimes \cdots \otimes P_d$ on $F_1 \times \cdots \times F_d$ and its image Q on a suitable class F of multivariate functions under the mapping

$$(f_1, \ldots, f_d) \mapsto f_1 \otimes \cdots \otimes f_d.$$

Obviously Q has mean zero and its covariance kernel is $K = \bigotimes_{\nu=1}^{d} K_\nu$. The corresponding random field is the *tensor product* of independent random functions.

Let P denote any zero mean measure on F with the same tensor product kernel K, and consider a linear problem with values in a Hilbert space G. From Section III.2.2 we know that

$$e_2(S_n, S, P) = e_2(S_n, S, Q)$$

for every linear method S_n. In the average case analysis we may therefore switch from P to Q and deal with a random field of tensor product form. We stress that this conclusion does not extend to nonlinear methods S_n or to merely normed spaces G, see Example VII.6.

EXAMPLE 22. We may apply the tensor product construction in particular to Gaussian measures P_ν. If Q were Gaussian then

$$\int_F f(t)^4 \, dQ(f) = 3 \cdot K(t,t)^2$$

for the fourth moments. However,

$$\int_F f(t)^4 \, dQ(f) = \prod_{\nu=1}^{d} \int_{F_\nu} f_\nu(\tau_\nu)^4 \, dP_\nu(f_\nu) = 3^d \cdot K(t,t)^2.$$

Thus Q is not Gaussian, except for the trivial cases $K = 0$ or $d = 1$. In particular, the Wiener sheet is not the tensor product of independent Brownian motions.

The Hilbert space with reproducing kernel $\bigotimes_{\nu=1}^{d} K_\nu$ is the *tensor product* of the Hilbert spaces $H(K_\nu)$ and characterized by the following properties. If $h_\nu \in H(K_\nu)$ for $\nu = 1, \ldots, d$ then $h_1 \otimes \cdots \otimes h_d \in H(K)$, and linear combinations of such tensor products are dense in $H(K)$. The norm of a linear combination is given by

$$\left\| \sum_{i=1}^{k} (h_{1,i} \otimes \cdots \otimes h_{d,i}) \right\|_K^2 = \sum_{i,j=1}^{k} \langle h_{1,i}, h_{1,j} \rangle_{K_1} \cdots \langle h_{d,i}, h_{d,j} \rangle_{K_d}.$$

In particular,

$$\| h_1 \otimes \cdots \otimes h_d \|_K = \| h_1 \|_{K_1} \cdots \| h_d \|_{K_d}.$$

2.2. Tensor Product Methods. The tensor product construction is also a well known and simple tool to build algorithms for multivariate problems from linear algorithms for univariate problems. This is attractive, for instance, since much more is known about good algorithms in the univariate case than in the multivariate case.

Let $\nu \in \{1, \ldots, d\}$ and suppose that linear methods

$$U^\nu(f) = \sum_{j=1}^{m_\nu} f(x_j^\nu) \cdot a_j^\nu$$

are given for integration or approximation of univariate functions. Here $m_\nu \in \mathbb{N}$ denotes the number of knots $x_j^\nu \in [0,1]$, and $a_j^\nu \in \mathbb{R}$ for integration while a_j^ν is a real-valued function on $[0, 1]$ for approximation. In the multivariate case we define the *tensor product method* by

$$(U^1 \otimes \cdots \otimes U^d)(f) = \sum_{j_1=1}^{m_1} \cdots \sum_{j_d=1}^{m_d} f(x_{j_1}^1, \ldots, x_{j_d}^d) \cdot (a_{j_1}^1 \otimes \cdots \otimes a_{j_d}^d).$$

On the right-hand side, $\otimes$ denotes either the tensor product of functions on $[0, 1]$ or ordinary multiplication of real numbers.

Obviously $U^1 \otimes \cdots \otimes U^d$ is a linear method for integration or approximation of functions on D. The tensor product method is based on functions values on the grid

$$X^1 \times \cdots \times X^d \subset D,$$

where

$$X^\nu = \{x_1^\nu, \ldots, x_{m_\nu}^\nu\} \subset [0, 1]$$

is the set of knots used by U^ν. We have already used tensor product methods in Remarks 3 and 9, namely, interpolation by (piecewise) polynomials of degree ℓ in each variable.

EXAMPLE 23. We illustrate how tensor product methods work for approximation in dimension $d = 2$. For every $(\tau_1, \tau_2) \in D$,

$$(U^1 \otimes U^2)(f)(\tau_1, \tau_2) = \sum_{j_1=1}^{m_1} \left(\sum_{j_2=1}^{m_2} f(x_{j_1}^1, x_{j_2}^2) \cdot a_{j_2}^2(\tau_2) \right) \cdot a_{j_1}^1(\tau_1).$$

Thus, in a first step, the values of f on $\{x_{j_1}^1\} \times X^2$ and the method U^2 are used to generate approximations to $f(x_{j_1}^1, \tau_2)$ for $j_1 = 1, \ldots, m_1$. Then U^1 is applied to this data and the output is evaluated at τ_1.

In general, $(U^1 \otimes \cdots \otimes U^d)(f)$ may be evaluated by iterated application of the methods U^ν in any order.

Suppose that $0 = x_1^\nu < \cdots < x_{m_\nu}^\nu = 1$ and that U^ν is defined by piecewise linear interpolation. Then $U^1 \otimes U^2$ is defined piecewise on the subcubes

$$[x_{j_1-1}^1, x_{j_1}^1] \times [x_{j_2-1}^2, x_{j_2}^2]$$

by interpolation of the data

$$f(x_{j_1-1}^1, x_{j_2-1}^2), \; f(x_{j_1}^1, x_{j_2-1}^2), \; f(x_{j_1}^1, x_{j_2}^2), \; f(x_{j_1-1}^1, x_{j_2}^2)$$

with a polynomial of degree at most two in both variables. This procedure is known as bilinear interpolation.

For covariance kernels of tensor product form, it seems reasonable to use a tensor product method with K_ν-spline algorithms U^ν. For approximation we obtain the K-spline algorithm that is based on the underlying grid. This fact is easy to verify; it is also an immediate consequence of Lemma 28 below.

The same conclusion holds true for integration, if the weight function ϱ is of tensor product form. Indeed, (B) implies

$$(21) \qquad \mathrm{Int}_\varrho \circ (U^1 \otimes \cdots \otimes U^d) = (\mathrm{Int}_{\varrho_1} \circ U^1) \otimes \cdots \otimes (\mathrm{Int}_{\varrho_d} \circ U^d)$$

for arbitrary linear methods U^ν for approximation. In particular, for K_ν-spline algorithms U^ν, the left-hand side is the K-spline algorithm for Int_ϱ and $\mathrm{Int}_{\varrho_\nu} \circ U^\nu$ is the K_ν-spline algorithm for $\mathrm{Int}_{\varrho_\nu}$.

In particular, we conclude for the Wiener sheet in dimension $d = 2$ that the spline algorithm for data from a grid is based on bilinear interpolation. See Remark II.11 and Examples 21 and 23.

At first glance one might think that tensor product methods are appropriate under conditions (A)–(C), or, more generally, for tensor product problems. Error bounds show that this is not true. In fact, take arbitrary sets X^ν and corresponding K_ν-spline algorithms U^ν. Put $S_n = U^1 \otimes \cdots \otimes U^d$, and take P_ν and S^ν as previously. Then

$$e_2(S_n, S, P)^2 = \prod_{\nu=1}^d e(0, \nu)^2 - \prod_{\nu=1}^d \left(e(0, \nu)^2 - e(1, \nu)^2 \right),$$

where

$$e(1, \nu) = e_2(U^\nu, S^\nu, P_\nu)$$

as well as

$$e(0, \nu)^2 = \int_{F_\nu} \mathrm{Int}_{\varrho_\nu}(f_\nu)^2 \, dP_\nu(f_\nu)$$

for integration and

$$e(0, \nu)^2 = \int_{F_\nu} \int_0^1 f_\nu(t)^2 \cdot \varrho_\nu(t) \, dt \, dP_\nu(f_\nu)$$

for L_2-approximation. This relation between errors for the multivariate and the univariate problems is an immediate consequence of Proposition 29 below. We conclude that

$$e_2(S_n, S, P) \geq \max_{\nu=1,\ldots,d} \left(e_2(U^\nu, S^\nu, P_\nu) \cdot \prod_{\tilde\nu \neq \nu} e(0, \tilde\nu) \right),$$

but the total number n of knots used by S_n satisfies

$$n = m_1 \cdots m_d \geq \min_{\nu=1,\ldots,d} m_\nu^d.$$

Hence

$$e_2(S_n, S, P) \geq e_2(\lceil n^{1/d} \rceil, \Lambda^{\mathrm{std}}, S^\nu, P_\nu) \cdot \prod_{\tilde\nu \neq \nu} e(0, \tilde\nu),$$

if m_ν is minimal among $m_1, \ldots, m_d$. The error of the tensor product algorithm with n knots is therefore bounded from below by a minimal error for one of the univariate problems where, however, only about $n^{1/d}$ knots are permitted. We will see that other methods lead to much better results.

The minimal errors for a tensor product problem are bounded from below by the minimal errors for the univariate problems.

PROPOSITION 24. *Suppose that (A)–(C) hold. Then*

$$e_2(n, \Lambda^{\text{std}}, S, P) \geq \max_{\nu=1,\ldots,d} \left(e_2(n, \Lambda^{\text{std}}, S^\nu, P_\nu) \cdot \prod_{\tilde{\nu} \neq \nu} e(0, \tilde{\nu}) \right).$$

PROOF. Take any n-point set $X \subset D$ and the corresponding K-spline algorithm S_n. Moreover, take n-point sets $X^1, \ldots, X^d \subset [0, 1]$ such that

$$X \subset X^1 \times \cdots \times X^d.$$

Let S_{n^d} denote the K-spline algorithm that is based on the grid $X^1 \times \cdots \times X^d$. The estimates for the error of S_{n^d} imply

$$e(S_n, S, P) \geq e(S_{n^d}, S, P) \geq e_2(n, \Lambda^{\text{std}}, S^\nu, P_\nu) \cdot \prod_{\tilde{\nu} \neq \nu} e(0, \tilde{\nu})$$

for every ν. $\square$

EXAMPLE 25. Specifically, consider the integration problem with the constant weight function $\varrho = 1$ and the Wiener sheet measure P. Recall that the minimal errors for integration with respect to the Wiener measure are of order n^{-1}, see Proposition II.9. Hence there exists a constant $c > 0$ such that

$$e_2(S_n, \text{Int}, P) \geq c \cdot n^{-1/d}$$

for every tensor product method using n knots. On the other hand, Proposition 2 already yields

$$e_2(n, \Lambda^{\text{std}}, \text{Int}, P) = O(n^{-(1/(2d)+1/2)}),$$

since the covariance kernel K from Example 21 satisfies a Hölder condition of order $\left(0, \frac{1}{2}\right)$. Observe that Proposition 24 yields a lower bound of order n^{-1} for the minimal errors.

We will present even better upper bounds in Example 33, and finally determine the order of minimal errors in Example 40.

2.3. Smolyak Formulas. Smolyak (1963) introduced a general construction of efficient algorithms for tensor product problems. The construction leads to almost optimal methods for every dimension $d > 1$ from order optimal methods in the univariate case $d = 1$. We use the construction for integration and L_2-approximation.

Let $\nu \in \{1, \ldots, d\}$ and suppose that for every component ν a sequence of linear methods

$$U^{i,\nu}(f) = \sum_{j=1}^{m_{i,\nu}} f(x_j^{i,\nu}) \cdot a_j^{i,\nu}, \qquad i \in \mathbb{N},$$

is given for the univariate problem S^ν, cf. Section 2.2. To simplify the notation we let

$$m_{i,\nu} = m_i$$

depend only on i in the sequel.

The tensor product methods $U^{i_1,1} \otimes \cdots \otimes U^{i_d,d}$ have a poor order of convergence for tensor product problems and only serve as building blocks for the more complicated construction of Smolyak. Put

$$|i| = i_1 + \cdots + i_d, \qquad i \in \mathbb{N}^d.$$

The *Smolyak formulas* are linear combinations of tensor products and given by

$$A(q,d) = \sum_{q-d+1 \leq |i| \leq q} (-1)^{q-|i|} \cdot \binom{d-1}{q-|i|} \cdot (U^{i_1,1} \otimes \cdots \otimes U^{i_d,d})$$

for integers $q \geq d$.

Clearly $A(q,d)$ is a linear method for the multivariate problem and $A(q,d)(f)$ depends on the function f only through its values at a finite number of knots. To describe these knots let

$$X^{i,\nu} = \{x_1^{i,\nu}, \ldots, x_{m_i}^{i,\nu}\} \subset [0,1]$$

denote the set of knots used by $U^{i,\nu}$. The tensor product algorithm $U^{i_1,1} \otimes \cdots \otimes U^{i_d,d}$ is based on the grid $X^{i_1,1} \times \cdots \times X^{i_d,d}$, and therefore $A(q,d)(f)$ depends on the values of f at the union

$$(22) \qquad H(q,d) = \bigcup_{q-d+1 \leq |i| \leq q} (X^{i_1,1} \times \cdots \times X^{i_d,d}) \subset D$$

of grids. Nested sets

$$(23) \qquad\qquad\qquad X^{i,\nu} \subset X^{i+1,\nu}$$

for all i and ν yield $H(q,d) \subset H(q+1,d)$ and

$$H(q,d) = \bigcup_{|i|=q} (X^{i_1,1} \times \cdots \times X^{i_d,d}).$$

Therefore nested sets seem to be the most economical choice; further reasons to use nested sets are mentioned in the sequel. The points $x \in H(q,d)$ are called *hyperbolic cross points* and $H(q,d)$ is called a *sparse grid*.

Sparse grids can be built from any sequence of sets $X^{i,\nu}$. From a practical point of view it is very important to choose $m_1 = 1$ for relatively large d, because otherwise the number of points in $H(q,d)$ increases too fast with d.

EXAMPLE 26. Let

$$(24) \qquad\qquad X^i = X^{i,\nu} = \{\tfrac{1}{2} + j\,2^{-i+1} : j \in \mathbb{Z}\} \cap [0,1]$$

for $i \in \mathbb{N}$ and every ν. Then

$$m_1 = 1, \qquad\qquad m_i = 2^{i-1} + 1, \quad i \geq 2.$$

Moreover, (23) holds. We obtain

$$H(q,d) = \left\{ \left(\tfrac{1}{2},\ldots,\tfrac{1}{2}\right) + \left(j_1 2^{-i_1+1},\ldots,j_d 2^{-i_d+1}\right) : |i| = q, \; j \in \mathbb{Z}^d \right\} \cap D.$$

Up to the shift by $\left(\tfrac{1}{2},\ldots,\tfrac{1}{2}\right)$ and the restriction to D, the sparse grid $H(q,d)$ is the union of all dyadic grids with product of mesh size given by 2^{d-q}.

For analysis it is often important to use the following representation of $A(q,d)$. See Wasilkowski and Woźniakowski (1995, p. 14) for the proof.

LEMMA 27. *Let*

$$\Delta^{i,\nu} = U^{i,\nu} - U^{i-1,\nu}$$

where $U^{0,\nu} = 0$. *Then*

$$A(q,d) = \sum_{|i| \leq q} (\Delta^{i_1,1} \otimes \cdots \otimes \Delta^{i_d,d}).$$

For given knots $x_j^{i,\nu}$, it seems reasonable to start with spline algorithms in the univariate case. The Smolyak formulas yield spline algorithms in the multivariate case, if the sets $X^{i,\nu}$ are nested.

LEMMA 28. *Suppose that (A)–(C) and (23) hold. If $U^{i,\nu}$ is the K_ν-spline algorithm that is based on $X^{i,\nu}$, then $A(q,d)$ is the K-spline algorithm that is based on $H(q,d)$.*

PROOF. Put

$$V^{i,\nu} = \operatorname{span}\{K_\nu(\cdot,x) : x \in X^{i,\nu}\},$$

assume $\dim V^{i,\nu} = m_i$ without loss of generality, and consider the case $S = \mathrm{App}_{2,\varrho}$. The coefficients of the K_ν-spline algorithm $U^{i,\nu}$ satisfy $a_j^{i,\nu} \in V^{i,\nu}$. Hence

$$a_{j_1}^{i_1,1} \otimes \cdots \otimes a_{j_d}^{i_d,d} \in \operatorname{span}\{K(\cdot,x) : x \in X^{i_1,1} \times \cdots \times X^{i_d,d}\}$$

$$\subset \operatorname{span}\{K(\cdot,x) : x \in H(q,d)\},$$

if $q - d + 1 \leq |i| \leq q$ and $1 \leq j_\nu \leq m_{i_\nu}$.

It remains to show that the interpolation property

(25) $$U^{i,\nu}(f)(x) = f(x), \qquad f \in C(D), \; x \in X^{i,\nu},$$

for $d = 1$ is preserved for $d > 1$ by the Smolyak construction with nested sets $X^{i,\nu}$. The proof is via induction over d. For $d = 1$ we have $A(q,1) = U^{q,1}$ and the statement follows. For the multivariate case we use the fact that $A(q,d+1)$ can be written in terms of $A(d,d),\ldots,A(q-1,d)$. Indeed, from Lemma 27 we get

$$A(q,d+1) = \sum_{\ell=d}^{q-1} A(\ell,d) \otimes (U^{q-\ell,d+1} - U^{q-\ell-1,d+1}).$$

The induction step is as follows. Assume $f = f_1 \otimes \cdots \otimes f_{d+1}$ without loss of generality. Let $x = (x_1,\ldots,x_{d+1}) \in H(q,d+1)$ and take $1 \leq k \leq q-d$ such that

$x_{d+1} \in X^{k,d+1} \setminus X^{k-1,d+1}$. Here $X^{0,d+1} = \emptyset$. Then $(x_1, \ldots, x_d) \in H(q - k, d)$ and therefore

$$A(q, d + 1)(f)(x)$$

$$= \sum_{\ell=q-k}^{q-1} A(\ell, d)(f_{i_1} \otimes \cdots \otimes f_{i_d})(x_1, \ldots, x_d) \cdot \Delta^{q-\ell,d+1}(f_{i_{d+1}})(x_{d+1})$$

$$= (f_{i_1} \otimes \cdots \otimes f_{i_d})(x_1, \ldots, x_d) \cdot \sum_{\ell=1}^{k}(U^{q,d+1} - U^{q-1,d+1})(f_{i_{d+1}})(x_{d+1})$$

$$= f(x),$$

as claimed.

For integration the spline algorithms are given as $\text{Int}_{\varrho_\nu} \circ U^{i,\nu}$ and $\text{Int}_\varrho \circ A(q, d)$ with the formulas $U^{i,\nu}$ and $A(q, d)$ as previously. Hence the Lemma follows from (21). □

The interpolation property of the Smolyak formulas holds if the univariate methods $U^{i,\nu}$ interpolate on nested sets. These methods need not be spline algorithms. By the proof above, (23) and (25) imply that $A(q, d)$ interpolates on the sparse grid $H(q, d)$, i.e.,

$$A(q, d)(f)(x) = f(x), \qquad f \in C(D), \ x \in H(q, d).$$

This property serves as a motivation for the binomial coefficients and the signs appearing in the definition of $A(q, d)$.

There are general techniques to obtain error bounds for Smolyak formulas from error bounds for the univariate formulas $U^{i,\nu}$. Here we consider the particular case of spline algorithms and nested sets. First we express the errors of the Smolyak formulas in terms of errors of the univariate formulas $U^{i,\nu}$. We put $U^{0,\nu} = 0$ and

$$e(i, \nu) = e_2(U^{i,\nu}, S^\nu, P_\nu),$$

where S^ν corresponds either to integration or L_2-approximation with weight function ϱ_ν and where P_ν is a zero mean measure with covariance kernel K_ν.

PROPOSITION 29. *Suppose that (A)–(C) and (23) hold. Moreover, let $A(q, d)$ be based on K_ν-spline algorithms $U^{i,\nu}$. Then*

$$e_2(A(q, d), S, P)^2 = \prod_{\nu=1}^{d} e(0, \nu)^2 - \sum_{|i| \leq q} \prod_{\nu=1}^{d} \left(e(i_\nu - 1, \nu)^2 - e(i_\nu, \nu)^2\right).$$

PROOF. Consider the integration problem. First we derive some facts for the univariate problems given by $S^\nu = \text{Int}_{\varrho_\nu}$. The spline algorithm $U^{i,\nu}$ is the orthogonal projection of S^ν onto the subspace of Λ^{P_ν} that is generated by function evaluation at the points $x \in X^{i,\nu}$, see Section III.3.2. Therefore

$$(26) \quad \int_{F_\nu} S^\nu(f) \cdot (U^{i,\nu} - U^{i-1,\nu})(f) \, dP_\nu(f)$$

$$= \int_{F_\nu} U^{i,\nu}(f)^2 \, dP_\nu(f) - \int_{F_\nu} U^{i-1,\nu}(f)^2 \, dP_\nu(f) = e(i-1,\nu)^2 - e(i,\nu)^2.$$

Orthogonality and $X^{i-1,\nu} \subset X^{i,\nu}$ yield

$$\int_{F_\nu} (S^\nu - U^{i,\nu})(f) \cdot (S^\nu - U^{i-1,\nu})(f) \, dP_\nu(f)$$

$$= \int_{F_\nu} (S^\nu - U^{i,\nu})(f) \cdot S^\nu(f) \, dP_\nu(f) = e(i,\nu)^2,$$

and hereby

$$(27) \quad \int_{F_\nu} (U^{i,\nu} - U^{i-1,\nu})(f)^2 \, dP_\nu(f)$$

$$= \int_{F_\nu} ((S^\nu - U^{i,\nu})(f) - (S^\nu - U^{i-1,\nu})(f))^2 \, dP_\nu(f)$$

$$= e(i-1,\nu)^2 - e(i,\nu)^2.$$

Moreover,

$$(28) \quad \int_{F_\nu} (U^{i,\nu} - U^{i-1,\nu})(f) \cdot (U^{j,\nu} - U^{j-1,\nu})(f) \, dP_\nu(f) = 0$$

for $i \neq j$.

In the multivariate case,

$$\int_F S(f)^2 \, dP(f) = \int_{F_1} \cdots \int_{F_d} S(f_1 \otimes \cdots \otimes f_d)^2 \, dP_1(f_1) \ldots dP_d(f_d)$$

$$= \prod_{\nu=1}^d \int_{F_\nu} S^\nu(f_\nu)^2 \, dP_\nu(f_\nu) = \prod_{\nu=1}^d e(0,\nu)^2,$$

see Section 2.1. We analogously get

$$\int_F S(f) \cdot A(q,d)(f) \, dP(f) = \sum_{|i| \leq q} \prod_{\nu=1}^d \int_{F_\nu} S^\nu(f_\nu) \cdot \Delta^{i_\nu,\nu}(f_\nu) \, dP_\nu(f_\nu)$$

$$= \sum_{|i| \leq q} \prod_{\nu=1}^d \left(e(i_\nu - 1, \nu)^2 - e(i_\nu, \nu)^2 \right)$$

and

$$\int_F A(q,d)(f)^2 \, dP(f) = \sum_{|i|,|j| \leq q} \prod_{\nu=1}^d \int_{F_\nu} \Delta^{i_\nu,\nu}(f_\nu) \cdot \Delta^{j_\nu,\nu}(f_\nu) \, dP_\nu(f_\nu)$$

$$= \sum_{|i| \leq q} \prod_{\nu=1}^d \left(e(i_\nu - 1, \nu)^2 - e(i_\nu, \nu)^2 \right).$$

from Lemma 27 and (26)–(28). This completes the proof for integration.

For L_2-approximation we use

$$e_2(A(q,d), S, P)^2 = \int_D \int_F (f(t) - A(q,d)(f)(t))^2 \, dP(f) \cdot \varrho(t) \, dt,$$

and we apply the consideration above for $S^\nu(f) = f(t)$ with fixed $t \in D$. Finally we take into account (B) when integrating with respect to t.   □

Often we have only got estimates for the univariate errors. Such estimates can be used in the following way.

PROPOSITION 30. *Suppose that constants $c_1 > 0$ and $0 < \kappa < 1$ exist such that*

$$e(i, \nu) \le c_1 \cdot \kappa^i, \qquad i \in \mathbb{N}_0,$$

for every ν. Under the assumptions of Proposition 29,

$$e_2(A(q,d), S, P) \le c_2 \cdot \kappa^q \cdot q^{(d-1)/2}$$

where $c_2 = c_1^d \cdot \kappa^{1-d} \cdot d^{1/2}$. If, additionally,

$$m_i \asymp 2^i, \qquad i \in \mathbb{N},$$

and

$$\kappa = 2^{-\gamma}$$

for some $\gamma > 0$, then

$$e_2(A(q,d), S, P) = O\left(n^{-\gamma} \cdot (\ln n)^{(d-1)\cdot(\gamma+1/2)}\right),$$

where $n = n(q,d)$ denotes the cardinality of $H(q,d)$.

PROOF. Put $E(q,d) = e_2(A(q,d), S, P)$. Clearly $E(q,1) = e(q,1)$. If $d > 1$ then Proposition 29 yields

$$E(q,d)^2 = \prod_{\nu=1}^d e(0,\nu)^2$$

$$- \sum_{|j| \le q-1} \prod_{\nu=1}^{d-1} (e(j_\nu - 1, \nu)^2 - e(j_\nu, \nu)^2) \cdot (e(0,d)^2 - e(q - |j|, d)^2)$$

$$= e(0,d)^2 \cdot E(q - 1, d - 1)^2$$

$$+ \sum_{|j| \le q-1} \prod_{\nu=1}^{d-1} (e(j_\nu - 1, \nu)^2 - e(j_\nu, \nu)^2) \cdot e(q - |j|, d)^2,$$

where $j \in \mathbb{N}^{d-1}$. Thus

$$E(q,d)^2 \le e(0,d)^2 \cdot E(q - 1, d - 1)^2 + \sum_{|j| \le q-1} \prod_{\nu=1}^{d-1} e(j_\nu - 1, \nu)^2 \cdot e(q - |j|, d)^2.$$

The upper bound for the univariate errors implies

$$E(q,d)^2 \leq c_1^2 \cdot E(q-1,d-1)^2 + c_1^{2d} \cdot \sum_{|j| \leq q-1} \kappa^{2(q-d+1)}$$

$$= c_1^2 \cdot E(q-1,d-1)^2 + c_1^{2d} \cdot \kappa^{2(q-d+1)} \cdot \binom{q-1}{d-1}$$

$$\leq c_1^2 \cdot E(q-1,d-1)^2 + c_1^{2d} \cdot \kappa^{2(q-d+1)} \cdot q^{d-1}.$$

Inductively we get

$$E(q,d)^2 \leq c_1^{2d} \cdot \kappa^{2(q-d+1)} \cdot \sum_{k=0}^{d-1} q^k \leq c_2^2 \cdot \kappa^{2q} \cdot q^{d-1},$$

as claimed.

By c we denote different positive constants, which depend only on c_1, κ, and d. From $m_i \asymp 2^i$ we get

$$n \leq \sum_{|i|=q} m_{i_1} \cdots m_{i_d} \leq c \cdot \#\{i \in \mathbb{N}^d : |i| = q\} \cdot 2^q$$

$$= c \cdot \binom{q-1}{d-1} \cdot 2^q \leq c \cdot q^{d-1} \cdot 2^q,$$

as well as

$$n \geq c \cdot 2^q.$$

Moreover, $\kappa = 2^{-\gamma}$ yields

$$E(q,d) \leq c \cdot \kappa^q \cdot q^{(d-1)/2} \leq c \cdot (n/q^{d-1})^{-\gamma} \cdot q^{(d-1)/2}$$

$$= c \cdot n^{-\gamma} \cdot q^{(d-1)\cdot(\gamma+1/2)},$$

which completes the proof. □

REMARK 31. Suppose that (A)–(C) and (23) hold. Moreover, let $A(q,d)$ be based on K_ν-spline algorithms $U^{i,\nu}$ that are order optimal with minimal errors of order $n^{-\gamma}$. By Proposition 24, we have a lower bound of the same order in the multivariate case. Then by Proposition 30, the Smolyak formulas $A(q,d)$ are almost optimal, up to a logarithmic factor, in the multivariate case.

2.4. Hölder Conditions. Now we consider tensor products of kernels that satisfy Hölder conditions. For simplicity we assume the same smoothness (r, β) in every component. In Remark 39 we indicate how to extend the results to kernels $K_1, \ldots, K_d$ of different smoothness. It is easy to check that the tensor product kernel also satisfies Hölder conditions of order (r, β), thus Proposition 2 is applicable. However, much better results are possible because of the tensor product structure. The following upper bound for approximation is achieved by Smolyak formulas. For integration the upper bound is proven non-constructively.

PROPOSITION 32. *Suppose that (A)–(C) hold with kernels K_ν that satisfy Hölder conditions of order (r, β). Then*

$$e_2(n, \Lambda^{\mathrm{std}}, \mathrm{App}_2, P) = O\left(n^{-(r+\beta)} \cdot (\ln n)^{(d-1)\cdot(r+\beta+1/2)}\right)$$

for the minimal error of L_2-approximation. Moreover, if additionally $\varrho \in L_2(D)$, then

$$e_2(n, \Lambda^{\mathrm{std}}, \mathrm{Int}, P) = O\left(n^{-(r+\beta+1/2)} \cdot (\ln n)^{(d-1)\cdot(r+\beta+1/2)}\right)$$

for the minimal error of integration.

PROOF. For L_2-approximation, take the sets X^i from Example 26 and the corresponding K_ν-spline algorithms $U^{i,\nu}$. Put $\kappa = 2^{-(r+\beta)}$. The univariate errors satisfy

$$e(i, \nu) = O\left(m_i^{-(r+\beta)}\right) = O\left(\kappa^i\right),$$

see Proposition 2 and Remark 3. Proposition 30 yields

$$e_2(A(q, d), \mathrm{App}_{2,\varrho}, P) = O\left(n^{-(r+\beta)} \cdot (\ln n)^{(d-1)\cdot(r+\beta+1/2)}\right).$$

The error bound for integration is now obtained by applying Proposition II.4. □

EXAMPLE 33. In Example 25 we have considered the integration problem with the constant weight function $\varrho = 1$ and the Wiener sheet measure P. This problem is a tensor product problem and the kernel $K_\nu(s, t) = \min(s, t)$ satisfies a Hölder condition of order $(0, \frac{1}{2})$. Hence Proposition 32 yields the improved upper bound

$$e_2(n, \Lambda^{\mathrm{std}}, \mathrm{Int}, P) = O\left(n^{-1} \cdot (\ln n)^{d-1}\right).$$

See Example 40 for a further improvement in term of powers of $\ln n$.

As in the isotropic case, Hölder conditions are too weak to determine the order of the minimal errors in the tensor product case. Therefore we strengthen the assumptions and require stationarity or Sacks-Ylvisaker conditions for the kernels K_ν. It turns out that the upper bounds for approximation in Proposition 32 cannot be improved in general for any $d \in \mathbb{N}$, $r \in \mathbb{N}_0$, and $0 < \beta < 1$. For integration no improvement is possible in terms of powers of n.

2.5. Stationary Random Fields. Suppose that the spectral density $\mathfrak{f}$ of a stationary covariance kernel on $(\mathbb{R}^d)^2$ is a tensor product

(29) $$\mathfrak{f} = \mathfrak{f}_1 \otimes \cdots \otimes \mathfrak{f}_d$$

of univariate spectral densities $\mathfrak{f}_\nu$. Then $\mathfrak{K}$ and its restriction K to D^2 are obviously of tensor product form. We study the approximation problem, assuming that all densities $\mathfrak{f}_\nu$ satisfy the decay conditions (7) and (8) with the same parameter $\gamma > \frac{1}{2}$. In Remark 39 we indicate how to extend the results to different values of γ.

Consider the integral operator with kernel $K = \bigotimes_{\nu=1}^{d} K_\nu$. The sequence of its eigenvalues $\mu_j(K)$ consists of all products of the eigenvalues $\mu_j(K_\nu)$ of the univariate integral operators with kernels K_ν. As previously, we assume that sequences of eigenvalues are decreasingly ordered.

LEMMA 34. *Assume that $\mu_j(K_\nu) \approx j^{-2\gamma}$ for $\nu = 1, \ldots, d$. Then*

$$\mu_j(K) \approx ((d-1)!)^{-2\gamma} \cdot ((\ln j)^{d-1}/j)^{2\gamma}$$

for every fixed $d \in \mathbb{N}$.

PROOF. Let $V(M, d)$ denote volume of the set of all $x \in [1, M]^d$ such that $x_1 \cdots x_d \leq M$. Inductively one gets

$$V(M, d) \approx 1/(d-1)! \cdot M \cdot (\ln M)^{d-1}$$

for fixed d as M tends to infinity. The same relation also holds if we additionally require $x \in \mathbb{N}^d$. Therefore

$$\#\{i \in \mathbb{N}^d : \mu_{i_1}(K_1) \cdots \mu_{i_d}(K_d) \geq 1/M\} \approx 1/((d-1)! \cdot 2\gamma) \cdot M^{1/(2\gamma)} \cdot (\ln M)^{d-1}.$$

Hereby the statement follows.   □

PROPOSITION 35. *Let P be a zero mean measure with spectral density (29), where all univariate densities $\mathfrak{f}_\nu$ satisfy (7) and (8) with the same parameter $\gamma > \frac{1}{2}$. Then*

$$e_2(n, \Lambda^{\mathrm{std}}, \mathrm{App}_{2,\varrho}, P) \asymp e_2(n, \Lambda^{\mathrm{all}}, \mathrm{App}_{2,\varrho}, P) \asymp n^{-(\gamma-1/2)} \cdot (\ln n)^{(d-1)\cdot\gamma}$$

for the minimal error of L_2-approximation.

PROOF. The upper bound for the class Λ^{std} follows from Proposition 8 with $d = 1$ and Proposition 30. From the proof of Proposition 11 we conclude that $\mu_j(K_\nu) \asymp j^{-2\gamma}$. We apply Lemma 34 and Proposition III.24 to obtain

$$e_2(n, \Lambda^{\mathrm{all}}, \mathrm{App}_2, P)^2 = \sum_{j=n+1}^{\infty} \mu_j(K) \asymp \sum_{j=n+1}^{\infty} ((\ln j)^{d-1}/j)^{2\gamma}$$
$$\asymp n^{-(2\gamma-1)} \cdot (\ln n)^{(d-1)\cdot 2\gamma}$$

for the class Λ^{all}.   □

As in the case of Hölder conditions, we may use Smolyak formulas based on K_ν-spline algorithms to achieve the upper bounds. To build the sparse grid we take equidistant knots $x_j^{i,\nu} = x_j^i$ with $x_1^1 = \frac{1}{2}$ and

$$x_j^i = (j-1) \cdot 2^{-(i-1)}, \qquad j = 1, \ldots, 1 + 2^{i-1},$$

for $i \geq 2$.

2.6. Sacks-Ylvisaker Conditions. In the univariate case the Sacks-Ylvisaker conditions lead to sharp bounds for integration and approximation, see Chapter IV. Therefore we now study tensor products of kernels satisfying these conditions. We determine the order of the minimal errors for integration and approximation, and we describe order optimal methods.

In dimension $d = 1$ the Sacks-Ylvisaker conditions determine the reproducing kernel Hilbert space uniquely up to a finite dimensional subspace of polynomials, see Section IV.1.2. We can use this fact, since the operation $\otimes$ preserves inclusions of the corresponding reproducing kernel Hilbert spaces.

LEMMA 36. *Let K_ν and L_ν be covariance kernels on $[0,1]^2$. If*

$$H(K_\nu) \subset H(L_\nu), \qquad \nu = 1, \ldots, d,$$

then

$$H\Big(\bigotimes_{\nu=1}^d K_\nu\Big) \subset H\Big(\bigotimes_{\nu=1}^d L_\nu\Big).$$

PROOF. Let $d = 2$. It is enough to show that $H(K_1 \otimes K_2) \subset H(K_1 \otimes L_2)$. Let $b_i \in \mathbb{R}, t_i = (u_i, v_i) \in [0,1]^2$, and let $(f_\ell)_\ell$ be an orthonormal basis in $H(K_1)$. Then

$$K_1(u_i, u_j) = \langle K_1(\cdot, u_i), K_1(\cdot, u_j)\rangle_{K_1} = \sum_\ell f_\ell(u_i) \cdot f_\ell(u_j).$$

Since $H(K_2) \subset H(L_2)$, there exists a positive constant c such that $cL_2 - K_2$ is nonnegative definite. We have

$$\sum_{i,j=1}^k a_i a_j \cdot (c \cdot K_1 \otimes L_2(t_i, t_j) - K_1 \otimes K_2(t_i, t_j))$$

$$= \sum_{i,j=1}^k a_i a_j \cdot K_1(u_i, u_j) \cdot (cL_2(v_i, v_j) - K_2(v_i, v_j))$$

$$= \sum_\ell \sum_{i,j=1}^k a_i f_\ell(u_i) \cdot a_j f_\ell(u_j) \cdot (cL_2(v_i, v_j) - K_2(v_i, v_j)) \geq 0.$$

Hence $H(K_1 \otimes K_2) \subset H(K_1 \otimes L_2)$.

For $d > 2$, let $K = \bigotimes_{\nu=1}^{d-1} K_\nu$ and $L = \bigotimes_{\nu=1}^{d-1} L_\nu$. By induction we can assume that $H(K) \subset H(L)$. Then $H(K \otimes K_d) \subset H(L \otimes L_d)$ follows as above. $\square$

Sacks-Ylvisaker conditions determine the order of minimal errors as follows.

PROPOSITION 37. *Suppose that (A)–(C) hold with kernels K_ν that satisfy Sacks-Ylvisaker conditions of the same order r. Then*

$$e_2(n, \Lambda^{\mathrm{std}}, \mathrm{Int}, P) \asymp n^{-(r+1)} \cdot (\ln n)^{(d-1)/2}$$

for the minimal error of integration and

$$e_2(n, \Lambda^{\mathrm{std}}, \mathrm{App}_2, P) \asymp n^{-(r+1/2)} \cdot (\ln n)^{(d-1)\cdot(r+1)}$$

as well as

$$e_2(n, \Lambda^{\text{all}}, \text{App}_2, P) \approx c_r \cdot n^{-(r+1/2)} \cdot (\ln n)^{(d-1)\cdot(r+1)}$$

for the minimal error of L_2-approximation, where

$$c_r = (\pi^d \cdot (d-1)!)^{-(r+1)} \cdot (2r+1)^{-1/2}.$$

PROOF. The upper bounds for $e_2(n, \Lambda^{\text{std}}, \text{App}_2, P)$ follow from Proposition 32, since Sacks-Ylvisaker conditions of order r imply Hölder conditions of order $(r, \frac{1}{2})$, see Remark IV.1.

In the univariate case the ordered eigenvalues of the integral operator with kernel K_ν satisfy

$$\mu_j(K_\nu) \approx (\pi j)^{-(2r+2)},$$

see Proposition IV.10. We apply Proposition III.24 and Lemma 34 to obtain

$$e_2(n, \Lambda^{\text{all}}, \text{App}_2, P)^2$$

$$= \sum_{j=n+1}^{\infty} \mu_j(K) \approx (\pi^d \cdot (d-1)!)^{-(2r+2)} \cdot \sum_{j=n+1}^{\infty} ((\ln j)^{d-1}/j)^{2r+2}$$

$$\approx c_r^2 \cdot n^{-(2r+1)} \cdot (\ln n)^{(d-1)\cdot(2r+2)},$$

which completes the proof for the approximation problem.

For integration we use Proposition III.17 and study the worst case on the unit ball in $H(K)$. The spaces $H(K_\nu)$ satisfy

$$H(P_r) \subset H(K_\nu) \subset H(R_r)$$

according to Proposition IV.8. By Lemma 36,

$$H(P_r^d) \subset H(K) \subset H(R_r^d),$$

where $P_r^d = \bigotimes_{\nu=1}^{d} P_r$ and $R_r^d = \bigotimes_{\nu=1}^{d} R_r$. From Frolov (1976) and Bykovskii (1985), see also Temlyakov (1994, Theorem IV.4.4), we get

$$(30) \qquad e_{\max}(n, \Lambda^{\text{std}}, \text{Int}, B(P_r^d)) \asymp n^{-(r+1)} \cdot (\ln n)^{(d-1)/2}.$$

Note that the spaces $H(P_r^d)$ and $H(R_r^d)$ differ only by the boundary conditions

$$h^{(\alpha)}(t) = 0, \qquad t \in \partial D, \ \alpha \in \{0, \ldots, r\}^d.$$

By a well known periodization technique, see Remark 38, the upper bound in (30) is also valid for $B(R_r^d)$, and therefore (30) also holds for $B(K)$.   □

We describe linear methods that are order optimal.

REMARK 38. For L_2-approximation the upper bound in Proposition 37 is achieved by Smolyak formulas $A(q, d)$. As univariate ingredients, we take K_ν-spline algorithms that are based on equidistant knots from X^i, given by (24). The formulas $A(q, d)$ remain order optimal under Sacks-Ylvisaker conditions if we drop the boundary condition (D) in our assumptions on the covariance kernels K_ν. The proof relies on Proposition IV.6, which says in particular that $H(K_\nu)$ is a closed and finite-codimensional subspace in $H(R_r)$.

Now we turn to the integration problem. Under Sacks-Ylvisaker conditions the upper bound from Proposition 32 is improved in terms of powers of $\ln n$, and order optimal cubature formulas are known explicitly. The formulas are constructed and analyzed by Frolov (1976), see also Temlyakov (1994, Section IV.4). Consider the polynomial

$$p(t) = \prod_{\nu=1}^{d} (t - (2\nu - 1)) - 1$$

and let $t_1, \ldots, t_d \in \mathbb{R}$ denote its pairwise different roots. Define

$$A = \begin{pmatrix} 1 & t_1 & \cdots & t_1^{d-1} \\ \vdots & \vdots & & \vdots \\ 1 & t_d & \cdots & t_d^{d-1} \end{pmatrix}$$

and the lattices

$$L(q) = \{1/q \cdot A^{-1} m : m \in \mathbb{Z}^d\}$$

for integers $q > 1$. The cubature formula

$$B(q)(g) = \frac{1}{q^d \cdot |\det A|} \cdot \sum_{x \in L(q) \cap D} g(x)$$

uses $n \asymp q^d$ knots and yields maximal errors of order $n^{-(r+1)} \cdot (\ln n)^{(d-1)/2}$ on $B(P_r^d)$.

For general K, where we do not have zero boundary conditions, we apply periodization. Let

$$\psi(\tau) = \frac{\int_0^\tau (u(1-u))^{r+1} \, du}{\int_0^1 (u(1-u))^{r+1} \, du}.$$

The polynomial ψ is strictly increasing on $[0,1]$ with $\psi(0) = 0$ and $\psi(1) = 1$. Moreover, $\psi^{(k)}$ vanishes at $\tau = 0$ and 1 for $k = 1, \ldots, r+1$. Put

$$\Psi = \psi \otimes \cdots \otimes \psi, \qquad \Psi' = \psi' \otimes \cdots \otimes \psi'$$

and consider the cubature formulas

(31) $$\widetilde{B}(q)(h) = \frac{1}{q^d \cdot |\det A|} \cdot \sum_{x \in L(q) \cap D} h(\Psi(x)) \cdot \Psi'(x).$$

We have

$$\int_D h(t) \, dt - \widetilde{B}(q)(h) = \int_D g(t) \, dt - B(q)(g)$$

where

$$g = h \circ \Psi \cdot \Psi'.$$

Note that $h \mapsto g$ defines a bounded linear operator from $H(R_r^d)$ to $H(P_r^d)$, and therefore $\widetilde{B}(q)$ is order optimal on $H(R_r^d)$ and $H(K)$.

REMARK 39. For tensor products of kernels of different smoothness the bounds in Propositions 32, 35, and 37 have to be modified as follows. If K_ν satisfies Sacks-Ylvisaker conditions of order r_ν, then Proposition 37 holds with r denoting the minimal smoothness,

$$r = \min\{r_\nu : 1 \leq \nu \leq d\},$$

and d being replaced by the number

$$\widetilde{d} = \#\{\nu : r_\nu = r\}$$

of components of minimal smoothness r, see Ritter, Wasilkowski, and Woźniakowski (1995). Propositions 32 and 35 are modified analogously.

EXAMPLE 40. Proposition 37 is applicable in particular for tensor products of kernels

$$Q_r(s,t) = \int_0^1 \frac{(s-u)_+^r \cdot (t-u)_+^r}{(r!)^2} \, du,$$

which correspond to the r-fold integrated Wiener measure, see Example IV.3. The zero mean Gaussian measure with kernel $\bigotimes_{\nu=1}^d Q_r$ is called the r-fold integrated Wiener sheet. For $r = 0$ this is the classical Wiener or Brownian sheet, see Example 21.

Proposition 37 gives, in particular, the order of the minimal errors for the integration problem with the Wiener sheet measure. We have $r = 0$ and therefore

$$e_2(n, \Lambda^{\mathrm{std}}, \mathrm{Int}, P) \asymp n^{-1} \cdot (\ln n)^{(d-1)/2}.$$

REMARK 41. The Smolyak construction can be used with regular sequences of knots in the univariate case, see Section IV.2.1. Hereby the sparse grid can be adjusted to the weight function. Moreover, one obtains asymptotic constants for the errors of respective methods. It is unknown, however, if we get asymptotically optimal methods in this way.

We briefly report on results from Müller-Gronbach (1998), who considers the approximation problem with Sacks-Ylvisaker conditions of order $r = 0$. Assume that the univariate weight functions ϱ_ν in (B) are positive and continuous. Moreover, assume that the same properties hold for the densities $\psi_1, \ldots, \psi_d$. Put

$$c = \left((d-1)! \cdot (\ln 2)^{d-1} \cdot 2^{d-1/2} \cdot 3^{d/2} \right)^{-1},$$

and recall the definition

$$I_{\varrho_\nu, 0}(\psi_\nu) = \left(\int_0^1 \varrho_\nu(t)/\psi_\nu(t) \, dt \right)^{1/2} \cdot \left(\int_0^1 \psi_\nu(t) \, dt \right)^{1/2}$$

from Section IV.3.1.

Choose $m_1 = 1$ and $m_i = 2^{i-1} + 1$ for $i \geq 2$. Define knots $x_j^{i,\nu} \in [0,1]$ by

$$\int_0^{x_j^{i,\nu}} \psi_\nu(t)\,dt = \frac{j-1}{m_i - 1} \cdot \int_0^1 \psi_\nu(t)\,dt, \qquad j = 1, \ldots, m_i,$$

for $i \geq 2$ and $x_1^{1,\nu} = x_2^{2,\nu}$. Finally, take the respective K_ν-spline algorithms $U^{i,\nu}$. A slight and straightforward modification of the proof in Müller-Gronbach (1998) yields

$$e(A(q,d), \mathrm{App}_{2,\varrho}, P) \approx c \cdot \prod_{\nu=1}^d I_{\varrho_\nu,0}(\psi_\nu) \cdot n^{-1/2} \cdot (\ln n)^{d-1}.$$

The best choice of the densities in this asymptotic sense is

$$\psi_\nu = \varrho_\nu^{1/2},$$

and then we have

$$e(A(q,d), \mathrm{App}_{2,\varrho}, P) \approx c \cdot \|\varrho^{1/2}\|_1 \cdot n^{-1/2} \cdot (\ln n)^{d-1}.$$

The dependence of the method on the kernel can be eliminated, if all weight functions ϱ_ν are Lipschitz continuous. Then the same results also hold if we take piecewise linear interpolation for $d = 1$ instead of K_ν-spline algorithms. Cf. Section IV.4.1.

REMARK 42. Consider the order optimal Smolyak formulas $A(q,d)$ for approximation under Sacks-Ylvisaker conditions that are defined in the proof of Proposition 32. By integration we get a Smolyak cubature formula

$$S_n(f) = \int_D A(q,d)(f)(t)\,dt$$

of the form

$$S_n(f) = \sum_{q-d+1 \leq |i| \leq q} (-1)^{q-|i|} \cdot \binom{d-1}{q-|i|} \cdot (V^{i_1,1} \otimes \cdots \otimes V^{i_d,d})(f).$$

Here $V^{i,\nu}$ is the K_ν-spline algorithm for integration based on X^i, given by (24). By Proposition 30,

$$e_2(S_n, \mathrm{Int}, P) = O\left(n^{-(r+1)} \cdot (\ln n)^{(d-1)\cdot(r+3/2)}\right),$$

which differs only by a factor of order $(\ln n)^{(d-1)\cdot(r+1)}$ from the minimal error.

Paskov (1993) analyzes the computational cost of constructing order optimal formulas $\tilde{B}(q)$, see (31), and Smolyak cubature formulas. The results show an advantage of the Smolyak formulas, taking into account the error and the cost of construction.

2.7. L_2-Discrepancy. The integration problem with the Wiener sheet is equivalent to the search for points $x_i \in D$ and coefficients $a_i \in \mathbb{R}$ with small L_2-discrepancy. In fact, one can introduce a general notion of L_2-discrepancy such that the average error of an arbitrary cubature formula and the L_2-discrepancy of its knots and coefficients coincide. The link is provided by an integral representation of the covariance kernel.

Consider a measurable mapping $B : D^2 \to \mathbb{R}$ such that $t \mapsto B(t, \cdot)$ is a well defined and continuous mapping from D to $L_2(D)$. By

$$(32) \qquad K(s,t) = \int_D B(s,v) \cdot B(t,v) \, dv$$

we get a continuous and nonnegative definite function on D^2. For every quadrature formula $S_n(f) = \sum_{i=1}^n f(x_i) \cdot a_i$ and $v \in D$ we put

$$\Delta(v; S_n, B) = \mathrm{Int}(B(\cdot, v)) - S_n(B(\cdot, v)).$$

It is easily verified that

$$\int_D (\mathrm{Int}(B(\cdot, v)))^2 \, dv = \int_D \int_D K(s,t) \, ds \, dt,$$

$$\int_D \mathrm{Int}(B(\cdot, v)) \cdot S_n(B(\cdot, v)) \, dv = \sum_{i=1}^n a_i \cdot \int_D K(s, x_i) \, ds$$

and

$$\int_D (S_n(B(\cdot, v)))^2 \, dv = \sum_{i,j=1}^n a_i \, a_j \cdot K(x_i, x_j).$$

Suppose that P is a zero mean measure with covariance kernel K. Then the quantities above are the variance of Int, the covariance of Int and S_n, and the variance of S_n, respectively. Therefore

$$(33) \qquad \left(\int_D \Delta^2(v; S_n, B) \, dv \right)^{1/2} = e_2(S_n, \mathrm{Int}, P).$$

Thus, for $q = 2$, the average error of S_n with respect to P and with respect to the Lebesgue measure on the parametric class of functions $B(\cdot, v)$ coincide.

In (32) we have defined the kernel K in terms of a given function B. Conversely, one may also start with a continuous covariance kernel K. Let $\mu_1 \geq \mu_2 \geq \cdots > 0$ denote the nonzero eigenvalues of the integral operator with kernel K, repeated according to their multiplicity, and let $\xi_1, \xi_2, \ldots$ denote a corresponding orthonormal system of eigenfunctions in $L_2(D)$. By Mercer's Theorem, K has the uniformly and absolutely convergent representation

$$K(s,t) = \sum_j \mu_j \cdot \xi_j(s) \, \xi_j(t).$$

Then

$$B(x,t) = \sum_j \mu_j^{1/2} \cdot \xi_j(s) \, \xi_j(t)$$

satisfies (32), see Lemma III.22. The integral representation (32) is known explicitly, for instance, for the kernel of the r-fold integrated Wiener measure, see Example IV.3.

The left-hand side in (33) is used to define the L_2-*discrepancy* of the points x_i and the coefficients a_i with respect to B. We stress that no tensor product properties were required so far.

To obtain the classical L_2-discrepancy as a particular instance, we take

$$(34) \qquad B(s, v) = 1_{[0,v[}(s),$$

where $1_{[0,v[}$ denotes the indicator function of $[0, v[= [0, v_1[\times \cdots \times [0, v_d[$. Moreover, we consider only quasi-Monte Carlo formulas S_n. By definition, the coefficients of S_n are given by $1/n$, i.e.,

$$S_n(f) = \frac{1}{n} \cdot \sum_{i=1}^{n} f(x_i)$$

for arbitrary knots $x_1, \ldots, x_n \in D$. Then

$$\Delta(v; S_n, B) = v_1 \cdots v_d - \frac{1}{n} \cdot \sum_{i=1}^{n} 1_{[0,v[}(x_i),$$

and

$$\mathrm{disc}_2(x_1, \ldots, x_n) = \left(\int_D \Delta^2(v; S_n, B)\, dv \right)^{1/2}$$

is called the L_2-discrepancy of the points $x_1, \ldots, x_n$. This quantity says how well, on the average, the quasi-Monte Carlo formula S_n with knots x_i approximates the volume of the boxes $[0, v[$.

For B given by (34) we get the tensor product kernel

$$K = \bigotimes_{\nu=1}^{d} K_\nu$$

in (32) with

$$K_\nu(\sigma, \tau) = 1 - \max(\sigma, \tau) = \min(1 - \sigma, 1 - \tau).$$

Therefore K is the covariance kernel of the Wiener sheet, up to the reflection at the midpoint $\left(\frac{1}{2}, \ldots, \frac{1}{2}\right)$ of D. Summarizing we see that the average error of an cubature formula with respect to the Wiener sheet coincides with the L_2-discrepancy of its knots (and coefficients), up to a reflection.

It is known that suitably shifted Hammersley points have L_2-discrepancy of order $n^{-1} \cdot (\ln n)^{(d-1)/2}$. We conclude that these points, together with equal weights $1/n$, yield order optimal cubature formulas with respect to the Wiener sheet.

2.8. Notes and References. 1. The first average case results for multivariate functions were obtained for tensor product problems. Tensor product cubature formulas, i.e., cubature formulas that are based on grids, were studied in Ylvisaker (1975) and Wittwer (1978) in dimension $d = 2$. Ylvisaker already demonstrated that tensor product formulas are not order optimal for tensor product kernels. Further results were obtained in Micchelli and Wahba (1981).

The first results on order optimality in the average case setting are due to Papageorgiou and Wasilkowski (1990) for L_2-approximation with $\Lambda = \Lambda^{\text{all}}$ and Woźniakowski (1991) for integration. Again, tensor product kernels are used in both papers.

2. See Aronszajn (1950) and Delvos and Schempp (1989) for a detailed treatment of tensor product kernels and tensor product spaces.

3. The facts from Section 2.2 are well known.

4. The formulas $A(q,d)$ as given in Lemma 27, together with a fundamental worst case error bound for general tensor product problems, are due to Smolyak (1963). Smolyak's technique can also be used in the average case setting, but the bounds are improved significantly by means of a different analysis. This analysis, which we have partially presented in Section 2.3, is due to Wasilkowski and Woźniakowski (1995). In the latter paper explicit bounds with strong emphasis on their dependence on d are obtained for the error of $A(q,d)$, as well as for the number of points in $H(q,d)$.

5. Smolyak's construction has been reinvented and used in many papers for specific problems. It is known under different names, including Biermann interpolation, *discrete blending method*, *Boolean method*, hyperbolic cross points, and sparse grid method. We present a partial list of references, which deal with the worst case setting.

If in the univariate case the operators are polynomial interpolation operators, then one obtains discrete versions of polynomial blending interpolation, which occur in computer aided design and in the finite element method. See Gordon (1971), Wahba (1978b) and Delvos and Schempp (1989). Integration of periodic functions is studied in Delvos (1990). The integration of non-periodic functions is analyzed in Novak and Ritter (1996a, 1999), and sparse grids that are based on non-equidistant sets in dimension $d = 1$ turn out to be very effective. The fully symmetric cubature formulas from Genz (1986) are a particular instance of Smolyak's construction.

For applications to integral equations we refer to Pereverzev (1986) and Frank, Heinrich, and Pereverzev (1996). Finite element methods based on sparse grids are investigated in Zenger (1991), Griebel, Schneider, and Zenger (1992), Werschulz (1996), and Bungartz and Griebel (1999). Global optimization is analyzed in Novak and Ritter (1996b). Further results and references for cubature and approximation are given in Temlyakov (1994).

The Smolyak construction, and modifications thereof, are used to compute path integrals, i.e., integrals over infinite-dimensional function spaces, in Novak,

Ritter, and Steinbauer (1998), Plaskota, Wasilkowski, and Woźniakowski (1999), and Steinbauer (1999).

6. For weighted tensor product problems Wasilkowski and Woźniakowski (1999) introduce weighted tensor product algorithms. These algorithms allow a different and much more flexible selection of multi-indices in the formula given in Lemma 27. We will mention a particular application in the following section.

7. A version of Proposition 32 with larger exponents for $\ln n$ is obtained in Ritter *et al.* (1993). A more general version of Lemma 34 is due to Papageorgiou and Wasilkowski (1990). Proposition 35 is new.

8. Proposition 37 is due to Ritter *et al.* (1995). The proposition was already known for r-fold integrated Wiener sheets; see Woźniakowski (1991) and Paskov (1993) for integration, Papageorgiou and Wasilkowski (1990) for approximation with $\Lambda = \Lambda^{\text{all}}$, and Woźniakowski (1992) for approximation with $\Lambda = \Lambda^{\text{std}}$.

9. Woźniakowski (1991) discovered the relation between the classical L_2-discrepancy and average errors of cubature formulas with respect to the Wiener sheet P. Hereby he could use results of Roth (1954, 1980) on L_2-discrepancy to determine the order of $e_2(n, \Lambda^{\text{std}}, \text{Int}, P)$. The general results from Section 2.7 are due to Frank and Heinrich (1996). We refer to Niederreiter (1992), Drmota and Tichy (1997), and Matoušek (1998) for a comprehensive account on discrepancy.

3. Tractability or Curse of Dimensionality?

In this section we discuss the dependence of the errors for multivariate integration and L_2-approximation with $\Lambda = \Lambda^{\text{std}}$ on the dimension d. The bounds that were presented in Sections 1 and 2 are of one of the following types,

$$(35) \qquad c_d \cdot n^{-\kappa/d},$$

$$(36) \qquad c_d \cdot n^{-(\kappa/d+1/2)},$$

$$(37) \qquad c_d \cdot n^{-\kappa} \cdot (\ln n)^{(d-1)\cdot\eta}.$$

Here $\kappa, \eta > 0$ capture the underlying smoothness and the constant $c_d > 0$ may depend on d. Recall that c_d was unspecified in most of the results.

For isotropic smoothness conditions we have established bounds of the form (35) for approximation and (36) for integration. Specifically,

$$\kappa = r + \beta \qquad \text{under Hölder conditions,}$$
$$\kappa = \gamma - \tfrac{1}{2} d \qquad \text{for stationary random fields,}$$
$$\kappa = 2\theta + \beta \qquad \text{for fractional Brownian fields.}$$

Note that the dependence of κ on d vanishes for stationary random fields, if we express κ in terms of the Hölder smoothness, see Lemma 5. With n tending to infinity, the order κ/d of the bounds for approximation depends strongly on d, and it is close to zero if d is large compared to the smoothness. This is not true for integration, since the order is at least $\frac{1}{2}$ in (36).

In the tensor product case the bounds are of the form (37) with

$$\kappa = r + \beta + \tfrac{1}{2}, \qquad \eta = r + \beta + \tfrac{1}{2} \qquad \text{under Hölder conditions,}$$
$$\kappa = r + 1, \qquad\qquad \eta = \tfrac{1}{2} \qquad\qquad \text{under Sacks-Ylvisaker conditions}$$

for integration and

$$\kappa = r + \beta, \qquad \eta = r + \beta + \tfrac{1}{2} \qquad \text{under Hölder conditions,}$$
$$\kappa = \gamma - \tfrac{1}{2}, \qquad \eta = \gamma \qquad\qquad\quad \text{for stationary random fields,}$$
$$\kappa = r + \tfrac{1}{2}, \qquad \eta = r + 1 \qquad\qquad \text{under Sacks-Ylvisaker conditions}$$

for approximation. Now the dimension only mildly affects the order of the minimal errors, namely, through a power of $(\ln n)^{d-1}$. Observe, however, that the bound (37) is strictly increasing for $n \leq \exp((d-1) \cdot \eta/\kappa)$, which limits its practical relevance for large d.

To fully understand the dependence on the dimension d we consider a sequence of problems, indexed by $d \in \mathbb{N}$. Then we have zero mean measures $P^{(d)}$ on classes $F^{(d)}$ of functions on $[0,1]^d$. Let $S^{(d)} = \text{Int}^{(d)}$ denote the d-dimensional integral and let $S^{(d)} = \text{App}_2^{(d)}$ correspond to the d-dimensional L_2-approximation problem. For simplicity we consider the constant weight function $\varrho = 1$ in both cases. Then one can study the behavior of the minimal errors as a function of n and d, simultaneously. Equivalently, one can study the minimal number of knots that is needed to obtain an average error ε in dimension d. To this end we put

$$m(\varepsilon, d) = \min\{n \in \mathbb{N} : e_2(n, \Lambda^{\text{std}}, S^{(d)}, P^{(d)}) \leq \varepsilon\}.$$

The sequence of problems is called *tractable* if there exist constants $a, b, c \geq 0$ such that

$$m(\varepsilon, d) \leq c \cdot \varepsilon^{-a} \cdot d^b$$

for all $d \in \mathbb{N}$ and $\varepsilon \in \,]0,1]$. It is called *strongly tractable* if this bound holds with $b = 0$, i.e.,

$$m(\varepsilon, d) \leq c \cdot \varepsilon^{-a}$$

for all $d \in \mathbb{N}$ and $\varepsilon \in \,]0,1]$. For obvious reasons one wants to know the infima of the exponents a and b, given tractability, and b, given strong tractability. These infima are called exponents of tractability.

We state two general observations concerning tractability, and we then turn to the integration problem with Wiener sheets $P^{(d)}$.

It is easily seen that a lower bound (35) for the minimal errors with fixed $\kappa > 0$ contradicts tractability, regardless of the behavior of the constants c_d on d. We conclude, in particular, that the approximation problem for fractional Brownian fields, as well as for stationary random fields with isotropic smoothness, is intractable.

From Proposition II.4 we know that

$$e_2(n, \Lambda^{\text{std}}, \text{Int}^{(d)}, P^{(d)}) \leq c_d \cdot n^{-1/2},$$

where

$$c_d = \int_{[0,1]^d} K^{(d)}(t,t)\, dt$$

and $K^{(d)}$ denotes the covariance kernel of $P^{(d)}$. Therefore the integration problem is tractable if the sequence of constants c_d is polynomially bounded, and boundedness by a constant implies strong tractability. Recall, however, that the proof of Proposition II.4 is not constructive. Under the very mild assumption of continuity of all kernels $K^{(d)}$ we can enforce strong tractability of integration by a suitable scaling. Furthermore, without scaling, integration is strongly tractable for Lévy's Brownian motion and for the Wiener sheet. In the latter case,

$$(38) \qquad\qquad c_d = 2^{-d}.$$

In the remaining part of this section, we survey recent results on tractability for integration with the Wiener sheet and closely related measures. From Proposition II.9, which deals with the case $d = 1$, we immediately get

$$a \geq 1$$

for the exponent of strong tractability. The error analysis for Smolyak formulas usually contains unspecified constants, cf. Propositions 32, 37, and 35. In contrast, Wasilkowski and Woźniakowski (1995) study general tensor product problems in the worst and average case setting, and they obtain explicit bounds for the error of $A(q, d)$ and as well as for the corresponding number of points. In particular they show that

$$a \leq 2.454,$$

which is the best known upper bound that is derived constructively. By nonconstructive arguments, we know that the exponent a of strong tractability is at most two. Wasilkowski and Woźniakowski (1997) establish the improved upper bound

$$a \leq 1.4778\ldots .$$

The proof uses their analysis of Smolyak formulas for approximation as well as Proposition II.4.

Furthermore, lower bounds for the minimal errors in smaller classes of cubature formulas are known. Modifying the definition of $m(\varepsilon, d)$ accordingly, one gets

$$a \geq 1.0669\ldots,$$

if only cubature formulas whose coefficients sum up to one are permitted, see Matoušek (1998). Analogously,

$$a \geq 2.1933$$

for arbitrary methods that are based on nested sparse grids, see Plaskota (1999).

The variance of the d-dimensional integral with respect to the Wiener sheet is given by

$$e(0, d)^2 = \int_{F^{(d)}} \mathrm{Int}^{(d)}(f)^2 \, dP^{(d)}(f) = \int_{[0,1]^d} \int_{[0,1]^d} K^{(d)}(s, t) \, ds \, dt = 3^{-d}.$$

Equivalently, $3^{-d/2}$ is the average error of the zero algorithm. Allowing $n \in \mathbb{N}_0$ in the definition of $m(\varepsilon, d)$ we therefore have

$$m(\varepsilon, d) = 0 \qquad \text{if} \qquad \varepsilon \geq 3^{-d/2},$$

so that integration with fixed accuracy ε eventually gets trivial if d increases.

This suggests modifying the definition of $m(\varepsilon, d)$ according to

(39) $$m(\varepsilon, d) = \min\{n \in \mathbb{N} : e_2(n, \Lambda^{\text{std}}, S^{(d)}, P^{(d)}) \leq \varepsilon \cdot e(0, d)\}.$$

Now, $m(\varepsilon, d)$ is the smallest number of knots that are needed to reduce the error of the zero algorithm by the factor ε. In different terms, one has normalized the problem such that the variance of the integral is one in every dimension d. In the sequel, tractability and strong tractability are understood with respect to (39).

Novak and Woźniakowski (1999) determine sufficient conditions for intractability of tensor product problems with linear functionals. Their results imply that integration with the normalized Wiener sheet is intractable. Therefore we also have intractability of the normalized L_2-discrepancy, see Section 2.7.

In some applications with large d, different variables have a different significance. Weighted tensor product kernels are appropriate for modeling such situations. We do not discuss the general results, but focus on kernels

$$K^{(d)}(s, t) = \prod_{\nu=1}^{d} (1 + \gamma_\nu \cdot \min(\sigma_\nu, \tau_\nu))$$

with a given sequence of weights

$$\gamma_1 \geq \gamma_2 \geq \cdots \geq 0.$$

In this product, the νth factor corresponds to a Brownian motion with variance parameter γ_ν and independent initial value with mean zero and variance one, see Example IV.3. In the analysis of the d-dimensional integration problem one may assume, without loss of generality, that the associated random field is the tensor product of independent random function of the previous type, see Section 2.1.

The decay of the weights determines whether (strong) tractability holds. In the nontrivial case $\gamma_1 > 0$ we clearly have

$$a \geq 1$$

for the exponent of strong tractability, if the problem is strongly tractable at all. Suppose that

(40) $$\sum_{\nu=1}^{\infty} \gamma_\nu^{1/w} < \infty$$

for some $w \geq 1$. Then strong tractability holds with exponent

$$a \leq \frac{2}{\min(w, 2)}.$$

In particular, (40) with $w = 2$ yields $a = 1$. Sloan and Woźniakowski (1998), for $w = 1$, and Hickernell and Woźniakowski (1999), for $w \leq 2$, use non-constructive arguments to derive this relation between w and a. Wasilkowski and Woźniakowski (1997) introduce weighted tensor product algorithms and provide constructive proofs of the following relations. If (40) holds with $w = 2$ or $w = 3$, then $a \leq 2$ or $a = 1$, respectively. On the other hand, (40) with $w = 1$ is also a necessary condition for strong tractability, see Novak and Woźniakowski (1999a). Furthermore, tractability is equivalent to

$$\limsup_{d \to \infty} \frac{1}{\ln d} \cdot \sum_{\nu=1}^{d} \gamma_\nu = \infty,$$

see Sloan and Woźniakowski (1998) and Novak and Woźniakowski (1999a).

3.1. Notes and References. 1. The notion of 'curse of dimensionality' dates back at least to Bellman (1961).

2. The definitions of tractability and strong tractability of linear multivariate problems are due to Woźniakowski (1992, 1994a). There are general results as well as results for specific problems in several settings. See, e.g., Woźniakowski (1992, 1994a, 1994b), Wasilkowski and Woźniakowski (1995, 1997), Sloan and Woźniakowski (1998), Hickernell and Woźniakowski (1999), and Novak and Woźniakowski (1999a). For a recent survey we refer to Novak and Woźniakowski (1999b).

3. Heinrich, Novak, Wasilkowski, and Woźniakowski (1999) show tractability of L_∞-discrepancy.

CHAPTER VII

Nonlinear Methods for Linear Problems

So far we have considered (affine) linear methods for linear problems. Because of this restriction, we only had to specify the covariance kernel and the mean of the measure in the average case analysis of the integration and L_2-approximation problems. Linear methods are easy to implement, if the knots and coefficients are precomputed, and they are well suited for parallel processing. Moreover, the search for asymptotically or order optimal linear method is often successful. On the other hand, nonlinear methods are frequently used in practice for linear problems. It is therefore natural to ask and important to know whether linear methods are optimal for linear problems.

In Section 1 we define the concepts of nonlinear algorithms, adaptive and nonadaptive information operators, and varying cardinality. For the corresponding classes of methods, we introduce minimal errors in the average case setting. The minimal errors allow us to compare the power of different classes of methods.

Gaussian measures are used to analyze linear problems in Section 2. It turns out that adaption with fixed cardinality does not help and linear methods are optimal. Moreover, varying cardinality does not help in many cases. Besides linearity and continuity of the operator to be approximated and the functionals to be used for this purpose, geometrical properties of Gaussian measures lead to the conclusions above. This is similar to the worst case setting where, for instance, symmetry and convexity of the function class implies that adaption may only help by a multiplicative factor of at most two. Furthermore, it is crucial that the measure P or the function class F is completely known.

Adaption may be very powerful for linear problems if any of the assumptions above does not hold. In Section 3 we study integration and approximation of functions with a priori unknown local smoothness. Adaptive methods, which explore the local smoothness and select further knots accordingly, are superior to all nonadaptive methods. In Section 4 we deal with integration of monotone functions. The Dubins-Freedman or Ulam measure, which is not Gaussian, is used in the average case setting and adaption helps significantly. We note that adaption does not help for integration of monotone functions in the worst case setting.

We are implicitly using a crude measure of the computational cost of numerical methods when studying minimal errors, namely, the number of functionals that are applied to the unknown function f. In Section 5 we sketch a proper definition of the computational cost, based on the real number model with an oracle. For the problems that are studied in these notes minimal errors turn out to be still the crucial quantity. The concepts from Sections 1 and 5 are fundamental in information-based complexity for the analysis of linear as well as nonlinear problems. Error, cost, and complexity are defined in various settings, including the worst case and average case setting.

1. Nonlinear Algorithms, Adaptive Information, and Varying Cardinality

Consider a bounded linear operator

$$(1) \qquad\qquad S : X \to G,$$

where X and G are normed linear spaces, and assume that a class

$$(2) \qquad\qquad \Lambda \subset X^*$$

of permissible functionals is given. Here X^* denotes the dual space of X, and we usually have $X = C^r(D)$ in these notes. A linear method S_n^{lin} for approximation of S may be written as

$$S_n^{\text{lin}} = \phi_n^{\text{lin}} \circ N_n^{\text{non}},$$

where

$$N_n^{\text{non}}(f) = (\lambda_1(f), \dots, \lambda_n(f))$$

with fixed functionals $\lambda_i \in \Lambda$ and

$$\phi_n^{\text{lin}}(y_1, \dots, y_n) = \sum_{i=1}^{n} y_i \cdot a_i$$

with fixed coefficients $a_i \in G$. Clearly $N_n^{\text{non}} : X \to \mathbb{R}^n$ is bounded and linear, and any such mapping is called a *nonadaptive information operator*. It is nonadaptive in the sense that the same set of functionals λ_i is applied to every $f \in X$. A linear mapping $\phi_n^{\text{lin}} : \mathbb{R}^n \to G$ is called a *linear algorithm*. We get a much larger class of methods for approximation of S, if we allow the adaptive selection of functionals $\lambda_i \in \Lambda$ and the nonlinear combination of the data $\lambda_1(f), \dots, \lambda_n(f)$. Moreover, the total number of functionals may depend on f via some adaptive termination criterion.

Adaptive information operators choose the functionals λ_i sequentially, depending on the previously computed data. They are formally defined by a fixed functional

$$\lambda_1 \in \Lambda,$$

which is applied in the first step to $f \in X$, and by mappings

$$\lambda_i : \mathbb{R}^{i-1} \to \Lambda, \qquad i = 2, \dots, n,$$

which determine the selection of functionals in the following steps. After n steps, the data

$$N_n^{\text{ad}}(f) = (y_1, \ldots, y_n) \tag{3}$$

with

$$y_1 = \lambda_1(f)$$

and

$$y_i = \lambda_i(y_1, \ldots, y_{i-1})(f), \qquad i = 2, \ldots, n,$$

are at hand. A mapping $\phi_n : \mathbb{R}^n \to G$ is called an (idealized) *algorithm*. The information operator and the algorithm define the *adaptive method*

$$S_n^{\text{ad}} = \phi_n \circ N_n^{\text{ad}}.$$

Constant mappings λ_i clearly yield nonadaptive information operators N_n^{non}, and *nonadaptive methods* are denoted by

$$S_n^{\text{non}} = \phi_n \circ N_n^{\text{non}}.$$

The total amount of information about $f \in X$ that is used to construct an approximation to $S(f)$ is often determined by an *adaptive termination criterion*. For instance, one may use more function values for integrating an (apparently) hard function f while stopping earlier when a good approximation to $\text{Int}(f)$ is already computed. A decision to stop or to select an additional knot can be made after each evaluation. This is formally described by mappings

$$\chi_i : \mathbb{R}^i \to \{0, 1\}, \qquad i \in \mathbb{N},$$

in addition to $\lambda_1 \in \Lambda$ and mappings

$$\lambda_i : \mathbb{R}^{i-1} \to \Lambda, \qquad i \geq 2.$$

The data $N_n^{\text{ad}}(f)$ from (3) are known after n steps, and the total number $\nu(f)$ of functionals that are used for f is determined by

$$\nu(f) = \min\{n \in \mathbb{N} : \chi_n(N_n^{\text{ad}}(f)) = 1\}.$$

For simplicity we assume that $\nu(f) < \infty$ for all $f \in X$. We obtain the information operator

$$N_\nu^{\text{var}}(f) = N_{\nu(f)}^{\text{ad}}(f),$$

which is called adaptive with *varying cardinality*. Constant mappings $\chi_1 = \cdots = \chi_{n-1} = 0$ and $\chi_n = 1$ yield adaptive operators N_n^{ad} with *fixed cardinality* as considered previously. An approximation $S_\nu^{\text{var}}(f)$ to $S(f)$ is constructed as

$$S_\nu^{\text{var}}(f) = \phi_{\nu(f)}(N_\nu^{\text{var}}(f))$$

with algorithms $\phi_i : \mathbb{R}^i \to G$. For simplicity we always assume that the mappings ϕ_i, χ_i, and λ_{i+1} are defined on the whole space $\mathbb{R}^i$.

Assume that a measure P on $F \subset X = C^r(D)$ is given. The definition

$$e_q(\widetilde{S}, S, P) = \left(\int_F \|S(f) - \widetilde{S}(f)\|_G^q \, dP(f) \right)^{1/q}$$

of the *q-average error* is extended to arbitrary measurable methods $\widetilde{S} : F \to G$. We always assume that the mappings ϕ_i, χ_i, and $(\lambda_i(\cdot))(f)$ for every fixed $f \in C^r(D)$ are measurable. Then the integrals we are using in the sequel are well defined.

So far we have studied minimal errors $e_q = e_q^{\text{lin}}$ where

$$e_q^{\text{lin}}(n, \Lambda, S, P) = \inf_{S_n^{\text{lin}}} e_q(S_n^{\text{lin}}, S, P)$$

in the class of all linear methods using n functionals from Λ, or, slightly more general, affine linear methods. Additionally we now study

$$e_q^{\text{non}}(n, \Lambda, S, P) = \inf_{S_n^{\text{non}}} e_q(S_n^{\text{non}}, S, P)$$

and

$$e_q^{\text{ad}}(n, \Lambda, S, P) = \inf_{S_n^{\text{ad}}} e_q(S_n^{\text{ad}}, S, P).$$

The latter two quantities are the *minimal errors* in the classes of all nonadaptive and adaptive methods, respectively, that use n functionals from Λ.

The average case setting provides a natural way to analyze methods with varying cardinality. We compare different methods by their average error and by their *average cardinality*

$$\text{card}(S_\nu^{\text{var}}, P) = \int_F \nu(f) \, dP(f).$$

This leads to the definition of the *minimal error*

$$e_q^{\text{var}}(n, \Lambda, S, P) = \inf\{e_q(S_\nu^{\text{var}}, S, P) : \text{card}(S_\nu^{\text{var}}, P) \leq n\}$$

in the class of all adaptive methods that use at most n functionals from Λ on the average.

Clearly

$$e_q^{\text{var}}(n, \Lambda, S, P) \leq e_q^{\text{ad}}(n, \Lambda, S, P) \leq e_q^{\text{non}}(n, \Lambda, S, P) \leq e_q^{\text{lin}}(n, \Lambda, S, P),$$

and it is important to know whether these quantities (almost) coincide.

REMARK 1. The concepts above have canonical counterparts in the worst case setting, where the average error is replaced by the maximal error over a function class F, see Section II.2.5. Instead of the average cardinality, the maximal cardinality is studied. It is obvious that a worst case analysis cannot justify the use of methods with varying cardinality.

1.1. Notes and References. Traub, Wasilkowski, and Woźniakowski (1988) give a comprehensive presentation and analysis of the concepts from this section.

2. Linear Problems with Gaussian Measures

In this section we report on results that hold for zero mean Gaussian measures P on $F \subset X = C^r(D)$. We assume that a linear problem is given with an operator S and a class Λ of permissible functionals, see (1) and (2), and we use K to denote the covariance kernel of P. The results from this section hold for q-average errors with arbitrary $q \in [1, \infty[$.

First we determine the conditional measures $P(\cdot \mid N_n^{\mathrm{ad}} = y)$. These measures are Gaussian, their means are the corresponding interpolating splines in the reproducing kernel Hilbert space $H(K)$, and their covariance kernels depend on y only through the functionals that are used in the case $N_n^{\mathrm{ad}}(f) = y$. Consequently, the K-spline algorithm, which applies S to the conditional mean, is optimal and its conditional error only depends on the functionals that are used. In particular, for a nonadaptive information operator the K-spline algorithm is linear and its conditional error does not dependent on y. Hence there are no favorable or unfavorable data. It follows that the adaptive selection of functionals cannot be superior to a proper nonadaptive selection. In many cases the minimal errors for linear methods enjoy convexity properties which imply that varying cardinality does not help.

According to the main results from this section the lower bounds from Chapters IV and VI are valid for adaptive methods with varying cardinality, if P is Gaussian. In Section IV.3 we have already studied specific linear problems with Gaussian measures, namely, L_p-approximation of univariate functions with $p \neq 2$. Since $G = L_p([0, 1])$ is not a Hilbert space in this case, the first two moments of an arbitrary measure are not even sufficient to analyze linear methods.

We end the section by a brief comparison of minimal average errors and average n-widths. In this comparison, we focus on L_p-approximation of univariate functions.

2.1. Desintegration. Let N_n^{ad} denote an adaptive information operator with fixed cardinality, which is defined by $\lambda_1, \ldots, \lambda_n$. Desintegration, i.e., the construction of the conditional measures $P(\cdot \mid N_n^{\mathrm{ad}} = y)$, is the key tool for analyzing linear problems with Gaussian measures.

Consider an arbitrary measure P on $C^r(D)$ at first, and let $Q = N_n^{\mathrm{ad}} P$ denote the image of P under the mapping N_n^{ad}. A family $(P(\cdot \mid N_n^{\mathrm{ad}} = y))_{y \in \mathbb{R}^n}$ of measures on $C^r(D)$ is called a *regular conditional probability distribution*, if the following properties hold. For Q-almost every $y \in \mathbb{R}^n$,

$$(4) \qquad P((N_n^{\mathrm{ad}})^{-1}\{y\} \mid N_n^{\mathrm{ad}} = y) = 1.$$

For every measurable set $A \subset C^r(D)$ the mapping $y \mapsto P(A \mid N_n^{\mathrm{ad}} = y)$ is measurable and

$$(5) \qquad P(A) = \int_{\mathbb{R}^n} P(A \mid N_n^{\mathrm{ad}} = y) \, dQ(y).$$

Parthasarathy (1967, p. 147) gives a basic theorem on existence and uniqueness Q-almost everywhere of regular conditional probabilities in Polish spaces, which applies here. In the sequel we identify regular conditional probabilities that coincide Q-almost everywhere.

Assume now that P is a zero mean Gaussian measure on $F \subset C^r(D)$. The conditional measures turn out to be Gaussian, too, and their means and covariance kernels can be described in terms of interpolating splines and reproducing kernels of subspaces of $H(K)$. From Proposition III.11 and Remark III.16 we get

$$H(K) \subset C^r(D)$$

for the Hilbert space with reproducing kernel K, if P is Gaussian. For every $y = (y_1, \ldots, y_n) \in \mathbb{R}^n$ we define

$$\xi_i^y = \mathfrak{J}(\lambda_i(y_1, \ldots, y_{i-1})), \qquad i = 1, \ldots, n.$$

Here $\mathfrak{J}$ denotes the fundamental isomorphism between Λ^P and $H(K)$, see Section III.1.2. Put

$$V^y = \operatorname{span}\{\xi_i^y : i = 1, \ldots, n\}.$$

Without loss of generality we may assume that $\dim V^y = n$ for every $y \in \mathbb{R}^n$, which is equivalent to the linear independence in Λ^P of the functionals that are applied by N_n^{ad} to yield the data y. Let K_1^y denote the reproducing kernel of the subspace $V^y \subset H(K)$ and let

$$K_0^y = K - K_1^y$$

denote the reproducing kernel of the orthogonal complement

$$(V^y)^\perp = \{h \in H(K) : \lambda_i(y_1, \ldots, y_{i-1})(h) = 0 \text{ for } i = 1, \ldots, n\}.$$

Finally, let m^y denote the spline in $H(K)$ that interpolates the data $y_1, \ldots, y_n$ given $\xi_1^y, \ldots, \xi_n^y$.

LEMMA 2. *Let P be a zero mean Gaussian measure. Then $P(\cdot \mid N_n^{\mathrm{ad}} = y)$ is the Gaussian measure on $C^r(D)$ with mean m^y and covariance kernel K_0^y.*

PROOF. We only consider the special case of a nonadaptive information operator. See Lee and Wasilkowski (1986) for a proof by induction on n in the general case.

For a nonadaptive information operator N_n^{non} the functions $\xi_i = \xi_i^y$, the subspace $V = V^y$, and the kernels $K_0 = K_0^y$ and $K_1 = K_1^y$ do not depend on y. Therefore we get $P(\cdot \mid N_n^{\mathrm{non}} = y)$ from $P(\cdot \mid N_n^{\mathrm{non}} = 0)$ by shifting with m^y. The support of a zero mean Gaussian measure on $C^r(D)$ is given as the closure of the associated reproducing kernel Hilbert space in $C^r(D)$, see Vakhania (1975). Since $V^\perp \subset \ker N_n^{\mathrm{non}}$, we conclude that

$$P(\ker N_n^{\mathrm{non}} \mid N_n^{\mathrm{non}} = 0) = 1,$$

which immediately implies (4) for every $y \in \mathbb{R}^n$.

Let P_1 denote the zero mean Gaussian measure on F with covariance kernel K_1. Observe that $y \mapsto m^y$ defines a bounded linear mapping between $\mathbb{R}^n$ and $C^r(D)$. We claim that $P_1 = m^y Q$. To verify this identity of zero mean Gaussian measures we have to show that the covariance kernels of P_1 and $m^y Q$ coincide. The covariance kernel K_1 of P_1 is given by

$$(6) \qquad K_1(s,t) = (\xi_1(s), \ldots, \xi_n(s)) \cdot \Sigma^{-1} \cdot (\xi_1(t), \ldots, \xi_n(t))',$$

where $\Sigma = (\langle \xi_i, \xi_j \rangle_K)_{1 \le i,j \le n}$, see Lemma III.30. Observe that Σ is the covariance matrix of the measure Q, and

$$m^y(t) = (y_1, \ldots, y_n) \cdot \Sigma^{-1} \cdot (\xi_1(t), \ldots, \xi_n(t))'$$

by Lemma III.29. Therefore

$$m^y(s) \cdot m^y(t)$$
$$= (\xi_1(s), \ldots, \xi_n(s)) \cdot \Sigma^{-1} \cdot (y_1, \ldots, y_n)' \cdot (y_1, \ldots, y_n) \cdot \Sigma^{-1} \cdot (\xi_1(t), \ldots, \xi_n(t))'$$

and the covariance kernel of $m^y Q$ satisfies

$$\int_F m^y(s) \cdot m^y(t) \, dQ(y) = K_1(s,t).$$

Clearly P is the convolution of $P(\cdot \mid N_n^{\text{non}} = 0)$ and P_1, since $K = K_0 + K_1$. This may be written as

$$P(A) = \int_F P(A - f_1 \mid N_n^{\text{non}} = 0) \, dP_1(f_1) = \int_{\mathbb{R}^n} P(A - m^y \mid N_n^{\text{non}} = 0) \, dQ(y)$$
$$= \int_{\mathbb{R}^n} P(A \mid N_n^{\text{non}} = y) \, dQ(y)$$

for every measurable set $A \subset F$. Hence we have established (5). $\qquad \square$

EXAMPLE 3. For illustration we consider the Wiener measure w on the class of functions $f \in C([0,1])$ with $f(0) = 0$ and a nonadaptive information operator

$$N_n^{\text{non}}(f) = (f(x_1), \ldots, f(x_n)).$$

Assume that $0 < x_1 < \cdots < x_n = 1$ and put $x_0 = y_0 = 0$. The spline and mean m^y of the conditional measure $w(\cdot \mid N_n^{\text{non}} = y)$ is given by piecewise linear interpolation of the values y_i in the knots x_i, see Example III.33. Moreover, Q is the zero mean Gaussian measure on $\mathbb{R}^n$ whose covariance matrix is given by (II.2).

Let $s, t \in [0,1]$ with $s \in [x_{i-1}, x_i]$ and $s \le t$. Using (II.5) and (6) we easily get

$$K_1(s,t) = x_{i-1} + \frac{(s - x_{i-1}) \cdot (\min(t, x_i) - x_{i-1})}{x_i - x_{i-1}}.$$

Therefore

$$K_0(s,t) = \min(s,t) - K_1(s,t) = \frac{(s - x_{i-1}) \cdot \max(x_i - t, 0)}{x_i - x_{i-1}}$$

is the covariance kernel of every conditional measure $w(\cdot \mid N_n^{\text{non}} = y)$.

Observe that $K_0(s,t) = 0$ if s and t belong to different subintervals, i.e., function values from different subintervals are conditionally independent. On the other hand, for $s, t \in [x_{i-1}, x_i]$ we get

$$K_0(s,t) = (x_i - x_{i-1}) \cdot L(Ts, Tt),$$

where $Tu = (u - x_{i-1})/(x_i - x_{i-1})$ and L denotes the covariance kernel of the Brownian bridge, see Section II.3.7. In this sense $w(\cdot \mid N_n^{\mathrm{non}} = y)$ is given by shifted and scaled independent Brownian bridges on the subintervals $[x_{i-1}, x_i]$. This particular property of the Wiener measure considerably simplifies the analysis of many problems.

2.2. Optimality of K-Spline Algorithms. Desintegration can be immediately applied in the average case analysis, since

$$e_q(\phi_n \circ N_n^{\mathrm{ad}}, S, P) = \left(\int_{\mathbb{R}^n} \int_{C^r(D)} \|S(f) - \phi_n(y)\|_G^q \, dP(f \mid N_n^{\mathrm{ad}} = y) \, dQ(y) \right)^{1/q}$$

for every measurable algorithm ϕ_n. Linearity and boundedness of S are irrelevant to derive this formula, and no particular property of P is used. It suffices that S be measurable. Define

$$U_q(\widetilde{g}; M) = \int_G \|g - \widetilde{g}\|_G^q \, dM(g)$$

for $\widetilde{g} \in G$ and every measure M on G. This quantity indicates how well the point $\widetilde{g}$ approximates the measure M, and a minimizer of $U_q(\cdot \, ; M)$ is called a q-center of M. Consider the family $(SP(\cdot \mid N_n^{\mathrm{ad}} = y))_{y \in R^n}$ of image measures on G to obtain

$$e_q(\phi_n \circ N_n^{\mathrm{ad}}, S, P) = \left(\int_{\mathbb{R}^n} U_q(\phi_n(y); SP(\cdot \mid N_n^{\mathrm{ad}} = y)) \, dQ(y) \right)^{1/q}.$$

Good algorithms ϕ_n aim at minimizing $U_q(\cdot \, ; SP(\cdot \mid N_n^{\mathrm{ad}} = y))$. The best algorithm selects q-centers of the image measures $SP(\cdot \mid N_n^{\mathrm{ad}} = y)$, provided such centers exist. The relevance of these facts depends on the knowledge about the operator S and the conditional measures.

For linear problems with Gaussian measures, we obtain several interesting consequences. At first we assume that an adaptive information operator N_n^{ad} is given. For every $y \in \mathbb{R}^n$ we assume without loss of generality that $\lambda_1, \ldots, \lambda_n(y_1, \ldots, y_{n-1})$ are linearly independent functionals in Λ^P. In the sequel we identify algorithms that coincide Q-almost everywhere. The q-average *radius* of the information operator N_n^{ad} is defined by

(7) $$r_q(N_n^{\mathrm{ad}}, S, P) = \inf\{e_q(\phi_n \circ N_n^{\mathrm{ad}}, S, P) : \phi_n \text{ measurable}\}.$$

We already know that the K-spline algorithm

$$\phi_n(y) = S(m^y)$$

has minimal error in the class of all linear algorithms that are based on a given nonadaptive information operator, see Proposition III.34. By the following result, this optimality also holds in the class of all measurable algorithms, and the given information operator may be adaptive. In the latter case, the K-spline algorithm is nonlinear.

PROPOSITION 4. *Consider an arbitrary linear problem with a zero mean Gaussian measure P, and assume that an adaptive information operator N_n^{ad} is given. Then*

$$e_q(\phi_n \circ N_n^{\mathrm{ad}}, S, P) = r_q(N_n^{\mathrm{ad}}, S, P)$$

iff ϕ_n is the K-spline algorithm.

PROOF. We apply the following simple observation. If a measure M on G is symmetric with respect to $\tilde{g} \in G$, i.e., $M(\tilde{g} + A) = M(\tilde{g} - A)$ for every measurable set $A \subset G$, then $\tilde{g}$ is a q-center of M. See Remark II.3, which deals with symmetry with respect to the origin. If $\tilde{g}$ additionally belongs to the support of M, then $\tilde{g}$ is the unique q-center of M, see Novak and Ritter (1989).

By Lemma 2 the measure $P(\cdot \mid N_n^{\mathrm{ad}} = y)$ is Gaussian with mean m^y. Since S is bounded and linear, the measure $SP(\cdot \mid N_n^{\mathrm{ad}} = y)$ is Gaussian, too, with mean $S(m^y)$. Note that $S(m^y)$ is the output of the K-spline algorithm, given $N_n^{\mathrm{ad}}(f) = y$. Moreover, every Gaussian measure M is symmetric with respect to its mean, which is contained in the support of M. Hereby the statement follows.   □

COROLLARY 5. *For every linear problem with a zero mean Gaussian measure,*

$$e_q^{\mathrm{lin}}(n, \Lambda, S, P) = e_q^{\mathrm{non}}(n, \Lambda, S, P).$$

EXAMPLE 6. Let us show that Corollary 5 does not extend to non-Gaussian measures. Consider the tensor product kernel

$$K(s, t) = \min(\sigma_1, \tau_1) \cdot \min(\sigma_2, \tau_2)$$

where $s = (\sigma_1, \sigma_2) \in [0, 1]^2$ and $t = (\tau_1, \tau_2) \in [0, 1]^2$. The zero mean Gaussian measure $\widetilde{P}$ on $C([0, 1]^2)$ with covariance kernel K is the two-dimensional Wiener sheet, see Example VI.21. A non-Gaussian measure P with the same covariance kernel and mean is obtained in the following way. Let w denote the Wiener measure on $C([0, 1])$ and let P denote the image of $w \otimes w$ under the mapping $(f, g) \mapsto f \otimes g$. Then P is a zero mean measure on $C([0, 1]^2)$ with covariance kernel K.

Consider the L_2-approximation problem. As long as we study linear methods, both the Wiener sheet and the measure P yield the same errors, see Section III.2.2. Therefore

$$e_2^{\mathrm{lin}}(n, \Lambda^{\mathrm{std}}, \mathrm{App}_2, P) = e_2^{\mathrm{lin}}(n, \Lambda^{\mathrm{std}}, \mathrm{App}_2, \widetilde{P}).$$

From Proposition VI.37 and Corollary 5 we know that

$$e_2^{\mathrm{lin}}(n, \Lambda^{\mathrm{std}}, \mathrm{App}_2, \widetilde{P}) = e_2^{\mathrm{non}}(n, \Lambda^{\mathrm{std}}, \mathrm{App}_2, \widetilde{P}) \asymp n^{-1/2} \cdot \ln n.$$

However, simple nonlinear algorithms and nonadaptive information operators turn out to be better than linear methods with respect to P. Evaluation of $f \otimes g$ on

$$X_n = \{(k/n, 1/n) : k = 1, \ldots, n\} \cup \{(1/n, k/n) : k = 1, \ldots, n\}$$

allows to recover $f \otimes g$ exactly on the grid

$$\widetilde{X}_n = \{(k/n, \ell/n) : k, \ell = 1, \ldots, n\}$$

with probability $P = 1$, since

$$(f \otimes g)(k/n, \ell/n) = \frac{(f \otimes g)(k/n, 1/n) \cdot (f \otimes g)(1/n, \ell/n)}{(f \otimes g)(1/n, 1/n)}.$$

Note that $(f \otimes g)(s, t) = 0$ with probability $P = 1$ if $s = 0$ or $t = 0$. It is easy to check that piecewise bilinear interpolation at the grid $\widetilde{X}_n$ leads to errors of order $n^{-1/2}$ with respect to P. Based on evaluations at X_n, this is a nonlinear algorithm that uses $2n - 1$ function values. Hence we get

$$e_2^{\mathrm{non}}(n, \Lambda^{\mathrm{std}}, \mathrm{App}_2, P) = O(n^{-1/2}).$$

See Müller-Gronbach and Schwabe (1996) for further results on approximation of tensor product random functions.

2.3. Adaptive versus Nonadaptive Information. Along with every adaptive information operator N_n^{ad}, we have a family of nonadaptive information operators

$$(8) \qquad N_n^{\mathrm{non}, y}(f) = (\lambda_1(f), \lambda_2(y_1)(f), \ldots, \lambda_n(y_1, \ldots, y_{n-1})(f))$$

with $y \in \mathbb{R}^n$. By definition $N_n^{\mathrm{non}, y}$ applies those functionals to all $f \in F$ that are used by N_n^{ad} for $f \in (N_n^{\mathrm{ad}})^{-1}\{y\}$.

PROPOSITION 7. *Consider an arbitrary linear problem with a zero mean Gaussian measure P, and assume that an adaptive information operator N_n^{ad} is given. Then there exist data $y \in \mathbb{R}^n$ such that*

$$r_q(N_n^{\mathrm{non}, y}, S, P) \leq r_q(N_n^{\mathrm{ad}}, S, P).$$

PROOF. Lemma 2 and the symmetry argument from the proof of Proposition 4 imply

$$(9) \qquad \int_{C^r(D)} \|S(f) - \phi_n(y)\|_G^q \, dP(f \mid N_n^{\mathrm{ad}} = y)$$

$$= \int_{C^r(D)} \|S(f + m^y) - \phi_n(y)\|_G^q \, dP(f \mid N_n^{\mathrm{non}, y} = 0)$$

$$\geq \int_{C^r(D)} \|S(f)\|_G^q \, dP(f \mid N_n^{\mathrm{non}, y} = 0)$$

for every algorithm ϕ_n. We have equality iff $\phi_n(y) = S(m^y)$. For nonadaptive information operators we obtain

$$(10) \qquad r_q(N_n^{\mathrm{non}}, S, P) = \left(\int_{C^r(D)} \|S(f)\|_G^q \, dP(f \mid N_n^{\mathrm{non}} = 0) \right)^{1/q},$$

so that

$$r_q(N_n^{\mathrm{ad}}, S, P) = \left(\int_{\mathbb{R}^n} (r_q(N_n^{\mathrm{non},y}, S, P))^q \, dN_n^{\mathrm{ad}} P(y) \right)^{1/q}$$

for adaptive information operators. Hereby the statement follows. □

Proposition 7 and Corollary 5 immediately yield the following fact.

COROLLARY 8. *For every linear problem with a zero mean Gaussian measure,*

$$e_q^{\mathrm{lin}}(n, \Lambda, S, P) = e_q^{\mathrm{ad}}(n, \Lambda, S, P).$$

According to the previous result adaption with fixed cardinality does not help. More precisely, for every adaptive method $S_n^{\mathrm{ad}} = \phi_n \circ N_n^{\mathrm{ad}}$, there exists a linear method $S_n^{\mathrm{lin}} = \phi_n^{\mathrm{lin}} \circ N_n^{\mathrm{non}}$ that uses the same number of functionals and whose error does not exceed the error of S_n^{ad}. The information operator N_n^{non} is already contained in N_n^{ad} in the sense that it uses n functionals for all $f \in F$ that are used by N_n^{ad} for some $f \in F$. Furthermore, the best choice of ϕ_n^{lin} is the K-spline algorithm based on N_n^{non}.

The search for optimal methods with fixed cardinality is therefore reduced to the optimal nonadaptive selection of functionals from Λ. For this problem no general solution is available, except if G is a Hilbert space, $q = 2$, and $\Lambda = \Lambda^{\mathrm{all}}$, see Section III.2.

A general negative result on superiority of adaptive methods for linear problems is also known in the worst case setting. In this setting a class F of functions is given and the maximal radius

$$r_{\max}(N_n^{\mathrm{ad}}, S, F) = \inf \left\{ \sup_{f \in F} \|S(f) - \phi_n \circ N_n^{\mathrm{ad}}(f)\|_G : \phi_n : \mathbb{R}^n \to G \right\}$$

is studied, cf. (7). The proof of the following result may be found in Traub and Woźniakowski (1980); see also Notes and References 2.6.9.

PROPOSITION 9. *Consider an arbitrary linear problem on a convex and symmetric class F, and assume that an adaptive information operator N_n^{ad} is given. Then*

$$r_{\max}(N_n^{\mathrm{non},0}, S, F) \leq 2 \cdot r_{\max}(N_n^{\mathrm{ad}}, S, F).$$

Thus we can simply take the functionals that N_n^{ad} uses for the zero function to obtain a nonadaptive information operator that is essentially as good as N_n^{ad}. We add that the factor two can sometimes be removed, e.g., if $G = \mathbb{R}$. Observe that Propositions 7 and 9 both rely on geometrical properties of P and F, respectively.

2.4. Varying Cardinality. Now we turn to information operators N_ν^{ad} with varying cardinality. Associated with N_ν^{ad} are two families of information operators, namely, the adaptive operators N_n^{ad} with fixed cardinality given in (3) and the nonadaptive operators $N_n^{\mathrm{non},y}$ given in (8). Let $n, k \in \mathbb{N}$, $y \in \mathbb{R}^n$, $z \in \mathbb{R}^k$ and $c \in \mathbb{R}$. We define

$$N_\mu^{\mathrm{var}}(f) = \begin{cases} N_n^{\mathrm{non},y}(f) & \text{if } \lambda_1(f) \leq c, \\ N_k^{\mathrm{non},z}(f) & \text{if } \lambda_1(f) > c. \end{cases}$$

Then N_μ^{var} selects one out of two nonadaptive operators that are contained in N_ν^{var}. In the first step all these operators apply the same functional λ_1, and the selection is made after the first step. Observe that $\lambda_1(f)$ differs from y_1 and z_1, in general. Let S_μ^{var} consist of N_μ^{var} and the corresponding K-spline algorithms, and put

$$(11) \qquad\qquad u = P(\{f \in F : \lambda_1(f) \leq c\}).$$

Using (10) with the nonadaptive operators above we obtain

$$e_q(S_\mu^{\mathrm{var}}, S, P)^q = u \cdot r_q(N_n^{\mathrm{non},y}, S, P)^q + (1 - u) \cdot r_q(N_k^{\mathrm{non},z}, S, P)^q.$$

Moreover,

$$\mathrm{card}(S_\mu^{\mathrm{var}}, P) = u \cdot n + (1 - u) \cdot k.$$

We show that it suffices to study information operators with varying cardinality that are of the very simple form N_μ^{var}.

PROPOSITION 10. *Consider an arbitrary linear problem with a zero mean Gaussian measure P, and assume that a method S_ν^{var} with varying cardinality is given. Then there exists $n, k \in \mathbb{N}$, $y \in \mathbb{R}^n$, $z \in \mathbb{R}^k$ and $c \in \mathbb{R}$, such that*

$$e_q(S_\mu^{\mathrm{var}}, S, P) \leq e_q(S_\nu^{\mathrm{var}}, S, P)$$

and

$$\mathrm{card}(S_\mu^{\mathrm{var}}, P) = \mathrm{card}(S_\nu^{\mathrm{var}}, P).$$

PROOF. Let $A_n = \{f \in F : \nu(f) = n\}$ and observe that $A_n = (N_n^{\mathrm{ad}})^{-1}(B_n)$ with a measurable set $B_n \subset \mathbb{R}^n$. Thus

$$e_q(S_\nu^{\mathrm{var}}, S, P)^q = \sum_{n=1}^\infty \int_{A_n} \|S(f) - \phi_n \circ N_n^{\mathrm{ad}}(f)\|_G^q \, dP(f)$$

$$= \sum_{n=1}^\infty \int_{B_n} \int_{C^r(D)} \|S(f) - \phi_n(y)\|_G^q \, dP(f \mid N_n^{\mathrm{ad}} = y) \, dN_n^{\mathrm{ad}} P(y)$$

$$\geq \sum_{n=1}^\infty \int_{B_n} \int_{C^r(D)} \|S(f)\|_G^q \, dP(f \mid N_n^{\mathrm{non},y} = 0) \, dN_n^{\mathrm{ad}} P(y)$$

due to (9). For suitable $y(n) \in B_n$ we obtain

$$e_q(S_\nu^{\mathrm{var}}, S, P)^q \geq \sum_{n=1}^\infty P(A_n) \cdot r_q(N_n^{\mathrm{non},y(n)}, S, P)^q$$

using (10). Momentarily treating the probabilities $P(A_n)$ as variables, one can show that

(12) $e_q(S_\nu^{\mathrm{var}}, S, P)^q \geq u \cdot r_q(N_n^{\mathrm{non},y(n)}, S, P)^q + (1 - u) \cdot r_q(N_k^{\mathrm{non},y(k)}, S, P)^q$

for some $n, k \in \mathbb{N}$, where $u \in [0, 1]$ is defined by

$$u \cdot n + (1 - u) \cdot k = \mathrm{card}(S_\nu^{\mathrm{var}}, P).$$

See Wasilkowski (1986a). It remains to take $y = y(n)$, $z = y(k)$ and to define c by (11). This is possible if $\lambda_1 \neq 0$ with probability one, which we may assume without loss of generality. $\square$

A sequence of positive integers v_n is called convex if

$$v_n \leq \tfrac{1}{2}(v_{n-1} + v_{n+1}), \qquad n \geq 2.$$

Under convexity assumptions the minimal errors for methods with varying cardinality do not differ much from those of linear methods. First assume that $e_q^{\mathrm{lin}}(n, \Lambda, S, P)^q$ is actually convex. Moreover, assume that optimal linear methods S_n^{lin} exist for every n, i.e.,

$$e_q(S_n^{\mathrm{lin}}, S, P) = e_q^{\mathrm{lin}}(n, \Lambda, S, P)$$

and S_n^{lin} uses n functionals from Λ. These assumptions are satisfied, for instance, if G is a Hilbert space, $q = 2$, and $\Lambda = \Lambda^{\mathrm{all}}$, see Proposition III.24 and Notes and References III.2.4.3. In this case (12) may be replaced by

$$e_q(S_\nu^{\mathrm{var}}, S, P)^q \geq u \cdot e_q(S_n^{\mathrm{lin}}, S, P)^q + (1 - u) \cdot e_q(S_k^{\mathrm{lin}}, S, P)^q$$

with the particular choice $n = \lceil \mathrm{card}(S_\nu^{\mathrm{var}}, P) \rceil$ and $k = n - 1$. Therefore

$$e_q(S_n^{\mathrm{lin}}, S, P) \leq e_q(S_\nu^{\mathrm{var}}, S, P)$$

while

$$\mathrm{card}(S_n^{\mathrm{lin}}, S, P) \leq \mathrm{card}(S_\nu^{\mathrm{var}}, S, P) + 1,$$

and so varying cardinality does not help.

Since the minimal errors $e_q^{\mathrm{lin}}(n, \Lambda, S, P)$ are usually known only up to strong or weak equivalence, we have to relax the convexity assumption accordingly. In many cases the qth powers of these minimal errors are weakly equivalent to a convex sequence, i.e.,

(13) $c_1 \cdot v_n \leq e_q^{\mathrm{lin}}(n, \Lambda, S, P)^q \leq c_2 \cdot v_n$

with constants $c_1, c_2 > 0$. For instance, this property holds for all problems that were studied in Chapters IV and VI. Sometimes we even have strong equivalence to a convex sequence and the minimal errors do not decay too fast, i.e.,

(14) $\displaystyle \lim_{n \to \infty} \frac{e_q^{\mathrm{lin}}(n, \Lambda, S, P)^q}{v_n} = 1,$ and $\displaystyle \lim_{n \to \infty} \frac{e_q^{\mathrm{lin}}(n, \Lambda, S, P)}{e_q^{\mathrm{lin}}(n + 1, \Lambda, S, P)} = 1.$

In these cases varying cardinality may only help up to a multiplicative constant or not at all, asymptotically. See Wasilkowski (1986a) for the proof of the following result.

COROLLARY 11. *Consider an arbitrary linear problem with a zero mean Gaussian measure P. If* (13) *holds then*

$$e_q^{var}(n, \Lambda, S, P) \geq \left(\frac{c_1}{c_2}\right)^{1/q} \cdot e_q^{lin}(n+1, \Lambda, S, P).$$

If (14) *holds then*

$$e_q^{var}(n, \Lambda, S, P) \approx e_q^{lin}(n, \Lambda, S, P).$$

REMARK 12. The error $\|S(f) - S_n^{lin}(f)\|_G$ defines a random variable on the space $F \subset X$, which is equipped with a zero mean Gaussian measure P. Instead of merely studying average errors, which are defined by expectations, one can analyze the distribution of the error.

Furthermore, the support of P is a linear subspace of X and therefore F has to be unbounded, in contrast to the standard situation in a worst case analysis. Hence it is interesting to define errors also with respect to restrictions

$$P_r(A) = \frac{P(\{f \in A : \|f\|_X \leq r\})}{P(\{f \in F : \|f\|_X \leq r\})}$$

of P to balls with radius r.

Given an accuracy level $\varepsilon > 0$, a probability level $0 < \delta < 1$ of failure, and a radius $0 < r \leq \infty$ we are looking for methods S_n^{lin} that satisfy

$$P_r(\{f \in F : \|S(f) - S_n^{lin}(f)\|_G > \varepsilon\}) \leq \delta.$$

In particular, we wish to know the smallest n such that this relation can hold for any S_n^{lin}. This approach is called *probabilistic analysis on bounded domains*. See Notes and References 2.6.6.

REMARK 13. Unbounded operators S occur in the analysis of *ill-posed problems*. If P is Gaussian and if $S : X \to G$ is a weakly measurable linear operator, i.e., if S is linear on a set of full measure, then most of the results presented in this section still hold. See Notes and References 2.6.7.

2.5. Average n-Widths. Adaptive operators N_n^{ad} cannot be one-to-one if $n < \dim X$. Thus the information $N_n^{ad}(f)$ is only partial and leads to errors in the approximation of $S(f)$. Furthermore, linear methods S_n^{lin} permit approximations only from a linear subspace of dimension at most n, which is a second source of errors. We are led to *average n-widths*

$$d(n, S, P) = \inf\left\{\int_F \inf_{g \in G_n} \|S(f) - g\|_G \, dP(f) : G_n \subset G, \dim G_n \leq n\right\},$$

if we stick to the second restriction but assume complete information on f. Average n-widths are the average case analogue of the classical *Kolmogorov n-widths*

$$\inf\Big\{ \sup_{f \in F} \inf_{g \in G_n} \|S(f) - g\|_G : G_n \subset G, \ \dim G_n \leq n \Big\},$$

where the worst case on the class F is studied.

If G is a Hilbert space, then

(15) $d(n, S, P) = e_1^{\text{lin}}(n, \Lambda^{\text{all}}, S, P)$

for every measure P and every bounded linear operator S with values in G, see Mathé (1990). In this case linear methods that may use arbitrary bounded linear functionals are as powerful as best approximation from finite dimensional subspaces. A generalization, which covers errors and widths defined by qth moments for arbitrary $1 \leq q < \infty$, is due to Sun (1992).

Let us consider L_p-approximation of univariate functions, where $S = \text{App}_p$ denotes the embedding of $C^r([0,1])$ into $G = L_p([0,1])$. By definition,

$$d(n, \text{App}_p, P)$$

$$= \inf\Big\{ \int_{C^r([0,1])} \inf_{g \in G_n} \|f - g\|_p \, dP(f) : G_n \subset L_p([0,1]), \ \dim G_n \leq n \Big\}.$$

PROPOSITION 14. *Let P be a zero mean Gaussian measure that satisfies Sacks-Ylvisaker conditions of order $r \in \mathbb{N}_0$. Then*

$$d(n, \text{App}_p, P) \asymp n^{-(r+1/2)}$$

for every $1 \leq p \leq \infty$.

Analogous results for linear methods were obtained in Section IV.3; by Corollary 11 these error bounds are also valid in the class of adaptive methods with varying cardinality. For $1 \leq p < \infty$ we know that the upper bound in Proposition 14 is achieved by classical linear operators mapping into spaces of splines or polynomials, see Remark IV.33. For $2 \leq p \leq \infty$ the lower bound is obtained in the following way. We have

$$d(n, \text{App}_p, P) \geq d(n, \text{App}_2, P) = e_1^{\text{lin}}(n, \Lambda^{\text{all}}, \text{App}_2, P),$$

see (15). The latter quantity is bounded from below by $e_1^{\text{lin}}(n, \Lambda^{\text{all}}, \text{App}_1, P)$, whence Proposition IV.30 is applicable. For $1 \leq p < 2$ the lower bound is stated in Maiorov (1992).

The case $p = \infty$ is of particular interest since $d(n, \text{App}_\infty, P)$ and the minimal errors $e_1^{\text{lin}}(n, \Lambda^{\text{std}}, \text{App}_\infty, P)$ differ by a factor of order $(\ln n)^{1/2}$. A nonconstructive proof of the upper bound for $d(n, \text{App}_\infty, P)$ with the Wiener measure $P = w_0$ is given in Maiorov (1993). The proof uses a discretization technique and estimates for diameters of balls in $\mathbb{R}^k$. As a corollary we get the upper bound for $P = w_r$, and it remains to use Propositions IV.6 and IV.8 and Anderson's inequality, see Tong (1980, p. 55), to cover the general case.

We stress that the order $n^{-(r+1/2)}$ cannot be achieved by linear methods for L_∞-approximation, see Proposition IV.32.

2.6. Notes and References.

1. Lemma 2 is derived by Lee and Wasilkowski (1986) more generally for Gaussian measures on separable Banach spaces.

2. q-centers of probability measures, in particular their existence and uniqueness, are studied in Novak and Ritter (1989). The worst case counterpart of q-centers are Chebyshev centers of subsets of normed linear spaces.

3. Proposition 4 and Corollary 5 are due to Lee and Wasilkowski (1986) and Wasilkowski (1986b). The uniqueness in Proposition 4 is shown in Novak and Ritter (1989). These papers deal with Gaussian measures on separable Banach spaces X. The same result is obtained in several papers under different assumptions on the measure P and the spaces X and G. The symmetry argument of the proof works for classes of measures P, which include in particular the Gaussian measures. We mention orthogonally or unitarily invariant measures, see Micchelli (1984), Wasilkowski and Woźniakowski (1984, 1986), Papageorgiou and Wasilkowski (1990), and Munch (1990), as well as the larger class of reflectable measures, see Mathé (1990).

4. Proposition 7 and Corollary 8 are due to Lee and Wasilkowski (1986) and Wasilkowski (1986b) more generally for Gaussian measures on separable Banach spaces. The result also holds for orthogonally invariant measures, see Traub, Wasilkowski, and Woźniakowski (1984), Wasilkowski and Woźniakowski (1984), and Munch (1990).

5. Proposition 10 and Corollary 11 are due to Wasilkowski (1986a) for Gaussian measures on separable Banach spaces and extended to elliptically contoured measures in Wasilkowski (1989). See Plaskota (1993) for further results on the power of varying cardinality. See Paskov (1995) for the analysis of adaptive termination criteria that are based on differences of consecutive approximations; see also Gao (1993).

6. For the probabilistic analysis, which we have briefly discussed in Remark 12, we refer to Lee and Wasilkowski (1986), Wasilkowski (1986b), Woźniakowski (1987), Heinrich (1990a), and Kon (1994).

7. An average case analysis of ill-posed problems, which we have mentioned in Remark 13, is given in Werschulz (1991), Kon, Ritter, and Werschulz (1991), and Vakhania (1991).

8. In addition to the original papers we refer to Traub et al. (1988, Chapter 6) for a comprehensive presentation of results for linear problems with Gaussian measures.

9. Optimality of linear algorithms and of spline algorithms, as well as the power of adaption for linear problems is also studied in the worst case setting. We refer to Traub and Woźniakowski (1980), Micchelli and Rivlin (1985), Novak (1988, 1996a), and Traub et al. (1988) for results and numerous references. Proposition 9 is due to Gal and Micchelli (1980) and Traub and Woźniakowski (1980), see also Bakhvalov (1971). See Korneichuk (1994) and Novak (1995a,

1995b) for linear problems where adaption helps significantly in the worst case setting. Novak (1996a) gives a survey on the power of adaption.

10. A general theory of linear problems with noisy data is developed in Plaskota (1996). In particular the power of adaption and varying cardinality is studied, and optimality of smoothing spline algorithms is proven.

11. In a series of papers, Maiorov studies average n-widths of embeddings into L_p-spaces. He obtains results for Gaussian measures on Sobolev spaces of periodic functions and for r-fold integrated Wiener measures. Proposition 14 is due to Maiorov (1992, 1993, 1996b). See Micchelli (1984), Mathé (1990), Sun (1992), Maiorov (1994, 1996a), Sun and Wang (1995), and Maiorov and Wasilkowski (1996) for further results on average n-widths. We add that some of these papers cited here also contain results on probabilistic widths. The notion of average n-widths is also used in a different meaning in the literature, namely, as a worst case concept that allows to deal with non-compact classes F of functions, e.g., functions defined on the whole space $\mathbb{R}^d$. See Magaril-Ilyaev (1994).

3. Unknown Local Smoothness

According to Corollary 8, adaption does not help for linear problems with Gaussian measures. Basically this conclusion relies on the assumption that the Gaussian measure P is completely known, and therefore one can design non-adaptive methods that are adapted to P in an optimal way. Now we only assume partial knowledge about P, and we are looking for methods that deal with different measures automatically.

Actually we have taken this approach to some extent already in Chapters IV–VI, see Section IV.4 in particular. We did not determine optimal methods for specific measures, but were looking for order optimal or asymptotically optimal methods for classes of measures. However, the measures in a given class shared the same smoothness properties, and therefore it was sufficient to consider nonadaptive methods.

Here we want to model problems with unknown local smoothness and unknown mean. This may be done in the following way. We first choose a zero mean Gaussian measure P on some class F of real-valued functions on D. The measure P describes our basic smoothness assumption. Then we consider transformations

$$(16) \qquad T_{m,u,v}(f) = m + u \cdot f \circ v, \qquad f \in F,$$

with unknown functions

$$m : D \to \mathbb{R}, \qquad u : D \to \,]0, \infty[, \qquad v : D \to D \text{ one-to-one}.$$

We hereby get a family of image measures

$$P_{m,u,v} = T_{m,u,v} P,$$

that is,

$$P_{m,u,v}(A) = P(\{f \in F : T_{m,u,v}(f) \in A\})$$

for every measurable set $A \subset F$. Using u and v in this way describes a deformation of the domain D and the range $\mathbb{R}$ of the random function induced by P. The resulting random function is shifted by m. Obviously, $P_{m,u,v}$ is a Gaussian measure with mean m. Its covariance kernel is given by

$$\int_F (T_{m,u,v}(f) - m)(s) \cdot (T_{m,u,v}(f) - m)(t) \, dP(f) = u(s) \cdot u(t) \cdot K(v(s), v(t))$$

for $s, t \in D$, where K denotes the covariance kernel of P.

In the sequel we discuss integration, $S = \mathrm{Int}$, and L_p-approximation, $S = \mathrm{App}_p$, with the constant weight function one, and we study methods using only function values, i.e., $\Lambda = \Lambda^{\mathrm{std}}$. We wish to determine a single sequence of methods that works well with respect to all measures $P_{m,u,v}$, where (m, u, v) varies in some set $\mathfrak{T}$. Specifically, we want to find a sequence of methods S_n^{ad} such that

$$\forall \, (m, u, v) \in \mathfrak{T} : \quad e_q(S_n^{\mathrm{ad}}, S, P_{m,u,v}) \approx e_q(n, \Lambda^{\mathrm{std}}, S, P_{m,u,v}).$$

Recall that the right-hand side coincides with $e_q^{\mathrm{ad}}(n, \Lambda^{\mathrm{std}}, S, P_{m,u,v})$, since $P_{m,u,v}$ is Gaussian. Instead of asymptotic optimality of the methods S_n^{ad} for all measures $P_{m,u,v}$ one can also analyze order optimality or perform a worst case analysis with respect to $(m, u, v) \in \mathfrak{T}$.

EXAMPLE 15. Let us demonstrate how the construction of measures $P_{m,u,v}$ effects locally stationary random functions in the univariate case $D = [0, 1]$. Suppose that the covariance kernel K of P satisfies the Sacks-Ylvisaker conditions (A) and (B) with $r = 0$, see Section IV.1. The corresponding random function is locally stationary with exponent $\beta = \frac{1}{2}$ and local Hölder constant $\gamma(t) = 1$ at every point $t \in [0, 1]$, see Example IV.35.

We determine the smoothness of the random function corresponding to $P_{m,u,v}$ under the assumption of continuous differentiability of m, u, and v with $v' > 0$. Since P has mean zero, we have

$$\int_F (f(s) - f(t))^2 \, dP_{m,u,v}(f) = \Delta(s,t) + (m(s) - m(t))^2,$$

where

$$\Delta(s,t) = \int_F ((u \cdot f \circ v)(s) - (u \cdot f \circ v)(t))^2 \, dP(f)$$
$$= u(s) \cdot \big(K(v(s), v(s)) \cdot u(s) - K(v(t), v(s)) \cdot u(t)\big)$$
$$+ u(t) \cdot \big(K(v(t), v(t)) \cdot u(t) - K(v(s), v(t)) \cdot u(s)\big).$$

Let $s < t$. Then

$$\Delta(s,t) = u(s) \cdot (s-t) \cdot \left(K^{(1,0)}(v(\eta), v(s)) \cdot v'(\eta) \cdot u(\eta) + K(v(\eta), v(s)) \cdot u'(\eta) \right)$$
$$- u(t) \cdot (s-t) \cdot \left(K^{(1,0)}(v(\xi), v(t)) \cdot v'(\xi) \cdot u(\xi) + K(v(\xi), v(t)) \cdot u'(\xi) \right)$$

for some $\eta, \xi \in]s, t[$. Therefore

$$\lim_{s \to t-} \frac{\Delta(s,t)}{t-s} = \left(K_-^{(1,0)}(t,t) - K_+^{(1,0)}(t,t) \right) \cdot u(t)^2 \cdot v'(t) = u(t)^2 \cdot v'(t).$$

The same result is valid for $s > t$ as well, and the convergence holds uniformly in $|t-s|$. Furthermore, $|m(s) - m(t)| = O(|s-t|)$ uniformly in s and t. We conclude that $P_{m,u,v}$ induces a locally stationary random function with exponent $\beta = \frac{1}{2}$ and local Hölder constant

(17)
$$\gamma(t) = u(t) \cdot \sqrt{v'(t)}.$$

Similar relations hold for other basic smoothness assumptions, namely local stationarity of a derivative with some exponent $0 < \beta \leq 1$, see Seleznjev (2000).

REMARK 16. Assume for simplicity that $m = 0$ and $v = \mathrm{id}_D$ is the identity mapping, and write $P_u = P_{0,u,\mathrm{id}_D}$ for brevity. Suppose for a moment that u were known. Then it suffices to consider nonadaptive information operators

$$N_n^{\mathrm{non}}(f) = (f(x_1), \ldots, f(x_n))$$

with $x_i \in D$. Transforming the measure P to P_u while keeping the weight function one is equivalent to introducing a weight function while keeping the measure P fixed. More precisely, the q-average radius of N_n^{non} satisfies

$$r_q(N_n^{\mathrm{non}}, S, P_u) = r_q(N_n^{\mathrm{non}}, S_u, P)$$

with $S_u = \mathrm{Int}_u$ if $S = \mathrm{Int}$, $S_u = \mathrm{App}_{p,u^p}$ if $S = \mathrm{App}_p$ and $p < \infty$, and $S_u = \mathrm{App}_{p,u}$ if $S = \mathrm{App}_p$ and $p = \infty$, cf. Section IV.6. Most of the results from Chapter IV in the univariate case and some of the results from Chapter VI in the multivariate case yield knots x_i that enjoy certain optimality properties and depend on u. Nevertheless the corresponding information operators are nonadaptive, as they use the same set of knots for every function f.

We cannot proceed in this way if u is unknown. Instead the following approach is reasonable. As a subproblem of integration or approximation we estimate the local smoothness of f, where f is distributed according to P_u. We stress that no independent and identically distributed copies of f are available for this purpose, and u itself is not accessible. The results for known u indicate how to make use of the estimated local smoothness for choosing further knots. Obviously, this is the outline of an adaptive method.

In the sequel we study the univariate case with P satisfying Sacks-Ylvisaker conditions of order $r = 0$, and we consider

(18)
$$\mathfrak{T} = \{(m, u, v) \in \left(C^1([0,1]) \right)^2 \times C^2([0,1]) : v' \text{ positive and } v([0,1]) \subset [0,1]\}.$$

We present adaptive methods for integration and approximation.

Basically the methods work as follows. In the first stage, which is nonadaptive, we use a small number of knots to estimate a certain power γ^λ of the local Hölder constant γ, see (17). We take

(19) $$\lambda = \begin{cases} \left(\frac{1}{2} + 1/p\right)^{-1} & \text{for } L_p\text{-approximation,} \\ \frac{2}{3} & \text{for integration.} \end{cases}$$

Then we select additional knots adaptively with "density" proportional to the estimate $\widehat{\gamma}$ of γ^λ. Finally, we use piecewise linear interpolation of the data for approximation. Analogously, we use a trapezoidal rule for integration. Obviously, the choice of λ is motivated by the results from Section IV.6, which deal with known local smoothness. Furthermore, Sections IV.4.1 and IV.4.2 motivate the choice of the algorithm, which is (based on) piecewise linear interpolation.

The nonadaptive part is determined by an integer

$$k \in \mathbb{N}$$

and a real number

$$0 < \delta < 1/k^2,$$

which define the knots

$$x_{i,j} = \frac{2i-1}{2k} - \frac{k\delta}{2} + j\delta$$

with $i = 1, \ldots, k$ and $j = 0, \ldots, k$. These knots are clustered around the points $(2i-1)/(2k)$ with distance δ within each cluster. Additionally, we use $x_{0,k} = 0$ and $x_{k+1,0} = 1$. The total number of nonadaptively selected knots $x_{0,k}, \ldots, x_{k+1,0}$ is given by $k \cdot (k+1) + 2$.

The nonadaptive part yields function values

$$y = (y_{0,k}, \ldots, y_{k+1,0})$$

at the respective knots, and y is used for estimating γ^λ. For every measure $P_{m,u,v}$ with P as above and $(m, u, v) \in \mathfrak{T}$, the differences $f(x_{i,j}) - f(x_{i,j-1})$ are normally distributed and weakly correlated with mean close to zero and second moment close to $\gamma((2i-1)/(2k))^2 \cdot \delta$, see Müller-Gronbach and Ritter (1998, Lemma 1). Therefore a natural choice for estimating γ^λ at $(2i-1)/(2k)$ is

$$\widehat{\gamma}_i(y) = \frac{1}{\nu_\lambda^\lambda \delta^{\lambda/2}} \cdot \frac{1}{k} \sum_{j=1}^{k} |y_{i,j} - y_{i,j-1}|^\lambda,$$

where ν_λ^λ denotes the absolute moment of order λ of the standard normal distribution. For technical reasons one must take care of $\widehat{\gamma}_i(y)$ getting too small, and therefore we use

$$\widetilde{\gamma}_i(y) = \max(\widehat{\gamma}_i(y), \varepsilon)$$

instead, where $0 < \varepsilon \leq 1$.

In the second stage additional knots $s_{i,j}(y)$ are adaptively placed in the subintervals $J_i = [x_{i,k}, x_{i+1,0}]$. These knots are determined by the values

$$\widetilde{\gamma}_1(y), \ldots, \widetilde{\gamma}_k(y)$$

together with an integer

$$n \in \mathbb{N},$$

which is roughly the total number of knots $s_{i,j}(y)$. We estimate γ^λ on $J_0 \cup \cdots \cup J_k$ by piecewise linear interpolation of the values $\widetilde{\gamma}_i(y)$ at the points $(2i-1)/(2k)$, and we use this estimate to construct an adaptive regular sequence.

More precisely,

$$r_i(y) = \left\lfloor n \cdot \frac{\widetilde{\gamma}_i(y) + \widetilde{\gamma}_{i+1}(y)}{2 \sum_{j=1}^{k} \widetilde{\gamma}_j(y)} \right\rfloor$$

is the number of knots that are placed in the subinterval J_i for $i = 1, \ldots, k-1$. Within each of these subintervals, the spacing is proportional to the linear density with boundary values $\widetilde{\gamma}_i(y)$ and $\widetilde{\gamma}_{i+1}(y)$. Formally,

$$\int_{x_{i,k}}^{s_{i,j}(y)} \left(\widetilde{\gamma}_i(y)\,(x_{i+1,0} - t) + \widetilde{\gamma}_{i+1}(y)\,(t - x_{i,k}) \right) dt$$

$$= j \cdot \frac{(\widetilde{\gamma}_i(y) + \widetilde{\gamma}_{i+1}(y)) \cdot (1/k - k\delta)^2}{2\,(r_i(y) + 1)}$$

for $j = 1, \ldots, r_i(y)$. We use equidistant knots in the subintervals J_0 and J_k. Here the numbers of knots are given by

$$r_0(y) = \left\lfloor n \cdot \frac{\widetilde{\gamma}_1(y)}{2 \sum_{j=1}^{k} \widetilde{\gamma}_j(y)} \right\rfloor \qquad \text{and} \qquad r_k(y) = \left\lfloor n \cdot \frac{\widetilde{\gamma}_k(y)}{2 \sum_{j=1}^{k} \widetilde{\gamma}_j(y)} \right\rfloor.$$

Therefore

$$s_{0,j}(y) = j \cdot \frac{1/k - k\delta}{2\,(r_0(y) + 1)}$$

with $j = 1, \ldots, r_0(y)$ and

$$s_{k,j}(y) = 1 - (r_k(y) + 1 - j) \cdot \frac{1/k - k\delta}{2\,(r_k(y) + 1)}$$

with $j = 1, \ldots, r_k(y)$.

Summarizing, the first stage depends on the parameters k and δ, and the second stage depends on the parameters ε, n, and λ. The total number of knots is bounded from above by

$$N = n + k \cdot (k + 1) + 2.$$

For approximation the algorithm consists of piecewise linear interpolation of the whole data from stages one and two. For integration we apply the trapezoidal rule to these data.

For the asymptotic analysis of the method we choose $k = k_n$, $\delta = \delta_n$, and $\varepsilon = \varepsilon_n$ in the following way. For any $0 < \alpha < \frac{1}{2}$ take

$$k_n = \lceil n^{1/2-\alpha} \rceil.$$

Hereby $N = N_n$ satisfies

$$N_n \approx n.$$

Moreover, take δ_n such that

$$\delta_n \leq c \cdot \frac{1}{n}$$

with a constant $c > 0$. Finally, take

$$\varepsilon_n = n^{-\tau}$$

for any $0 < \tau < \frac{1}{4} - \frac{1}{2}\alpha$. We denote the resulting sequence of adaptive methods by S_n^λ, where λ is chosen according to (19). See Müller-Gronbach and Ritter (1998) for the proof of the following result.

PROPOSITION 17. *Let P be a zero mean Gaussian measure whose covariance kernel satisfies Sacks-Ylvisaker conditions of order $r = 0$. Then*

$$\forall\, (m, u, v) \in \mathfrak{T}: \quad e_q(S_n^\lambda, S, P_{m,u,v}) \approx e_q(n, \Lambda^{\mathrm{std}}, S, P_{m,u,v}).$$

On the other hand, for every sequence of nonadaptive methods S_n^{non} and every $c \geq 1$, there exists $(m, u, v) \in \mathfrak{T}$ such that

$$\limsup_{n \to \infty} \frac{e_q(S_n^{\mathrm{non}}, S, P_{m,u.v})}{e_q(n, \Lambda^{\mathrm{std}}, S, P_{m,u,v})} \geq c.$$

Here $S = \mathrm{Int}$ with $q = 2$ or $S = \mathrm{App}_p$ with $1 \leq p = q < \infty$, and $\mathfrak{T}$ is defined by (18).

We therefore have a single sequence of adaptive methods that are simultaneously asymptotically optimal for integration or approximation of random functions with different local smoothness. Moreover, this kind of optimality cannot be achieved by any sequence of nonadaptive methods, i.e., adaption helps for problems with unknown local smoothness in the average case setting. We add that

$$e_2(S_n^\lambda, \mathrm{Int}, P_{m,u,v}) \approx \frac{1}{2\sqrt{3}} \cdot \left(\int_0^1 \gamma(t)^{2/3}\, dt \right)^{3/2} \cdot n^{-1}$$

for integration and

$$e_p(S_n^\lambda, \mathrm{App}_p, P_{m,u,v}) \approx \nu_p \cdot d_p \cdot \left(\int_0^1 \gamma(t)^\lambda\, dt \right)^{1/\lambda} \cdot n^{-1/2}$$

for L_p-approximation with finite p. Here γ is given by (17) and

$$d_p = \left(\int_0^1 (t \cdot (1-t))^{p/2}\, dt \right)^{1/p}.$$

We conjecture that Proposition 17 extends to the case $p = \infty$. Müller-Gronbach and Ritter (1998, Remark 4) sketch how the methods S_n^λ might be modified to deal with other and in particular higher degrees of smoothness. See Notes and References 3.1.2 for results on estimating the local or global smoothness of univariate Gaussian random functions; this task is a natural subproblem in the present setting.

REMARK 18. Let us consider random functions with singularities. One possible construction is based on transformations $T_{m,u,v}$, see (16), with functions u having singularities. Here we report on a different approach in the case $D = [0, 1]$, which is due to Wasilkowski and Gao (1992).

Let g_1 and g_2 denote independent random functions, take $z \in]0, 1[$, and define

$$g(t) = \begin{cases} g_1(t) & \text{if } t \leq z, \\ g_2(t) & \text{if } t > z. \end{cases}$$

In general, g lacks continuity at z. By r-fold integration, see (IV.3), we then get a random function f with a singularity in its rth derivative. Specifically, Wasilkowski and Gao assume that g_1 and g_2 correspond to k-fold integrated Wiener measures, where g_1 has reversed time. Then $r + k + \frac{1}{2}$ is the smoothness in quadratic mean sense of f at all points $t \neq z$. To obtain a measure on $C^r([0, 1])$, say, one could place a distribution on z. The following results are robust with respect to such distributions and actually hold in a worst case sense with respect to $z \in [0, 1]$.

Wasilkowski and Gao construct a method that locates the singularity z from a small number of evaluations of f. The output of the method is a small interval that contains z with high probability. This method is then applied for integration of the random function f, as it allows full use of the additional smoothness k away from the singularity. Furthermore, the probabilistic analysis shows that the adaptive method is far superior to any nonadaptive method. The latter can only exploit the global smoothness, which is essentially $r + \frac{1}{2}$.

REMARK 19. We feel that an average case formulation as well as the basic idea of an adaptive method is outlined in Davis and Rabinowitz (1984, p. 425). These authors suggest the following strategy for automatic integration: "After a preliminary scan of certain function values and based upon certain characteristic behavior within each class F_i, the program makes a decision as to which class F_j the integrand is most likely to be in and then the control is shifted to the F_j portion of the program."

REMARK 20. Cohen and d'Ales (1997) construct random functions f with discontinuities given by a Poisson process on $[0, 1]$. Between successive discontinuities, f is stationary and smooth. The random function may be represented as

$$f = \sum_{i=1}^{\infty} \int_0^1 f(t) \cdot e_i(t) \, dt \cdot e_i$$

for any orthonormal basis e_i of $L_2([0, 1])$. If the first n terms are used to approximate f in the L_2-norm, then the optimal basis corresponds to the Karhunen-Loève representation of f, see Remark III.25. If different n terms may be chosen for different realizations of f, then much smaller errors can be obtained. To this end, a wavelet basis is selected and the coordinates that correspond to the largest coefficients in absolute value are used. In general, this approach is called nonlinear approximation or n-term approximation.

Let us stress the following conceptual difference between the approximation problem, as studied for instance in these notes, and nonlinear approximation. We compare methods that use the same number of linear functionals, e.g., function evaluations. In nonlinear approximation methods are compared that produce approximating functions of the same 'size'. Formally an infinite number of linear functionals, namely the inner products with all basis functions e_i, is needed. Alternatively, one might think of an oracle that on input k yields the value of the kth largest inner product and its index i. This oracle is obviously nonlinear in f.

3.1. Notes and References. 1. Proposition 17 is due to Müller-Gronbach and Ritter (1998), who also report on numerical experiments for the adaptive method.

2. Istas (1996) considers Sacks-Ylvisaker conditions of order $r \in \mathbb{N}_0$, and he studies estimation of the local Hölder constant of the r-th derivative of Gaussian processes. See Istas and Lang (1994) and Benassi, Cohen, Istas, and Jaffard (1998) for estimating the Hölder exponent and the local Hölder constant of Gaussian processes. The analogous problem for multifractional Gaussian processes is studied in Benassi, Cohen, and Istas (1998). The estimators in these papers use generalized quadratic variations. See Gao and Wasilkowski (1993) for estimation of the global smoothness of r-fold integrated Wiener measure with unknown r.

3. The approach and results from Remark 18 are due to Wasilkowski and Gao (1992).

4. In the worst case setting uncertainty regarding the function class F is sometimes called the fat F problem, see Woźniakowski (1986).

4. Integration of Monotone Functions

Consider the class

$$F = \{f \in C([0, 1]) : f(0) = 0, \ f(1) = 1, \ f \text{ increasing}\}.$$

A Gaussian measure with support F does not exists, because Gaussian measures are supported by affine linear subspaces. We use the Dubins-Freedman or *Ulam measure* on F to study integration with the constant weight function $\varrho = 1$. We consider methods that use function values only, i.e., we take $\Lambda = \Lambda^{\text{std}}$. Worst and average case results for integration of monotone functions differ significantly.

A detailed analysis of the Ulam measure P is given in Graf, Mauldin, and Williams (1986). We briefly describe its construction. The value $f(\frac{1}{2})$ is uniformly distributed on $[0,1]$ with respect to P, and a P-random function can be constructed on the dyadic rationals inductively. First put $f(0) = 0$ and $f(1) = 1$. Knowing $f(i/2^k)$ for $i = 0,\dots,2^k$, choose $f((2i+1)/2^{k+1})$ for $i = 0,\dots,2^k - 1$ independently according to the uniform distribution on $[f(i/2^k), f((i+1)/2^k)]$. With probability one, these discrete data belong to a strictly increasing function $f \in F$. Furthermore, F is the support and $m(t) = t$ is the mean of P.

Let us compare the Ulam measure P on F and the Brownian bridge Q on the class of all continuous functions f with $f(0) = 0$ and $f(1) = 1$. In the latter case Q is the Gaussian measure on $C([0,1])$ with mean $m(t) = t$ and covariance kernel $K(s,t) = \min(s,t) - s \cdot t$, see Example 3. Both measures are symmetric with respect to their common mean m, and the random functions are rather irregular in both cases. The functions f are nowhere differentiable with Q-probability one, and strictly singular with P-probability one. The Brownian bridge Q has the quasi-Markov property, i.e., the restrictions of f to $[0,s] \cup [t,1]$ and to $[s,t]$ are independent given the function values $f(s)$ and $f(t)$ at $s < t$. Furthermore, the conditional measure $Q(\cdot \mid f(s) = a, f(t) = b)$ is the image of Q under the scaling operation $f \mapsto (b-a)\cdot f((\cdot-s)/(t-s))+a$ with probability one. For the Ulam measure P we only have these properties for all dyadic rationals $s = i/2^k$ and $t = (i+1)/2^k$.

We first study nonadaptive methods

$$S_n^{\text{non}} = \phi_n \circ N_n^{\text{non}},$$

where

$$N_n^{\text{non}}(f) = (f(x_1),\dots,f(x_n))$$

with fixed knots x_i and $\phi_n : \mathbb{R}^n \to \mathbb{R}$ is measurable. In particular, we consider the trapezoidal rule

$$(20) \qquad S_n^{\text{trp}}(f) = \frac{1}{2n+2} + \frac{1}{n+1} \cdot \sum_{i=1}^{n} f\left(\frac{i}{n+1}\right)$$

Formally, by multiplying its first term by $f(1)$, the trapezoidal rule can be regarded as a linear method using $n+1$ knots. According to the construction of the Ulam measure, the number $n = 2^k - 1$ of knots in the trapezoidal rule is most suitable for analysis.

PROPOSITION 21. *Let P denote the Ulam measure. The average error of the trapezoidal rule S_n^{trp} with $n = 2^k - 1$ knots is given by*

$$e_2(S_n^{\text{trp}}, \text{Int}, P) = (40 \cdot 6^k)^{-1/2}.$$

Furthermore,

$$e_2^{\mathrm{non}}(n, \Lambda^{\mathrm{std}}, \mathrm{Int}, P) \asymp n^{-\ln 6/\ln 4},$$

so that trapezoidal rules are order optimal in the class of nonadaptive methods.

PROOF. We only sketch the proof, see Graf and Novak (1990) for details. The constant algorithm $S_0(f) = \frac{1}{2}$ yields

$$e_2(S_0, \mathrm{Int}, P)^2 = \int_F \mathrm{Int}(f)^2 \, dP(f) - \mathrm{Int}(m)^2 = \tfrac{11}{40} - \tfrac{1}{4} = \tfrac{1}{40}.$$

Using the conditional independence and scaling properties of P one can show

$$e_2(S_{2n+1}^{\mathrm{trp}}, \mathrm{Int}, P)^2 = \tfrac{1}{6} \cdot e_2(S_n^{\mathrm{trp}}, \mathrm{Int}, P)^2$$

for the trapezoidal rules with $n = 2^k - 1$ and $2n + 1 = 2^{k+1} - 1$ knots, respectively. Inductively we get the statement on $e_2(S_n^{\mathrm{trp}}, \mathrm{Int}, P)$.

The matching lower bound is obtained as follows. Let S_n^{non} use increasingly ordered knots x_i and assume $x_\ell \leq \frac{1}{2}$ and $x_{\ell+1} > \frac{1}{2}$. The conditional independence and scaling properties of P yield

$$e_2(S_n^{\mathrm{non}}, \mathrm{Int}, P)^2$$

$$\geq \tfrac{1}{12} \cdot \left(\int_F (\mathrm{Int}(f) - \phi_\ell(f(2x_1), \ldots, f(2x_\ell)))^2 \, dP(f) \right.$$

$$\left. + \int_F (\mathrm{Int}(f) - \phi_{n-\ell}(f(2x_{\ell+1} - 1), \ldots, f(2x_n - 1)))^2 \, dP(f) \right),$$

where ϕ_ℓ and $\phi_{n-\ell}$ are the conditional expectations of the integral, given the respective data. As a consequence, the minimal errors

$$e(n) = e_2^{\mathrm{non}}(n, \Lambda^{\mathrm{std}}, \mathrm{Int}, P)$$

satisfy

$$e(n)^2 \geq \tfrac{1}{12} \cdot \inf_{\ell = 0, \ldots, n} (e(\ell)^2 + e(n - \ell)^2).$$

This implies

$$e(n)^2 \geq c \cdot n^{-\ln 6/\ln 2}$$

with a constant $c > 0$. $\qquad\square$

Now we turn to adaptive methods with fixed cardinality. The following method S_n^{ad} is motivated by worst case considerations. In the first step we take $x_1 = \frac{1}{2}$. Let $f(x_1) = y_1, \ldots, f(x_{i-1}) = y_{i-1}$ denote the function values that are known at pairwise different knots $x_j \in]0, 1[$ after $i - 1$ steps. Let

$$0 = z_0 < z_1 < \cdots < z_{i-1} < z_i = 1,$$

with

$$\{z_1, \ldots, z_{i-1}\} = \{x_1, \ldots, x_{i-1}\}$$

and put $d(s,t) = (t - s) \cdot (f(t) - f(s))$. Suppose that $z \in]z_{\ell-1}, z_\ell[$ is chosen in step i. The trapezoidal rule, applied to the data after i steps, is worst case optimal and yields an error bound

$$e(z, f(z)) = \tfrac{1}{2} \cdot \Big(\sum_{j \neq \ell} d(z_{j-1}, z_j) + d(z_{\ell-1}, z) + d(z, z_\ell) \Big).$$

A general strategy, which is known as optimal-in-one-step method or greedy method, is to choose z (worst case) optimally under the assumption that the next evaluation is the final evaluation of f. For our particular problem we get

$$\sup\{e(z, f(z)) : f \in F, \ f(x_j) = y_j \text{ for } j = 1, \dots, i-1\}$$
$$= \tfrac{1}{2} \cdot \Big(\sum_{j \neq \ell} d(z_{j-1}, z_j) + \max(z - z_{\ell-1}, z_\ell - z) \cdot (f(z_\ell) - f(z_{\ell-1})) \Big).$$

Hence, in the greedy algorithm, we take z as the midpoint of a subinterval $[z_{j-1}, z_j]$ where $(z_j - z_{j-1}) \cdot (f(z_j) - f(z_{j-1}))$ is maximal.

PROPOSITION 22. *Let P denote the Ulam measure. The average error of the greedy method S_n^{ad} satisfies*

$$e_2(S_n^{\mathrm{ad}}, \mathrm{Int}, P) \leq 5 \cdot n^{-3/2}.$$

See Novak (1992) for the proof. Combining Propositions 21 and 22 and observing $\ln 6 / \ln 4 = 1.29\ldots$, we conclude that adaption helps significantly for integration of monotone functions in an average case setting. Lower bounds for $e_2^{\mathrm{ad}}(n, \Lambda^{\mathrm{std}}, \mathrm{Int}, P)$ or results on varying cardinality are not known.

Let us compare these average case results with worst case results, which are proven in Kiefer (1957).

PROPOSITION 23. *Let S_n^{trp} denote the trapezoidal rule, defined by (20). Then*

$$e_{\max}(S_n^{\mathrm{trp}}, \mathrm{Int}, F) = \inf_{S_n^{\mathrm{ad}}} e_{\max}(S_n^{\mathrm{ad}}, \mathrm{Int}, F) = (2n + 2)^{-1}.$$

Thus the trapezoidal rule is worst case optimal and adaption does not help in the worst case setting. The superiority of adaptive methods can only be proven in the average case setting.

4.1. Notes and References. Propositions 21 and 22 are due to Graf and Novak (1990) and Novak (1992), respectively. Proposition 23 is due to Kiefer (1957).

5. Computational Complexity of Continuous Problems

Consider a linear method S_n^{lin}. After n evaluations of functionals $\lambda_i \in \Lambda$, at most n scalar multiplications and $n - 1$ additions are needed to compute the approximation to $S(f)$ for any $f \in F$. Therefore the number n is a reasonable measure of the computational cost of every linear method S_n^{lin}. This is no longer true if we consider nonlinear algorithms, adaptive information operators, and varying cardinality. The numbers n or $\nu(f)$, respectively, do not capture the computational cost for selecting the functionals via λ_i, checking the termination criteria χ_i, and applying ϕ_n or $\phi_{\nu(f)}$, respectively.

Of course a proper definition of computation and computational cost is needed at this point. We do not give a rigorous definition here; instead we refer to the literature that is quoted in Notes and References 5.1.

The *real number model* of computation is reasonable for problems from continuous mathematics, which are studied in these notes. This model permits the arithmetic operations, comparisons, and copy instructions. Its basic entities are real numbers, and the instructions are carried out exactly at unit cost. Furthermore, if $G \neq \mathbb{R}$, one either permits computations also with the elements of G or these elements are described in terms of real coefficients with respect to a given family in G. A similar extension is needed to deal with the dual space X^* of X, unless X is a function space, say $X = C^r(D)$, and Λ consists of point evaluations.

An *oracle* that yields values $\lambda(f)$ for $f \in X$ and $\lambda \in \Lambda \subset X^*$ must be incorporated in the model. Each call of this oracle costs $c > 0$. In many cases an evaluation of $\lambda(f)$ is much more expensive than arithmetic operations etc., so that $c \gg 1$.

The total *computational cost* for obtaining $S_\nu^{\mathrm{var}}(f)$, say, consists of the number of arithmetic operations etc. plus c times the number of oracle calls for the particular $f \in X$. We consider the average cost of S_ν^{var} with respect to P, which clearly satisfies

$$(21) \qquad c \cdot \mathrm{card}(S_\nu^{\mathrm{var}}, P) \leq \mathrm{cost}(S_\nu^{\mathrm{var}}, P).$$

Recall that the notion of minimal errors is based on $\mathrm{card}(S_\nu^{\mathrm{var}}, P)$. Therefore lower bounds for the minimal errors also hold, up to the constant c, if we study the average cost instead of the average cardinality. In fact, we strengthen the lower bounds by ignoring all cost that is caused by arithmetic and similar operations.

On the other hand, care is needed with respect to the upper bounds, since the gap in (21) is huge for some methods. Fortunately, this gap is small for the methods that yield the upper bounds in these notes. For instance, we have

$$\mathrm{cost}(S_n^{\mathrm{lin}}, P) \leq (2 + c) \cdot \mathrm{card}(S_n^{\mathrm{lin}}, P) - 1$$

for linear methods S_n^{lin} with $\mathrm{card}(S_n^{\mathrm{lin}}, P) = n$. Similar estimates, which say that the average cost and the average cardinality are of the same order, hold for the methods studied in Sections 3, 4, and Chapter VIII. The only exception is the P-algorithm for global optimization, whose cost depends quadratically on the cardinality with respect to the Wiener measure.

Instead of fixing a cost bound and asking for a small error one can also fix an accuracy $\varepsilon > 0$ and ask for the minimal average cost that is needed to obtain an average error of at most ε. Any method that uses functionals from Λ is permitted. This leads to the definition

$$\mathrm{comp}_q(\varepsilon, \Lambda, S, P) = \inf\{\mathrm{cost}(S_\nu^{\mathrm{var}}, P) : e_q(S_\nu^{\mathrm{var}}, S, P) \leq \varepsilon\}$$

of the ε-complexity of the given problem. Clearly

$$\mathrm{comp}_q(\varepsilon, \Lambda, S, P) \geq c \cdot \inf\{\mathrm{card}(S_\nu^{\mathrm{var}}, P) : e_q(S_\nu^{\mathrm{var}}, S, P) \leq \varepsilon\},$$

but, as explained above, this lower bound is sharp for the problems that are studied in these notes.

5.1. Notes and References. 1. The complexity theory of continuous problems, whose basic terms are presented in this section, is called *information-based complexity*. The key features are the use of the real number model of computation and the fact that only partial and priced information is available. Several settings, including the average case setting, are studied. We refer to the monographs by Traub and Woźniakowski (1980) and Traub *et al.* (1988), as well as to the recent introductory book by Traub and Werschulz (1998). See Blum, Shub, and Smale (1989) and Blum, Cucker, Shub, and Smale (1998) for a complexity theory over the real numbers for problems with complete information; here, a principal example is polynomial zero finding.

2. See Novak (1995c) for the precise definition and analysis of the model of computation that is sketched in this section.

CHAPTER VIII

Nonlinear Problems

This chapter provides an outlook on average case results for nonlinear problems of numerical analysis. So far the number of average case results for nonlinear problems is small, compared to what is known in the linear case. In particular, lower bounds seem to be known only for zero finding and global optimization in the univariate case. We consider zero finding in Section 1 and global optimization in Section 2.

Worst case and average case results for nonlinear problems sometimes differ significantly. Moreover, the worst case approach also leads to new numerical methods. Adaptive termination criteria, which yield methods with varying cardinality, cannot be adequately analyzed in the worst case setting. On the other hand, varying cardinality turns out to be very powerful for zero finding in the average case setting. For global optimization adaption does not help much in the worst case on convex and symmetric classes. In an average case setting, however, adaption is very powerful for global optimization.

Upper bounds for nonlinear problems are known for several other problems in the average case setting.

Heinrich (1990b, 1991) studies Fredholm integral equations of the second kind with variable kernels and right-hand sides. Clearly the solution depends nonlinearly on the right-hand side. Classes of kernels and right-hand sides of different smoothness are equipped with Gaussian measures, and the Galerkin and iterated Galerkin method are analyzed.

Average case results are also known for linear programming and polynomial zero finding. These problems are usually analyzed for fixed dimension and degree, so that the problem elements belong to finite dimensional spaces. This fact has two important consequences. First, the Lebesgue measure can be used to define the average case setting. Second, a finite number of linear functional evaluations is sufficient to specify a problem instance completely. Hence this number of evaluations does not lead to interesting results. Instead, the number of arithmetic operations and branches is crucial in the definition of the cost, cf. Section VII.5. In a series of papers, Borgwardt constructs a variant of the simplex algorithm and proves a polynomial bound for the average number of steps. In contrast, an exponential number of steps is necessary in the worst case. See Borgwardt (1981, 1987, 1994). Average case results are also obtained for other linear programming methods and with respect to different measures, see, e.g., Smale (1983) and Todd (1991). Polynomial zero finding is studied in Smale (1981, 1985, 1987). Systems of polynomials are studied in Renegar (1987) and in a series of papers by Shub and Smale, see Shub and Smale (1994). In the latter papers, a path-following method, which is based on projective Newton steps, is constructed and a polynomial bound for the average number of steps is established. See Blum, Cucker, Shub, and Smale (1998) for an overview and new results on polynomial zero finding.

1. Zero Finding

Let

$$F_r = \{f \in C^r([0,1]) : f(0) < 0 < f(1)\},$$

where $r \in \mathbb{N}_0$ or $r = \infty$, and consider a class

$$F \subset F_r$$

of univariate and at least continuous functions with a change of sign. We wish to approximate a zero $x^* = x^*(f)$ of $f \in F$. There is no linear dependence of $x^*(f)$ on f, and several zeros of f may exist.

Methods for zero finding usually evaluate f or one of its derivatives adaptively, and they often apply adaptive termination criteria, so that the total number $\nu(f)$ of evaluations depends on f. The output is a point from the domain of f, which is the unit interval in our case. We formally define the class of adaptive methods S_ν^{var} with varying cardinality as in Section VII.1, with $G = [0,1]$ and Λ consisting of all functionals

$$\lambda(f) = f^{(k)}(x)$$

for $x \in [0,1]$ and $k \in \mathbb{N}_0$ with $k \leq r$. In particular, we get a method S_n^{ad} with fixed cardinality n if $\nu(f) = n$ for every function $f \in F$.

The error of the method S_ν^{var} for $f \in F$ is defined either in the root sense by

$$\Delta^{\mathrm{ro}}(S_\nu^{\mathrm{var}}, f) = \inf\{|x^* - S_\nu^{\mathrm{var}}(f)| : f(x^*) = 0\}$$

or in the residual sense by

$$\Delta^{\mathrm{re}}(S_\nu^{\mathrm{var}}, f) = |f(S_\nu^{\mathrm{var}}(f))|.$$

Assume that P is a measure on F. For simplicity we only study *average errors* that are defined by first moments,

$$e_1^{\mathrm{ro}}(S_\nu^{\mathrm{var}}, \mathrm{Zero}, P) = \int_F \Delta^{\mathrm{ro}}(S_\nu^{\mathrm{var}}, f) \, dP(f)$$

or

$$e_1^{\mathrm{re}}(S_\nu^{\mathrm{var}}, \mathrm{Zero}, P) = \int_F \Delta^{\mathrm{re}}(S_\nu^{\mathrm{var}}, f) \, dP(f).$$

We write e_1 in statements that hold for e_1^{ro} as well as for e_1^{re}. The definition of e_1 corresponds to q-average errors with $q = 1$ for linear problems.

First we discuss methods with fixed cardinality n, and we begin with a worst case result. The *maximal error* of S_n^{ad} on F is defined by

$$e_{max}^{ro}(S_n^{ad}, \text{Zero}, F) = \sup_{f \in F} \Delta^{ro}(S_n^{ad}, f)$$

or

$$e_{max}^{re}(S_n^{ad}, \text{Zero}, F) = \sup_{f \in F} \Delta^{re}(S_n^{ad}, f),$$

and we write e_{max} in statements that hold for e_{max}^{ro} as well as for e_{max}^{re}.

Clearly, the bisection method S_n^{bis} yields an error

$$\Delta^{ro}(S_n^{bis}, f) \le \frac{1}{2^{n+1}}$$

after n steps for every $f \in F_0$. This bound is not only sharp for bisection on F_0, but best possible in the worst case setting for any method and even for classes of smooth functions that have only simple zeros. Let

$$e_{max}^{ro,ad}(n, \Lambda, \text{Zero}, F) = \inf_{S_n^{ad}} e_{max}^{ro}(S_n^{ad}, \text{Zero}, F)$$

denote the minimal maximal root error on F in the class of all adaptive methods using n function or derivative values.

PROPOSITION 1. *If*

$$\{f \in F_\infty : f' \ge 1\} \subset F \subset F_0,$$

then

$$e_{max}^{ro,ad}(n, \Lambda, \text{Zero}, F) = \frac{1}{2^{n+1}}.$$

See Kung (1976) for the proof. Worst case optimality of bisection may be surprising, since iterative methods are known that are globally and superlinearly convergent on

$$F \subset \{f \in F_2 : f'(x^*) \ne 0 \text{ if } f(x^*) = 0\},$$

i.e., on classes of smooth functions having only simple zeros. Hence, after several steps, the number of correct digits of the approximation to $x^*(f)$ is roughly improved by a factor $c > 1$ with each additional step. However, the number of steps that is required before this fast convergence occurs depends on f. We see that asymptotic results for iterative methods do not imply error bounds that hold after a finite number of steps. Due to the significant difference between worst case results and asymptotic results, an average case analysis of zero finding is of interest.

Average case results are known for the classes

$$(1) \qquad F = \{f \in F_r : f^{(k)}(0) = a_k,\ f^{(k)}(1) = b_k \text{ for } k = 0, \dots, r\}$$

with $a_0 < 0 < b_0$, equipped with a *conditional r-fold integrated Wiener measure* P, as well as for

$$F = \{f \in F_0 : f(0) = -1, \ f(1) = 1, \ f \text{ increasing}\},$$

equipped with the Ulam measure. We have used the Ulam measure for the analysis of integration of monotone functions in Section VII.4.

The r-fold integrated Wiener measure w_r is constructed in Example IV.3 and used for the analysis of integration, approximation, and differentiation of smooth functions in Chapters IV and V. To match the boundary conditions in (1), a translation by polynomials is used. More precisely, let $g \in C^r([0,1])$ and let $p(g)$ be the polynomial of degree at most $2r + 1$ such that $g - p(g) \in F$. Then the measure P on F is the image of w_r under the translation $g \mapsto g - p(g)$. For $r \geq 1$ almost every function has only simple zeros and the number of zeros is finite, see Novak, Ritter, and Woźniakowski (1995). We add that P is called a conditional measure because of the following fact. Let $a_k = b_k = 0$ for $k = 0, \ldots, r$ and put

$$N^{\mathrm{non}}_{n \cdot (r+1)}(f) = (f(x_1), \ldots, f^{(r)}(x_1), \ldots, f(x_n), \ldots, f^{(r)}(x_n)).$$

Then $w_r(\cdot \mid N^{\mathrm{non}}_{n \cdot (r+1)} = y)$ is given by shifted and scaled independent copies of P. See Example VII.3 for the case $r = 0$, where P corresponds to a Brownian bridge.

The worst case bound $1/2^{n+1}$ from Proposition 1 can be improved even for functions of low regularity in the average case setting. See Notes and References 1.1.2 for references concerning the proof of the following result.

PROPOSITION 2. *Let P be a conditional r-fold integrated Wiener measure or the Ulam measure. There exists a constant $\gamma < \frac{1}{2}$ and a sequence of methods S_n^{ad} that only use function values with*

$$e_1^{\mathrm{ro}}(S_n^{\mathrm{ad}}, \mathrm{Zero}, P) \leq \gamma^n.$$

This upper bound only guarantees linear convergence of the minimal average case errors, and in fact one cannot achieve superlinear convergence with methods of fixed cardinality. In the root and residual sense we have the following lower bound on the nth minimal average errors

$$e_1^{\mathrm{ad}}(n, \Lambda, \mathrm{Zero}, P) = \inf_{S_n^{\mathrm{ad}}} e_1(S_n^{\mathrm{ad}}, \mathrm{Zero}, P)$$

for adaptive methods that use function or derivative values.

PROPOSITION 3. *Let P be as in Proposition 2. There exists a constant $\theta > 0$ such that*

$$e_1^{\mathrm{ad}}(n, \Lambda, \mathrm{Zero}, P) \geq \theta^n.$$

We describe the basic idea of the proof, see Ritter (1994) for details. For an arbitrary adaptive method S_n^{ad}, a set $U_n \subset F$ of unfavorable functions is defined. The functions $f \in U_n$ have the following properties. There are points $s, t \in [0,1]$ with $t - s = 1/4^n$ such that f is negative on $[0, s]$ and positive on $[t, 1]$. Moreover the absolute values of f are sufficiently large on $[0, s] \cup [t, 1]$ and the boundary values $(f(s), \ldots, f^{(r)}(s))$ and $(f(t), \ldots, f^{(r)}(t))$ are in certain prescribed subsets of $\mathbb{R}^{r+1}$. Finally all knots that are used by S_n^{ad} for the function f are outside the interval $[s, t]$. One can show that the conditional expectation of the error given U_n, as well as the probability of U_n, have lower bounds of the form θ^n.

The lower bound from Proposition 3 implies the following. The number of bisection steps that is necessary to guarantee a root error ε for all functions $f \in F$ differs from the number of function or derivative evaluations that is necessary to obtain an average error ε at most by a multiplicative factor. The number of steps is proportional to $\ln(1/\varepsilon)$, and bisection is almost optimal also with respect to the average error.

So far we have considered zero finding methods that use the same number of function or derivative evaluations for all functions $f \in F$. In practice, however, the number of evaluations is often determined by an adaptive termination criterion, such as $|f(x_i)| \leq \varepsilon$, and may therefore vary over the class F. A worst case analysis cannot justify the use of methods with varying cardinality for a numerical problem. In an average case approach, however, we can study the average cardinality

$$\text{card}(S_\nu^{\text{var}}, P) = \int_F \nu(f) \, dP(f).$$

Varying cardinality does not help for many linear problems with respect to Gaussian measures P, see Section VII.2. For zero finding, we will see that varying cardinality is very powerful.

Novak et al. (1995) construct a zero finding method that yields a worst case error ε on the class F_0 but uses only about $\ln \ln(1/\varepsilon)$ function evaluations on the average for smooth functions. The error ε is guaranteed in the residual sense and, in particular, in the root sense, so that an enclosing method has to be used. The method chooses knots x_i, either by secant or by bisection steps. From the corresponding data $f(x_i)$, a sequence of subintervals $[s_i, t_i] \subset [0, 1]$ is constructed that contain a zero of f.

The method depends on the error criterion and on the required accuracy $\varepsilon > 0$ via its adaptive termination criterion and the formula for the output. These components are very simple and given by the test $t_i - s_i \leq 2\varepsilon$ and the output $\frac{1}{2}(s_i + t_i)$, if we consider the error in the root sense, and by $|f(x_i)| \leq \varepsilon$ and x_i, if we consider the error in the residual sense.

A secant step is denoted by

$$\text{SEC}(x, y) = \begin{cases} x - \dfrac{x - y}{f(x) - f(y)} \cdot f(x) & \text{if } f(x) \neq f(y), \\ \text{undefined} & \text{otherwise} \end{cases}$$

with $x, y \in [0, 1]$, and for a bisection step we compute

$$x_i = \tfrac{1}{2}(s_{i-1} + t_{i-1}).$$

After each step, a new subinterval $[s_i, t_i]$ is constructed according to

$$[s_i, t_i] = \begin{cases} [s_{i-1}, x_i] & \text{if } f(x_i) > 0, \\ [x_i, t_{i-1}] & \text{if } f(x_i) < 0, \\ [x_i, x_i] & \text{otherwise.} \end{cases}$$

The *hybrid secant-bisection method* is defined as follows.

> $i := 0$;
> $[s_0, t_0] := [0, 1]$; compute $f(0)$, $f(1)$;
> do
>   for $j := 1, 2$ do   (regula falsi)
> $i := i + 1$;
> $x_i := \text{SEC}(s_{i-1}, t_{i-1})$; compute $f(x_i), [s_i, t_i]$;
>   od;
>   repeat   (secant)
> $q := \text{SEC}(x_i, x_{i-1})$;
>   if $q \notin\,]s_i, t_i[$ then break fi;
> $i := i + 1$;
> $x_i := q$; compute $f(x_i), [s_i, t_i]$;
>   until $t_i - s_i > (t_{i-3} - s_{i-3})/2$;
> $i := i + 1$;   (bisection)
> $x_i := (s_i + t_i)/2$; compute $f(x_i), [s_i, t_i]$;
> od;

Additionally, each evaluation of f is followed by a check of the termination criterion. Note that the computational cost per step is bounded by a small constant.

We mention some properties of the hybrid method that are also important for the proof of average case results. The method uses steps of the regula falsi (R), the secant method (S), and the bisection method (B). A typical iteration pattern is

$$\text{R R S} \ldots \text{S B R R S} \ldots \text{S B R R S S S S S S} \ldots,$$

and we always have a length reduction of the subintervals $[s_i, t_i]$ by a factor of at least 2 after at most 4 steps. The following result shows that, on the average, the reduction is much faster on classes $F \subset F_2$.

PROPOSITION 4. *Let P be a conditional r-fold integrated Wiener measure with $r \geq 2$, and let S_ν^{var} denote the hybrid method with required accuracy $0 < \varepsilon < \frac{1}{2}$. Then*

$$e_{\max}(S_\nu^{\text{var}}, \text{Zero}, F_0) \leq \varepsilon$$

for the maximal error and

$$\text{card}(S_\nu^{\text{var}}, P) \leq \frac{1}{\ln \beta} \cdot \ln \ln(1/\varepsilon) + c$$

for the average number of steps. Here c depends on P and

$$\beta = \frac{1 + \sqrt{5}}{2}.$$

We sketch the basic idea of the proof and refer to Novak *et al.* (1995) for details. For $\gamma_i > 0$ we define a subset G of F_r by

$$G = \{g \in F_r : |g''(x) - g''(y)| \leq \gamma_1 |x - y|^{1/3},$$
$$|g''(x^*)| \geq \gamma_2 \text{ and } |g'(x^*)| \geq \gamma_3 \text{ if } g(x^*) = 0\}.$$

Let $g \in G$. After $k = k(G)$ steps we can guarantee the following: If there is a further bisection step then this is the last one and we obtain the pattern B R R S S S S ... It is possible to give explicit estimates of $k(G)$. It is also possible to estimate the probability $P(G)$ and so, finally, the result can be proven.

One should note that such an average case analysis includes a worst case analysis for many different subsets G of F_r. It is crucial, of course, that a membership $f \in G$ can not be checked by any algorithm that is based on a finite number of function or derivative evaluations. Hence the constants γ_i cannot be used in a program.

According to the following lower bound, the hybrid method is almost optimal, even if we do not require the error bound ε for all $f \in F$ but only on the average. See Novak *et al.* (1995) for the proof.

PROPOSITION 5. *Let P be a conditional r-fold integrated Wiener measure with $r \geq 2$, and let*

$$\alpha > r + \tfrac{1}{2}.$$

There exists a constant c, which depends only on P and α, with the following property. For every $0 < \varepsilon < \frac{1}{2}$ and every zero finding method S_ν^{var} with average error

$$e_1(S_\nu^{\text{var}}, \text{Zero}, P) \leq \varepsilon$$

the average cardinality satisfies

$$\text{card}(S_\nu^{\text{var}}, P) \geq \frac{1}{\ln \alpha} \cdot \ln \ln(1/\varepsilon) + c.$$

The value of β in Proposition 4 is due to the asymptotic properties of the secant method. We conjecture, however, that the lower bound $r + \frac{1}{2}$ from Proposition 5 is optimal.

Propositions 4 and 5 determine the ε-complexity

$$\text{comp}_1(\varepsilon, \Lambda, \text{Zero}, P) = \inf\{\text{cost}(S_\nu^{\text{var}}, P) : e_1(S_\nu^{\text{var}}, \text{Zero}, P) \leq \varepsilon\}$$

of zero finding with respect to conditional r-fold integrated Wiener measures P if $r \geq 2$. We refer to Section VII.5 for a discussion of the average cost of S_ν^{var} and its relation to the average cardinality. Clearly $\text{cost}(S_\nu^{\text{var}}, P)$ and $\text{card}(S_\nu^{\text{var}}, P)$

are of the same order for the hybrid secant-bisection method S_ν^{var}, and therefore we conclude that

$$\text{comp}_1(\varepsilon, \Lambda, \text{Zero}, P) \asymp \ln\ln(1/\varepsilon)$$

as ε tends to zero. Proposition 1 implies that the corresponding worst case quantity is only of order $\ln(1/\varepsilon)$. According to the previous results, this huge difference is caused by switching from worst case cardinality or cost to average cardinality or cost, while the corresponding change for the errors does not effect the orders.

1.1. Notes and References. 1. Proposition 1 is due to Kung (1976) and extended by Sikorski (1982, 1989) to smaller classes F and to methods that may use arbitrary linear functionals λ_i.

2. Proposition 2 is due to Graf, Novak, and Papageorgiou (1989) in case of the Ulam measure and due to Novak (1989) in case of the Brownian bridge. For the conditional r-fold integrated Wiener measure with $r \geq 1$ a proof could be given similarly.

3. Proposition 3 is due to Novak and Ritter (1992) for the case of a Brownian bridge and due to Ritter (1994) in the other cases.

4. Propositions 4 and 5 are due to Novak *et al.* (1995). Novak (1996b) presents numerical tests for the hybrid secant-bisection method and other zero finding methods.

5. See Novak and Ritter (1993) for a survey of asymptotic, worst case, and average case results on the complexity of zero finding.

2. Global Optimization

Another important nonlinear problem is global optimization. We discuss this problem for classes F of at least continuous functions on compact domains D, and we assume that function values may be used to approximately compute a global maximizer of $f \in F$. The formal definition of the respective classes of methods is taken from Section VII.1, with $G = D$ and $\Lambda = \Lambda^{\text{std}}$ consisting of the functionals

$$\lambda(f) = f(x)$$

for $x \in D$. We consider nonadaptive methods S_n^{non} and, more generally, adaptive methods S_n^{ad} with fixed cardinality and adaptive methods S_ν^{var} with varying cardinality. The error of S_ν^{var} for a function $f \in F$ is defined in the residual sense by

$$\Delta(S_\nu^{\text{var}}, f) = \max_{t \in D} f(t) - f(S_\nu^{\text{var}}(f)).$$

As for zero finding, we start with a negative worst case result. To this end let

$$e_{\max}(S_n^{\text{ad}}, \text{Opt}, F) = \sup_{f \in F} \Delta(S_n^{\text{ad}}, f)$$

denote the *maximal error* of S_n^{ad} on F. The *minimal maximal errors* in the classes of all nonadaptive and adaptive methods, respectively, that use n function values are given by

$$e_{\max}^{\mathrm{ad}}(n, \Lambda^{\mathrm{std}}, \mathrm{Opt}, F) = \inf_{S_n^{\mathrm{ad}}} e_{\max}(S_n^{\mathrm{ad}}, \mathrm{Opt}, F)$$

and

$$e_{\max}^{\mathrm{non}}(n, \Lambda^{\mathrm{std}}, \mathrm{Opt}, F) = \inf_{S_n^{\mathrm{non}}} e_{\max}(S_n^{\mathrm{non}}, \mathrm{Opt}, F).$$

It turns out that these quantities coincide, up to one function evaluation, under rather general assumptions on F, e.g., if F is a unit ball in some function space.

PROPOSITION 6. *Let F be a convex and symmetric class. Then*

$$e_{\max}^{\mathrm{ad}}(n, \Lambda^{\mathrm{std}}, \mathrm{Opt}, F) \geq e_{\max}^{\mathrm{non}}(n+1, \Lambda^{\mathrm{std}}, \mathrm{Opt}, F).$$

Let us mention the major steps in the proof. Wasilkowski (1984) and Novak (1988) show that

$$e_{\max}^{\mathrm{ad}}(n, \Lambda^{\mathrm{std}}, \mathrm{Opt}, F) \geq e_{\max}^{\mathrm{ad}}(n+1, \Lambda^{\mathrm{std}}, \mathrm{App}_\infty, F)$$

for convex and symmetric classes F, i.e., global optimization is almost as hard as L_∞-approximation on the whole domain. The latter problem is a linear one and, according to a general result, adaption does not help up to a factor of at most two on convex and symmetric classes. For L_∞-approximation one may drop this factor, see Notes and References VII.2.6.9. Thus

$$e_{\max}^{\mathrm{ad}}(n+1, \Lambda^{\mathrm{std}}, \mathrm{App}_\infty, F) = e_{\max}^{\mathrm{non}}(n+1, \Lambda^{\mathrm{std}}, \mathrm{App}_\infty, F).$$

It remains to observe that global optimization is not harder than L_∞-approximation, as far as minimal errors are concerned.

In computational practice we usually apply adaptive optimization methods. For instance, in the case of Lipschitz optimization, we hope that subregions may be excluded during the computation. This cannot be justified, however, by a worst case analysis, as shown in Proposition 6.

We now turn to the average case setting, where we study the *average error*

$$e_1(S_\nu^{\mathrm{var}}, \mathrm{Opt}, P) = \int_F \Delta(S_\nu^{\mathrm{var}}, f) \, dP(f)$$

of methods S_ν^{var} with respect to a measure P on F. Obviously, the definition of e_1 corresponds to q-average errors with $q = 1$ for linear problems. So far, error bounds have only been obtained in the univariate case with respect to the Wiener measure on the class of continuous functions f on $[0,1]$ with $f(0) = 0$. On the other hand, the average case setting is used to derive new global optimization algorithms in the univariate and multivariate case, see Notes and References 2.1.3.

The order of the *nth minimal errors*

$$e_1^{\mathrm{non}}(n, \Lambda^{\mathrm{std}}, \mathrm{Opt}, P) = \inf_{S_n^{\mathrm{non}}} e_1(S_n^{\mathrm{non}}, \mathrm{Opt}, P)$$

of nonadaptive methods is known when P is the Wiener measure.

PROPOSITION 7. *Let w denote the Wiener measure. Then*

$$e_1^{\mathrm{non}}(n, \Lambda^{\mathrm{std}}, \mathrm{Opt}, w) \asymp n^{-1/2}.$$

See Ritter (1990) for the proof. We add that the optimal order is obtained by function evaluation at equidistant knots $x_{i,n} = i/n$. Moreover, the best algorithm for any given adaptive or nonadaptive information operator simply chooses a point where the largest function value has been observed.

REMARK 8. We briefly discuss results on asymptotic constants for the errors of nonadaptive methods. Recall that a sequence of knots $x_{1,n}, \ldots, x_{n,n}$ is a regular sequence if these knots form quantiles with respect to a given density ψ, i.e.,

$$\int_0^{x_{i,n}} \psi(t)\,dt = \frac{i-1}{n-1} \cdot \int_0^1 \psi(t)\,dt,$$

see Section IV.2.1. Let

$$S_n^{\psi}(f) = \min\{x_{i,n} : f(x_{i,n}) = \max_{j=1,\ldots,n} f(x_{j,n})\},$$

where we take the minimum merely to obtain uniqueness of the output for every function f.

With probability one the trajectories of the Brownian motion have a single global maximum, the location of which is distributed according to the arcsine law. Let

$$\xi(t) = \frac{1}{\pi\sqrt{t \cdot (1-t)}}, \qquad t \in [0,1],$$

denote the density of the arcsine distribution. The corresponding regular sequence therefore seems to be a natural choice. Calvin (1996) proves that

$$e_1(S_n^{\psi}, \mathrm{Opt}, w) \approx c \cdot \int_0^1 \xi(t) \cdot \psi(t)^{-1/2}\,dt \cdot \left(\int_0^1 \psi(t)\,dt\right)^{1/2} \cdot n^{-1/2}$$

for some positive constant c. The numerical value of the asymptotic constant in the case $\psi = \xi$ is $c \cdot 0.956$, so that the arcsine distribution is slightly better than the uniform distribution, viz. equidistant knots. However, the best regular sequence is given by $\psi = \xi^{2/3}$, which corresponds to the $\beta(2/3, 2/3)$-distribution and yields the asymptotic constant $c \cdot 0.937$.

The following results suggest that the order $n^{-1/2}$ of average errors can be improved significantly by adaption and, in particular, by varying cardinality. The results deal with the distribution of suitably normalized errors, where the normalizing sequence tends to infinity much faster that $n^{1/2}$.

For every $0 < \delta < 1$, Calvin (1997) constructs adaptive Monte Carlo methods S_n^{ad} with the following properties. The normalized error $n^{1-\delta} \cdot \Delta(S_n^{\mathrm{ad}}, f)$ convergences in distribution under the Wiener measure, and the computational cost of S_n^{ad} is proportional to the cardinality n. It would be interesting to know whether the order of the average errors $e_1(S_n^{\mathrm{ad}}, \mathrm{Opt}, w)$ is also close to n^{-1}.

We now discuss an adaptive method with varying cardinality, which is likely to be far superior to the previously discussed methods with fixed cardinality. First we describe the method for arbitrary D and P, and then consider $D = [0, 1]$ and the Wiener measure as a particular case.

For every $n \in \mathbb{N}$ let $Z_n : \mathbb{R}^n \to [0, \infty[$ denote a measurable mapping. In the first step we compute $N_1^{\mathrm{ad}}(f) = f(x_1)$ for a fixed knot $x_1 \in D$. Suppose that N_n^{ad} has already been defined and put

$$M_n(y) = \max\{y_1, \ldots, y_n\}$$

for $y = (y_1, \ldots, y_n) \in \mathbb{R}^n$. Let

(2) $$h_n(x, y) = P(\{f \in F : f(x) \geq M_n(y) + Z_n(y)\} \mid N_n^{\mathrm{ad}} = y)$$

denote the conditional probability that the function value at x exceeds the adaptively chosen threshold $M_n(y) + Z_n(y)$. Define the next point $\lambda_{n+1}(y)$ of evaluation by maximizing this probability, assuming that the maximum exists, i.e.,

(3) $$h_n(\lambda_{n+1}(y), y) = \sup_{x \in D} h_n(x, y).$$

Note that an auxiliary optimization problem has to be solved in every step, and the associated computational cost may be arbitrarily large in general. If P is Gaussian, then the conditional measures are Gaussian, too, and the conditional mean m^y and covariance kernel K_0^y are given in Section VII.2.1. We obtain

$$h_n(x, y) = 1 - \Phi\left(\frac{M_n(y) + Z_n(y) - m^y(x)}{K_0^y(x, x)^{1/2}}\right),$$

where Φ denotes the distribution function of the standard normal distribution. Choosing the next point of evaluation is therefore equivalent to maximizing

$$\frac{K_0^y(x, x)}{(M_n(y) + Z_n(y) - m^y(x))^2}$$

over all $x \in D$.

The latter problem is rather easy to solve for the Wiener measure, due to the Markov property and the explicit formulas from Example VII.3. The computational cost in the nth step is proportional to n, and we therefore have quadratic dependence of the computational cost on the cardinality.

Now we describe the choice of the mappings Z_n for the Wiener measure. Trivially, Z_n must be strictly positive, since $\lambda_{n+1}(y)$ would otherwise coincide with a knot from the previous steps where the maximal value $M_n(y)$ was observed. Take a strictly increasing sequence of positive integers γ_n such that

$$\lim_{n \to \infty} \gamma_n = \infty,$$

$$\lim_{n \to \infty} \frac{\gamma_{n+1}}{\gamma_n} = 1,$$

$$\lim_{n \to \infty} \frac{\gamma_n^2 \cdot \ln \gamma_n}{n} = 0.$$

After n evaluations the respective knots form a partition of $[0, 1]$ into n non-overlapping subintervals, and we let $\tau_n(y)$ denote the length of the smallest such interval. Define

(4) $$Z_n(y) = \gamma_n \cdot \tau_n(y)^{1/2}.$$

It remains to define the termination criterion and the output of the method. Let $k \in \mathbb{N}$ be given. After every evaluation we check a simple condition, and we stop if the condition is satisfied for the kth time. After $n + 1$ evaluations the condition is as follows. Put $x_{n+1} = \lambda_{n+1}(y)$ and $\widetilde{y} = (y, f(x_{n+1}))$, where $y = N_n^{\mathrm{ad}}(f)$. Check whether

$$\tau_{n+1}(\widetilde{y}) < \tau_n(y)$$

and

$$h_{n+1}(\lambda_{n+2}(\widetilde{y}), \widetilde{y}) \leq h_n(x_{n+1}, y).$$

The output is a knot where the largest function value has been found. We use $S_{\nu(\cdot, k)}^{\mathrm{var}}$ to denote the method above, which also depends on the sequence $(\gamma_n)_{n \in \mathbb{N}}$ in addition to k.

A properly normalized error $\Delta(S_{\nu(\cdot, k)}^{\mathrm{var}}, \cdot)$ converges to zero in probability with respect to the Wiener measure w.

PROPOSITION 9. *Let $c < \frac{1}{16}$. Then*

$$\lim_{k \to \infty} w(\{f \in F : \exp(\nu(f, k) \cdot c / \gamma_{\nu(f, k)}) \cdot \Delta(S_{\nu(\cdot, k)}^{\mathrm{var}}, f) \geq \varepsilon\}) = 0$$

for every $\varepsilon > 0$.

Observe that the normalizing sequence depends almost exponentially on the cardinality $\nu(f, k)$, since γ_n may converge to infinity arbitrarily slow. This is in sharp contrast to the results discussed previously. To fully compare nonadaptive and adaptive methods, it would be interesting to have sharp upper bounds on average error of $S_{\nu(\cdot, k)}^{\mathrm{var}}$ in terms of the average cardinality. Moreover, lower bounds on the minimal errors for adaptive methods with fixed cardinality are needed. Proposition 9 suggests that the method $S_{\nu(\cdot, k)}^{\mathrm{var}}$ is far superior at least to all nonadaptive methods.

2.1. Notes and References. 1. Proposition 6 is due to Wasilkowski (1984) and Novak (1988).

2. Objective functions for global optimization are modeled as random functions in Kushner (1962) for the first time. Kushner suggests in particular (2) and (3), with a nonadaptive choice of Z_n, to define an adaptive method. The method was later called the P-algorithm. See Žilinskas (1985) for a proof of convergence.

3. A detailed study of the Bayesian approach to global optimization is given in Mockus (1989) and Törn and Žilinskas (1989), see also Boender and Romeijn (1995) for a survey. It seems that so far the focus is on the construction of algorithms.

4. Proposition 7 is due to Ritter (1990).

5. Calvin studies global optimization in the average case setting in a series of papers, including Calvin (1996, 1997, 1999). The method $S^{\mathrm{var}}_{\nu(\cdot,k)}$, with the adaptive choice (4) of Z_n, and Proposition 9 are due to Calvin (1999).

Bibliography

Abt, M. (1992), Some exact optimal designs for linear covariance functions in one dimension, Commun. Statist.-Theory Meth. **21**, 2059–2069.

Adler, R. J. (1981), The geometry of random fields, Wiley, New York.

Adler, R. J. (1990), An introduction to continuity, extrema, and related topics for general Gaussian processes, Lect. Notes Vol. **12**, IMS, Hayward.

Arestov, V. V. (1975), On the best approximation of the operators of differentiation and related questions, in: Approximation theory, Proc. Conf. Poznań 1972, Z. Ciesielski, J. Musielak, eds., pp. 1–9, Reidel, Dordrecht.

Aronszajn, N. (1950), Theory of reproducing kernels, Trans. Amer. Math. Soc. **68**, 337–404.

Atteia, M. (1992), Hilbertian kernels and spline functions, North-Holland, Amsterdam.

Bakhvalov, N. S. (1959), On approximate computation of multiple integrals (in Russian), Vestnik Moscow Univ. Ser. I Mat. Mekh. **4**, 3–18.

Bakhvalov, N. S. (1971), On the optimality of linear methods for operator approximation in convex classes of functions, USSR Comput. Math. Math. Phys. **11**, 244–249.

Barrow, D. L., and Smith, P. W. (1978), Asymptotic properties of best L_2-approximation by splines with variable knots, Quart. Appl. Math. **36**, 293–304.

Barrow, D. L., and Smith, P. W. (1979), Asymptotic properties of optimal quadrature formulas, in: Numerische Integration, G. Hämmerlin, ed., ISNM **45**, pp. 54–66, Birkhäuser Verlag, Basel.

Bellman, R. (1961), Adaptive Control Processes: a Guided Tour, Princeton University, Princeton.

Benassi, A., Cohen, S., and Istas, J. (1998), Identifying the multifractional function of a Gaussian process, Stat. Prob. Letters **39**, 337–345.

Benassi, A., Cohen, S., Istas, J., and Jaffard, S. (1998), Identification of filtered white noise, Stochastic Processes Appl. **75**, 31–49.

Benassi, A., Jaffard, S., and Roux, D. (1993), Analyse multi-échelle des processus gaussiens markoviens d'ordre p indexés par $(0, 1)$, Stochastic Processes Appl. **47**, 275–297.

Benhenni, K. (1997), Approximation d'intégrales de processus stochastiques, C. R. Acad. Sci. **325**, 659–663.

Benhenni, K. (1998), Approximating integrals of stochastic processes: extensions, J. Appl. Probab. **35**, 843–855.

Benhenni, K., and Cambanis, S. (1992a), Sampling designs for estimating integrals of stochastic processes, Ann. Statist. **20**, 161–194.

Benhenni, K., and Cambanis, S. (1992b), Sampling designs for estimating integrals of stochastic processes using quadratic mean derivatives, in: Approximation theory, G. A. Anastassiou, ed., pp. 93–123, Dekker, New York.

Benhenni, K., and Cambanis, S. (1996), The effect of quantization on the performance of sampling design, Tech. Rep. No. 481, Dept. of Statistics, Univ. of North Carolina, Chapel Hill.

Benhenni, K., and Istas, J. (1998), Minimax results for estimating integrals of analytic processes, ESAIM, Probab. Stat. **2**, 109–121.

Berman, S. M. (1974), Sojourns and extremes of Gaussian processes, Ann. Probab. **2**, 999–1026.

Birman, M. Š., and Solomjak, M. Z. (1967), Piecewise polynomial approximations of functions of the classes W_p^α, Math. USSR - Sbornik **2**, 295–317.

Blum, L., Cucker, F., Shub, M., and Smale, S. (1998), Complexity and real computation, Springer, New York.

Blum, L., Shub, M., and Smale, S. (1989), On a theory of computation and complexity over the real numbers: NP completeness, recursive functions and universal machines, Bull. Amer. Math. Soc. **21**, 1–46.

Boender, C. G. E., and Romeijn, H. E. (1995), Stochastic methods, in: Handbook of global optimization, R. Horst and P. M. Pardalos, eds., pp. 829–869, Kluwer, Dordrecht.

Bojanov, B. D., Hakopian, H. A., and Sahakian, A. A. (1993), Spline functions and multivariate interpolations, Kluwer, Dordrecht.

Borgwardt, K. H. (1981), The expected number of pivot steps required by a certain variant of the simplex method is polynomial, Methods Oper. Res. **43**, 35–41.

Borgwardt, K. H. (1987), The simplex method, Springer-Verlag, Berlin.

Borgwardt, K. H. (1994), Verschärfung des Polynomialitätsbeweises für die erwartete Anzahl von Schattenecken im Rotationssymmetrie-Modell, in: Beiträge zur Angewandten Analysis und Informatik, E. Schock, ed., pp. 13–33, Verlag Shaker, Aachen.

Braß, H. (1977), Quadraturverfahren, Vandenhoek & Ruprecht, Göttingen.

Bungartz, H.-J., and Griebel, M. (1999), A note on the complexity of solving Poisson's equation for spaces of bounded mixed derivatives, J. Complexity **15**, 167–199.

Buslaev, A. P., and Seleznjev, O. V. (1999), On certain extremal problems in the theory of approximation of random processes, East J. Approx. Theory **4**, 467–481.

Bykovskii, V. A. (1985), On exact order of optimal quadrature formulas for spaces of functions with bounded mixed derivatives (in Russian), Report of Dalnevostochnoi Center of Academy of Sciences, Vladivostok.

Calvin, J. M. (1996), An asymptotically optimal non-adaptive algorithm for minimization of Brownian motion, in: Mathematics of Numerical Analysis, J. Renegar, M. Shub, S. Smale, eds., Lectures in Appl. Math. **32**, pp. 157–163, AMS, Providence.

Calvin, J. M. (1997), Average performance of a class of adaptive algorithms for global optimization, Ann. Appl. Prob. **7**, 711–730.

Calvin, J. M. (1999), A one-dimensional optimization algorithm and its convergence rate under the Wiener measure, submitted for publication.

Cambanis, S. (1985), Sampling designs for time series, in: Time series in the time domain, Handbook of Statistics, Vol. 5, E. J. Hannan, P. R. Krishnaiah, and M. M. Rao, eds., pp. 337–362, North-Holland, Amsterdam.

Cambanis, S., and Masry, E. (1983), Sampling designs for the detection of signals in noise, IEEE Trans. Inform. Theory **IT-29**, 83–104.

Cambanis, S., and Masry, E. (1994), Wavelet approximation of deterministic and random signals: convergence properties and rates, IEEE Trans. Inform. Theory **IT-40**, 1013–1029.

Christensen, R. (1987), Plane answers to complex questions, Springer-Verlag, New York.

Christensen, R. (1991), Linear models for multivariate, time series, and spatial data, Springer-Verlag, New York.

Ciesielski, Z. (1975), On Lévy's Brownian motion with several-dimensional time, in: Probability-Winter School, Z. Ciesielski and K. Urbanik, eds., Lect. Notes in Math. **472**, pp. 29–56, Springer-Verlag, Berlin.

Cohen, A., and d'Ales, J.-P. (1997), Nonlinear approximation of random functions, SIAM J. Appl. Math. **57**, 518–540.

Cressie, N. A. C. (1978), Estimation of the integral of a stochastic process, Bull. Austr. Math. Soc. **18**, 83–93.

Cressie, N. A. C. (1993), Statistics for spatial data, Wiley, New York.

Currin, C., Mitchell, T., Morris, M., and Ylvisaker, D. (1991), Bayesian prediction of deterministic functions, with applications to the design and analysis of computer experiments, J. Amer. Statist. Assoc. **86**, 953–963.

Davis, P. J., and Rabinowitz, P. (1984), Methods of numerical integration, Academic Press, New York.

Delvos, F.-J. (1990), Boolean methods for double integration, Math. Comp. **55**, 683–692.

Delvos, F.-J., and Schempp, W. (1989), Boolean methods in interpolation and approximation, Pitman Research Notes in Mathematics Series 230, Longman, Essex.

DeVore, R. A., and Lorentz, G. G. (1993), Constructive approximation, Springer-Verlag, Berlin.

Diaconis, P. (1988), Bayesian numerical analysis, in: Statistical decision theory and related topics IV, Vol. 1, S. S. Gupta and J. O. Berger, eds., pp. 163–175, Springer-Verlag, New York.

Donoho, D. L. (1994), Asymptotic minimax risk for sup-norm loss: solution via optimal recovery, Probab. Theory Relat. Fields **99**, 145–170.

Drmota, M., and Tichy, R. F. (1997), Sequences, discrepancies, and applications, Lect. Notes in Math. **1651**, Springer-Verlag, Berlin.

Engels, H. (1980), Numerical quadrature and cubature, Academic Press, London.

Ermakov, S. M. (1975), Die Monte-Carlo-Methode und verwandte Fragen, Oldenbourg Verlag, München.

Eubank, R. L. (1988), Spline smoothing and nonparametric regression, Dekker, New York.

Eubank, R. L., Smith, P. L., and Smith, P. W. (1981), Uniqueness and eventual uniqueness of optimal designs in some time series models, Ann. Statist. **9**, 486–493.

Eubank, R. L., Smith, P. L., and Smith, P. W. (1982), A note on optimal and asymptotically optimal designs for certain time series models, Ann. Statist. **10**, 1295–1301.

Fedorov, V. V. (1972), Theory of optimal experiments, Academic Press, New York.

Frank, K., and Heinrich, S. (1996), Computing discrepancies of Smolyak quadrature rules, J. Complexity **12**, 287–314.

Frank, K., Heinrich, S., and Pereverzev, S. (1996), Information complexity of multivariate Fredholm integral equations in Sobolev classes, J. Complexity **12**, 17–34.

Frolov, K. K. (1976), Upper error bounds for quadrature formulas on function classes, Soviet Math. Dokl. **17**, 1665–1669.

Gabushin, V. N. (1967), Inequalities for the norms of a function and its derivatives in metric L_p, Math. Notes **1**, 194–198.

Gal, S., and Micchelli, C. (1980), Optimal sequential and non-sequential procedures for evaluating a functional, Appl. Anal. **10**, 105–120.

Gao, F. (1993), On the role of computable error estimates in the analysis of numerical approximation algorithms, in: Proceedings of the Smalefest, M. W. Hirsch, J. E. Marsden, and M. Shub, eds., pp. 387–394, Springer-Verlag, New York.

Gao, F., and Wasilkowski, G. W. (1993), On detecting regularity of functions: a probabilistic analysis, J. Complexity **9**, 373–386.

Genz, A. C. (1986), Fully symmetric interpolatory rules for multiple integrals, SIAM J. Numer. Anal. **23**, 1273–1283.

Gihman, I. I., and Skorohod, A. V. (1974), The theory of stochastic processes I, Springer-Verlag, Berlin.

Golomb, M., and Weinberger, H. F. (1959), Optimal approximation and error bounds, in: On numerical approximation, R. E. Langer, ed., pp. 117–190, Univ. of Wisconsin Press, Madison.

Gordon, W. J. (1971), Blending function methods of bivariate and multivariate interpolation and approximation, SIAM J. Numer. Anal. **8**, 158–177.

Graf, S., Mauldin, R. D., and Williams, S. C. (1986), Random homeomorphisms, Adv. Math. **60**, 239–359.

Graf, S., and Novak, E. (1990), The average error of quadrature formulas for functions of bounded variation, Rocky Mountain J. Math. **20**, 707–716.

Graf, S., Novak, E., and Papageorgiou, A. (1989), Bisection is not optimal on the average, Numer. Math. **55**, 481–491.

Griebel, M., Schneider, M., and Zenger, Ch. (1992), A combination technique for the solution of sparse grid problems, in: Iterative methods in linear algebra, R. Beauwens and P. de Groen, eds., pp. 263–281, Elsevier, North-Holland.

Hájek, J. (1962), On linear statistical problems in stochastic processes, Czech. Math. J. **12**, 404–440.

Hájek, J., and Kimeldorf, G. S. (1974), Regression designs in autoregressive stochastic processes, Ann. Statist. **2**, 520–527.

Heinrich, S. (1990a), Probabilistic complexity analysis for linear problems on bounded domains, J. Complexity **6**, 231–255.

Heinrich, S. (1990b), Invertibility of random Fredholm operators, Stochastic Anal. Appl. **8**, 1–59.

Heinrich, S. (1991), Probabilistic analysis of numerical methods for integral equations, J. Integral Eq. Appl. **3**, 289–319.

Heinrich, S., and Mathé, P. (1993), The Monte Carlo complexity of Fredholm integral equations, Math. Comp. **60**, 257–278.

Heinrich, S., Novak, E., Wasilkowski, G. W., and Woźniakowski, H. (1999), The star-discrepancy depends linearly on the dimension, submitted for publication.

Hickernell, F. J., and Woźniakowski, H. (1999), Integration and approximation in arbitrary dimension, to appear in Adv. Comput. Math.

Hjort, N. L., and Omre, H. (1994), Topics in spatial statistics, Scand. J. Statist. **21**, 289–357.

Hüsler, J. (1999), Extremes of Gaussian processes, on results of Piterbarg and Seleznjev, Stat. Prob. Letters **44**, 251–258.

Istas, J. (1992), Wavelet coefficients of a Gaussian process and applications, Ann. Inst. Henri Poincaré, Probab. Stat. **28**, 537–556.

Istas, J. (1996), Estimating the singularity functions of a Gaussian process with applications, Scand. J. Statist. **23**, 581–595.

Istas, J. (1997), Estimation d'intégrales de processus multi-fractionaires, C. R. Acad. Sci. **324**, 565–568.

Istas, J., and Lang, G. (1994), Variations quadratiques et estimation de l'exposant de Hölder local d'un processus gaussien, C. R. Acad. Sci. **319**, 201–206.

Istas, J., and Laredo, C. (1994), Estimation d'intégrales de processus aléatoires à partir d'observations discrétisées, C. R. Acad. Sci. **319**, 85–88.

Istas, J., and Laredo, C. (1997), Estimating functionals of a stochastic process, Adv. Appl. Prob. **29**, 249–270.

Ivanov, A. V., and Leonenko, N. N. (1989), Statistical analysis of random fields, Kluwer, Dordrecht.

Kadane, J. B., and Wasilkowski, G. W. (1985), Average case ε-complexity in computer science — a Bayesian view, in: Bayesian statistics, J. M. Bernardo, ed., pp. 361–374, Elsevier, North-Holland.

Kahane, J.-P. (1985), Some random series of functions, Cambridge Univ. Press, Cambridge.

Kiefer, J. (1957), Optimum sequential search and approximation methods under regularity assumptions, J. Soc. Indust. Appl. Math. **5**, 105–136.

Kimeldorf, G. S., and Wahba, G. (1970a), A correspondence between Bayesian estimation on stochastic processes and smoothing by splines, Ann. Math. Statist. **41**, 495–502.

Kimeldorf, G. S., and Wahba, G. (1970b), Spline functions and stochastic processes, Sankhyā Ser. A **32**, 173–180.

Koehler, J. R., and Owen, A. B. (1996), Computer experiments, in: Design and analysis of experiments, Handbook of Statistics, Vol. 13, S. Gosh and C. R. Rao, eds., pp. 261–308, North Holland, Amsterdam.

Kolmogorov, A. (1936), Über die beste Annäherung von Funktionen einer gegebenen Funktionenklasse, Ann. Math. **37**, 107–110.

Kon, M. A. (1994), On the $\delta \to 0$ limit in probabilistic complexity, J. Complexity **10**, 356–365.

Kon, M. A., Ritter, K., and Werschulz, A. G. (1991), On the average case solvability of ill-posed problems, J. Complexity **7**, 220–224.

Korneichuk, N. P. (1991), Exact constants in approximation theory, Cambridge Univ. Press, Cambridge.

Korneichuk, N. P. (1994), Optimization of active algorithms for recovery of monotonic functions from Hölder's class, J. Complexity **10**, 265–269.

Kung, H. T. (1976), The complexity of obtaining starting points for solving operator equations by Newton's method, in: Analytic computational complexity, J. F. Traub, ed., pp. 35–57, Academic Press, New York.

Kushner, H. J. (1962), A versatile stochastic model of a function of unknown and time varying form, J. Math. Anal. Appl. **5**, 150–167.

Kwong, Man Kam, and Zettl, A. (1992), Norm inequalities for derivatives and differences, Lect. Notes in Math. **1536**, Springer-Verlag, Berlin.

Larkin, F. M. (1972), Gaussian measure in Hilbert space and application in numerical analysis, Rocky Mountain J. Math. **2**, 379–421.

Lasinger, R. (1993), Integration of covariance kernels and stationarity, Stochastic Processes Appl. **45**, 309–318.

Lee, D. (1986), Approximation of linear operators on a Wiener space, Rocky Mountain J. Math. **16**, 641–659.

Lee, D., and Wasilkowski, G. W. (1986), Approximation of linear functionals on a Banach space with a Gaussian measure, J. Complexity **2**, 12–43.

Levin, M., and Girshovich, J. (1979), Optimal quadrature formulas, Teubner Verlagsgesellschaft, Leipzig.

Lifshits, M. A. (1995), Gaussian random functions, Kluwer, Dordrecht.

Loève, M. (1948), Fonctions aléatoires du second ordre, supplement to: Processus stochastiques et mouvement Brownien, Lévy, P., pp. 299–353, Gauthier-Villars, Paris.

Lorentz, G. G., v. Golitschek, M., and Makovoz, Y. (1996), Constructive Approximation, Advanced Problems, Springer-Verlag, Berlin.

Magaril-Ilyaev, G. G. (1994), Average widths of Sobolev classes on $\mathbb{R}^n$, J. Approx. Theory 76, 65–76.

Maiorov, V. E. (1992), Widths of spaces endowed with a Gaussian measure, Russian Acad. Sci. Dokl. Math. 45, 305–309.

Maiorov, V. E. (1993), Average n-widths of the Wiener space in the L_∞-norm, J. Complexity 9, 222–230.

Maiorov, V. E. (1994), Linear widths of function spaces equipped with the Gaussian measure, J. Approx. Theory 77, 74–88.

Maiorov, V. E. (1996a), About widths of Wiener space in L_q-norm, J. Complexity 12, 47–57.

Maiorov, V. E. (1996b), Widths and distributions of values of the approximation functional on the Sobolev space with measure, Constr. Approx. 12, 443–462.

Maiorov, V. E., and Wasilkowski, G. W. (1996), Probabilistic and average linear widths in L_∞-norm with respect to r-fold Wiener measure, J. Approx. Theory 84, 31–40.

Mardia, K. V., Kent, J. T., Goodall, C. R., and Little, J. A. (1996), Kriging and splines with derivative information, Biometrika 83, 207–221.

Mathé, P. (1990), s-numbers in information-based complexity, J. Complexity 6, 41–66.

Mathé, P. (1993), A minimax principle for the optimal error of Monte Carlo methods, Constr. Approx. 9, 23–39.

Matoušek, J. (1998), The exponent of discrepancy is at least $1,0669$, J. Complexity 14, 448–453.

Matoušek, J. (1999), Geometric discrepancy: an illustrated guide, Springer-Verlag, Berlin.

Mazja, V. G. (1985), Sobolev spaces, Springer-Verlag, Berlin.

Micchelli, C. A. (1976), On an optimal method for the numerical differentiation of smooth functions, J. Approx. Theory 18, 189–204.

Micchelli, C. A. (1984), Orthogonal projections are optimal algorithms, J. Approx. Theory 40, 101–110.

Micchelli, C. A., and Rivlin, T. J. (1977), A survey of optimal recovery, in: Optimal estimation in approximation theory, C. A. Micchelli and T. J. Rivlin, eds., pp. 1–54, Plenum, New York.

Micchelli, C. A., and Rivlin, T. J. (1985), Lectures on optimal recovery, in: Numerical analysis Lancaster 1984, P. R. Turner, ed., pp. 21–93, Lect. Notes in Math. 1129, Springer-Verlag, Berlin.

Micchelli, C. A., and Wahba, G. (1981), Design problems for optimal surface interpolation, in: Approximation theory and applications, Z. Ziegler, ed., pp. 329–347, Academic Press, New York.

Mitchell, T., Morris, M., and Ylvisaker, D. (1990), Existence of smoothed stationary processes on an interval, Stochastic Processes Appl. 35, 109–119.

Mockus, J. (1989), Bayesian approach to global optimization, Kluwer, Dordrecht.

Molchan, G. M. (1967), On some problems concerning Brownian motion in Lévy's sense, Theory Probab. Appl. 12, 682–690.

Müller-Gronbach, T. (1994), Asymptotically optimal designs for approximating the path of a stochastic process with respect to the L^∞-norm, in: Probastat'94, A. Paźman and J. Volaufová, eds., Tatra Mountains Math. Publ. 7, pp. 87–95, Bratislava.

Müller-Gronbach, T. (1996), Optimal designs for approximating the path of a stochastic process, J. Statist. Planning Inf. 49, 371–385.

Müller-Gronbach, T. (1998), Hyperbolic cross designs for approximation of random fields, J. Statist. Planning Inf. 66, 321–344.

Müller-Gronbach, T., and Ritter, K. (1997), Uniform reconstruction of Gaussian processes, Stochastic Processes Appl. 69, 55–70.

Müller-Gronbach, T., and Ritter, K. (1998), Spatial adaption for predicting random functions, Ann. Statist. 26, 2264–2288.

Müller-Gronbach, T., and Schwabe, R. (1996), On optimal allocations for estimating the surface of a random field, Metrika 44, 239–258.

Munch, N. J. (1990), Orthogonally invariant measures and best approximation of linear operators, J. Approx. Theory 61, 158–177.

Näther, W. (1985), Effective observation of random fields, Teubner Verlagsgesellschaft, Leipzig.

Niederreiter, H. (1992), Random number generation and quasi-Monte Carlo methods, CBSM-NSF Regional Conf. Ser. Appl. Math. 63, SIAM, Philadelphia.

Nikolskij, S. M. (1950), On the problem of approximation estimate by quadrature formulas (in Russian), Usp. Mat. Nauk 5, 165–177.

Nikolskij, S. M. (1975), Approximation of functions of several variables and imbedding theorems, Springer-Verlag, Berlin.

Nikolskij, S. M. (1979), Integration formulas (in Russian), Nauka, Moscow.

Novak, E. (1988), Deterministic and stochastic error bounds in numerical analysis, Lect. Notes in Math. **1349**, Springer-Verlag, Berlin.

Novak, E. (1989), Average-case results for zero finding, J. Complexity **5**, 489–501.

Novak, E. (1992), Quadrature formulas for monotone functions, Proc. Amer. Math. Soc. **115**, 59–68.

Novak, E. (1995a), Optimal recovery and n-widths for convex classes of functions, J. Approx. Theory **80**, 390–408.

Novak, E. (1995b), The adaption problem for nonsymmetric convex sets, J. Approx. Theory **82**, 123–134.

Novak, E. (1995c), The real number model in numerical analysis, J. Complexity **11**, 57–73.

Novak, E. (1996a), On the power of adaption, J. Complexity **12**, 199–237.

Novak, E. (1996b), The Bayesian approach to numerical problems: results for zero finding, in: Proc. IMACS-GAMM Int. Symp. Numerical Methods and Error Bounds, G. Alefeld, J. Herzberger, eds., pp. 164–171, Akademie Verlag, Berlin.

Novak, E., and Ritter, K. (1989), A stochastic analog to Chebyshev centers and optimal average case algorithms, J. Complexity **5**, 60–79.

Novak, E., and Ritter, K. (1992), Average errors for zero finding: lower bounds, Math. Z. **211**, 671–686.

Novak, E., and Ritter, K. (1993), Some complexity results for zero finding for univariate functions, J. Complexity **9**, 15–40.

Novak, E., and Ritter, K. (1996a), High dimensional integration of smooth functions over cubes, Numer. Math. **75**, 79–97.

Novak, E., and Ritter, K. (1996b), Global optimization using hyperbolic cross points, in: State of the Art in Global Optimization, C. A. Floudas, P. M. Pardalos, eds., pp. 19–33, Kluwer, Boston.

Novak, E., and Ritter, K. (1999), Simple cubature formulas with high polynomial exactness, Constr. Approx. **15**, 499–522.

Novak, E., Ritter, K., and Steinbauer, A. (1998), A multiscale method for the evaluation of Wiener integrals, in: Approximation Theory IX, Vol. 2, C. K. Chui and L. L. Schumaker, eds., pp. 251–258, Vanderbilt Univ. Press, Nashville.

Novak, E., Ritter, K., and Woźniakowski, H. (1995), Average case optimality of a hybrid secant-bisection method, Math. Comp. **64**, 1517–1539.

Novak, E., and Woźniakowski, H. (1999a), Intractability results for integration and discrepancy, submitted for publication.

Novak, E., and Woźniakowski, H. (1999b), When are integration and discrepancy tractable, in preparation.

O'Hagan, A. (1992), Some Bayesian numerical analysis, in: Bayesian statistics 4, J. M. Bernardo, J. O. Berger, A. P. Dawid, and A. F. M. Smith, eds., pp. 345–363, Oxford Univ. Press, Oxford.

Ossiander, M., and Waymire, E. C. (1989), Certain positive-definite kernels, Proc. Amer. Math. Soc. **107**, 487–492.

Papageorgiou, A., and Wasilkowski, G. W. (1990), On the average complexity of multivariate problems, J. Complexity **6**, 1–23.

Parthasarathy, K. R. (1967), Probability measures on metric spaces, Academic Press, New York.

Parzen, E. (1959), Statistical inference on time series by Hilbert space methods, I, in: Time series analysis papers, E. Parzen, ed., pp. 251–382, Holden-Day, San Francisco, 1967.

Parzen, E. (1962), An approach to time series analysis, Ann. Math. Statist. **32**, 951–989.

Paskov, S. H. (1993), Average case complexity of multivariate integration for smooth functions, J. Complexity **9**, 291–312.

Paskov, S. H. (1995), Termination criteria for linear problems, J. Complexity **11**, 105–137.

Pereverzev, S. V. (1986), On optimization of approximate methods of solving integral equations, Soviet Math. Dokl. **33**, 347–351.

Pereverzev, S. V. (1996), Optimization of methods for approximate solution of operator equations, Nova Science, New York.

Pilz, J. (1983), Bayesian estimation and experimental design in linear regression models, Teubner Verlagsgesellschaft, Leipzig.

Pinkus, A. (1985), n-widths in approximation theory, Springer-Verlag, Berlin.

Piterbarg, V., and Seleznjev, O. (1994), Linear interpolation of random processes and extremes of a sequence of Gaussian non-stationary processes, Univ. of North Carolina, Chapel Hill, Center of Stoch. Proc., Tech. Rep. No. **446**.

Pitt, L., Robeva, R., and Wang, D. Y. (1995), An error analysis for the numerical calculation of certain random integrals: Part 1, Ann. Appl. Prob. **5**, 171–197.

Plaskota, L. (1992), Function approximation and integration on the Wiener space, J. Complexity **8**, 301–321.

Plaskota, L. (1993), A note on varying cardinality in the average case setting, J. Complexity **9**, 458–470.

Plaskota, L. (1996), Noisy information and computational complexity, Cambridge Univ. Press, Cambridge.

Plaskota, L. (1998), Average case L_∞-approximation in the presence of Gaussian noise, J. Approx. Theory **93**, 501–515.

Plaskota, L. (2000), The exponent of discrepancy of sparse grids is at least 2.1933, Adv. Comput. Math. **12**, 3–24.

Plaskota, L., Wasilkowski, G. W., and Woźniakowski, H. (1999), A new algorithm and worst case complexity for Feynman-Kac path integration, submitted for publication.

Poincaré, H. (1896), Calcul des probabilités, Georges Carré, Paris.

Powell, M. J. D. (1992), The theory of radial basis function approximation in 1990, in: Advances in numerical analysis II, W. A. Light, ed., Oxford Univ. Press, Oxford.

Pukelsheim, F. (1993), Optimal design of experiments, Wiley, New York.

Ragozin, D. L., (1983), Error bounds for derivative estimates based on spline smoothing of exact or noisy data, J. Approx. Theory **37**, 335–355.

Renegar, J. (1987), On the efficiency of Newton's method in approximating all zeros of a system of complex polynomials, Math. Oper. Res. **12**, 121–148.

Richter, M. (1992), Approximation of Gaussian random elements and statistics, Teubner Verlagsgesellschaft, Stuttgart.

Ritter, K. (1990), Approximation and optimization on the Wiener space, J. Complexity **6**, 337–364.

Ritter, K. (1994), Average errors for zero finding: lower bounds for smooth or monotone functions, Aequationes Math. **48**, 194–219.

Ritter, K. (1996a), Asymptotic optimality of regular sequence designs, Ann. Statist. **24**, 2081–2096.

Ritter, K. (1996b), Almost optimal differentiation using noisy data, J. Approx. Theory bf 86, 293–309.

Ritter, K. (1996c), Average case analysis of numerical problems, Habilitationsschrift, Mathematisches Institut, Univ. Erlangen-Nürnberg.

Ritter, K., and Wasilkowski, G. W. (1996a), Integration and L_2-approximation: Average case setting with isotropic Wiener measure for smooth functions, Rocky Mountain J. Math. **26**, 1541–1557.

Ritter, K., and Wasilkowski, G. W. (1996b), On the average case complexity of solving Poisson equations, in: Mathematics of Numerical Analysis, J. Renegar, M. Shub, S. Smale, eds., Lectures in Appl. Math. **32**, pp. 677–687, AMS, Providence.

Ritter, K., Wasilkowski, G. W., and Woźniakowski, H. (1993), On multivariate integration for stochastic processes, in: Numerical integration, H. Braß and G. Hämmerlin, eds., ISNM **112**, pp. 331–347, Birkhäuser Verlag, Basel.

Ritter, K., Wasilkowski, G. W., and Woźniakowski, H. (1995), Multivariate integration and approximation for random fields satisfying Sacks-Ylvisaker conditions, Ann. Appl. Prob. **5**, 518–540.

Roth, K. (1954), On irregularities of distribution, Mathematika **1**, 73–79.

Roth, K. (1980), On irregularities of distribution, IV, Acta Arith. **37**, 67–75.

Sacks, J., Welch, W. J., Mitchell, T. J., and Wynn, H. P. (1989), Design and analysis of computer experiments, Statist. Sci. **4**, 409–435.

Sacks, J., and Ylvisaker, D. (1966), Designs for regression with correlated errors, Ann. Math. Statist. **37**, 68–89.

Sacks, J., and Ylvisaker, D. (1968), Designs for regression problems with correlated errors; many parameters, Ann. Math. Statist. **39**, 49–69.

Sacks, J., and Ylvisaker, D. (1970a), Design for regression problems with correlated errors III, Ann. Math. Statist. **41**, 2057–2074.

Sacks, J., and Ylvisaker, D. (1970b), Statistical design and integral approximation, in: Proc. 12th Bienn. Semin. Can. Math. Congr., R. Pyke, ed., pp. 115–136, Can. Math. Soc., Montreal.

Samaniego, F. J. (1976), The optimal sampling design for estimating the integral of a process with stationary independent increments, IEEE Trans. Inform. Theory **IT-22**, 375–376.

Sard, A. (1949), Best approximate integration formulas, best approximation formulas, Amer. J. Math. **71**, 80–91.

Sarma, V. L. N. (1968), Eberlein measure and mechanical quadrature formulae. I: Basic Theory, Math. Comp. **22**, 607–616.

Sarma, V. L. N., and Stroud, A. H. (1969), Eberlein measure and mechanical quadrature formulae. II: Numerical Results, Math. Comp. **23**, 781–784.

Schaback, R. (1997), Reconstruction of multivariate functions from scattered data, Manuscript, Göttingen.

Schoenberg, I. (1974), Cardinal interpolation and spline functions VI, semi-cardinal interpolation and quadrature formulae, J. Analyse Math. **27**, 159–204.

Seleznjev, O. V. (1991), Limit theorems for maxima and crossings of a sequence of Gaussian processes and approximation of random processes, J. Appl. Probab. **28**, 17–32.

Seleznjev, O. V. (1996), Large deviations in the piecewise linear approximation of Gaussian processes with stationary increments, Adv. Appl. Prob. **28**, 481–499.

Seleznjev, O. V. (2000), Spline approximation of random processes and design problems, J. Statist. Planning Inf. **84**, 249–262.

Shub, M., and Smale, S. (1994), Complexity of Bezout's theorem V: Polynomial time, Theor. Comput. Sci. **133**, 141–164.

Sikorski, K. (1982), Bisection is optimal, Numer. Math. **40**, 111-117.

Sikorski, K. (1989), Study of linear information for classes of polynomial equations, Aequationes Math. **37**, 1-14.

Singer, P. (1994), An integrated fractional Fourier transform, J. Comput. Appl. Math. **54**, 221–237.

Sloan, I. H., and Joe, S. (1994), Lattice methods for multiple integration, Clarendon Press, Oxford.

Sloan, I. H., and Woźniakowski, H. (1998), When are quasi-Monte Carlo algorithms efficient for high dimensional integrals, J. Complexity **14**, 1–33.

Smale, S. (1981), The fundamental theorem of algebra and complexity theory, Bull. Amer. Math. Soc. **4**, 1–36.

Smale, S. (1983), On the average speed of the simplex method, Math. Programming **27**, 241–262.

Smale, S. (1985), On the efficiency of algorithms of analysis, Bull. Amer. Math. Soc. **13**, 87–121.

Smale, S. (1987), Algorithms for solving equations, in: Proc. Int. Congress Math., Berkeley, A. M. Gleason, ed., pp. 172–195, Amer. Math. Soc., Providence.

Smolyak, S. A. (1963), Quadrature and interpolation formulas for tensor products of certain classes of functions, Soviet Math. Dokl. **4**, 240-243.

Sobolev, S. L. (1992), Cubature formulas and modern analysis, Gordon and Breach, Philadelphia.

Speckman, P. (1979), L_p approximation of autoregressive Gaussian processes, Tech. Rep., Dept. of Statistics, Univ. of Oregon, Eugene.

Stein, M. L. (1990a), Uniform asymptotic optimality of linear predictions of a random field using an incorrect second order structure , Ann. Statist. **18**, 850–872.

Stein, M. L. (1990b), Bounds on the efficiency of linear predictions using an incorrect covariance function, Ann. Statist. **18**, 1116–1138.

Stein, M. L. (1993), Asymptotic properties of centered systematic sampling for predicting integrals of spatial processes, Ann. Appl. Prob. **3**, 874–880.

Stein, M. L. (1995a), Predicting integrals of random fields using observations on a lattice, Ann. Statist. **23**, 1975–1990.

Stein, M. L. (1995b), Locally lattice sampling designs for isotropic random fields, Ann. Statist. **23**, 1991–2012. Correction Note: Ann. Statist. **27** (1999).

Stein, M. L. (1995c), Predicting integrals of stochastic processes, Ann. Appl. Prob. **5**, 158–170.

Stein, M. L. (1999), Interpolation of spatial data, Springer-Verlag, New York.

Steinbauer, A. (1999), Quadrature formulas for the Wiener measure, J. Complexity **15**, 476–498.

Stroud, A. H. (1971), Approximate calculation of multiple integrals, Prentice Hall, Englewood Cliff.

Su, Y., and Cambanis, S. (1993), Sampling designs for estimation of a random process, Stochastic Processes Appl. **46**, 47–89.

Suldin, A. V. (1959), Wiener measure and its applications to approximation methods I (in Russian), Izv. Vyssh. Ucheb. Zaved. Mat. **13**, 145–158.

Suldin, A. V. (1960), Wiener measure and its applications to approximation methods II (in Russian), Izv. Vyssh. Ucheb. Zaved. Mat. **18**, 165–179.

Sun, Yong-Sheng (1992), Average n-width of point set in Hilbert space, Chinese Sci. Bull. **37**, 1153–1157.

Sun, Yong-Sheng, and Wang, Chengyong (1994), μ-average n-widths on the Wiener space, J. Complexity **10**, 428–436.

Sun, Yong-Sheng, and Wang, Chengyong (1995), Average error bounds of best approximation of continuous functions on the Wiener space, J. Complexity **11**, 74–104.

Taylor, J. M. G., Cumberland, W. G., and Sy, J. P (1994), A stochastic model for analysis of longitudinal AIDS data, J. Amer. Statist. Assoc. **89**, 727–736.

Temirgaliev, N. (1988), On an application of infinitely divisible distributions to quadrature problems, Anal. Math. **14**, 253–258.

Temlyakov, V. N. (1994), Approximation of periodic functions, Nova Science, New York.

Tikhomirov, V. M. (1976), Some problems in approximation theory (in Russian), Moscow State Univ., Moscow.

Tikhomirov, V. M. (1990), Approximation theory, in: Analysis II, R. V. Gamkrelidze, ed., Encyclopaedia of mathematical sciences, Vol. **14**, Springer-Verlag, Berlin.

Todd, M. J. (1991), Probabilistic models for linear programming, Math. Oper. Res. **16**, 671–693.

Törn, A., and Žilinskas, A. (1989), Global optimization, Lect. Notes in Comp. Sci. **350**, Springer-Verlag, Berlin.

Tong, Y. L. (1980), Probability inequalities in multivariate distributions, Academic Press, New York.

Traub, J. F., Wasilkowski, G. W., and Woźniakowski, H. (1983), Information, uncertainty, complexity, Addison-Wesley, Reading.

Traub, J. F., Wasilkowski, G. W., and Woźniakowski, H. (1984), Average case optimality for linear problems, J. Theor. Comput. Sci. **29**, 1–25.

Traub, J. F., Wasilkowski, G. W., and Woźniakowski, H. (1988), Information-based complexity, Academic Press, New York.

Traub, J. F., and Werschulz, A. G. (1998), Complexity and information, Cambridge Univ. Press, Cambridge.

Traub, J. F., and Woźniakowski, H. (1980), A general theory of optimal algorithms, Academic Press, New York.

Triebel, H. (1983), Theory of function spaces, Birkhäuser Verlag , Basel.

Vakhania, N. N. (1975), The topological support of Gaussian measure in Banach space, Nagoya Math. J. **57**, 59–63.

Vakhania, N. N. (1991), Gaussian mean boundedness of densely defined linear operators, J. Complexity **7**, 225–231.

Vakhania, N. N., Tarieladze, V. I., and Chobanyan, S. A. (1987), Probability distributions on Banach spaces, Reidel, Dordrecht.

Voronin, S. M., and Skalyga, V. I. (1984), On quadrature formulas, Soviet Math. Dokl. **29**, 616–619.

Wahba, G. (1971), On the regression design problem of Sacks and Ylvisaker, Ann. Math. Statist. **42**, 1035–1043.

Wahba, G. (1974), Regression design for some equivalence classes of kernels, Ann. Statist. **2**, 925–934.

Wahba, G. (1978a), Improper priors, spline smoothing and the problem of guarding against model errors in regression, J. Roy. Statist. Soc. Ser. B **40**, 364–372.

Wahba, G. (1978b), Interpolating surfaces: high order convergence rates and their associated designs, with applications to X-ray image reconstruction, Tech. Rep. No. 523, Dept. of Statistics, Univ. of Wisconsin, Madison.

Wahba, G. (1990), Spline models for observational data, CBSM-NSF Regional Conf. Ser. Appl. Math. **59**, SIAM, Philadelphia.

Wang, Chengyong (1994), μ-average n-widths on the Wiener space, Approx. Theory Appl. **10**, 17–31.

Wasilkowski, G. W. (1983), Local average errors, Tech. Rep., Dept. of Computer Science, Columbia Univ., New York.

Wasilkowski, G. W. (1984), Some nonlinear problems are as easy as the approximation problem, Comput. Math. Appl. **10**, 351–363.

Wasilkowski, G. W. (1986a), Information of varying cardinality, J. Complexity **2**, 204–228.

Wasilkowski, G. W. (1986b), Optimal algorithms for linear problems with Gaussian measures, Rocky Mountain J. Math. **16**, 727–749.

Wasilkowski, G. W. (1989), On adaptive information with varying cardinality for linear problems with elliptically contoured measures, J. Complexity **5**, 363–368.

Wasilkowski, G. W. (1992), On average complexity of global optimization problems, Math. Programming **57**, 313–324.

Wasilkowski, G. W. (1993), Integration and approximation of multivariate functions: average case complexity with isotropic Wiener measure, Bull. Amer. Math. Soc. (N. S.) **28**, 308–314. Full version (1994), J. Approx. Theory **77**, 212–227.

Wasilkowski, G. W. (1996), Average case complexity of multivariate integration and function approximation, an overview, J. Complexity **12**, 257–272.

Wasilkowski, G. W., and Gao, F. (1992), On the power of adaptive information for functions with singularities, Math. Comp. **58**, 285–304.

Wasilkowski, G. W., and Woźniakowski, H. (1984), Can adaption help on the average?, Numer. Math. **44**, 169–190.

Wasilkowski, G. W., and Woźniakowski, H. (1986), Average case optimal algorithms in Hilbert spaces, J. Approx. Theory **47**, 17–25.

Wasilkowski, G. W., and Woźniakowski, H. (1995), Explicit cost bounds of algorithms for multivariate tensor product problems, J. Complexity **11**, 1–56.

Wasilkowski, G. W., and Woźniakowski, H. (1997), The exponent of discrepancy is at most 1.4778, Math. Comp. **66**, 1125–1132.

Wasilkowski, G. W., and Woźniakowski, H. (1999), Weighted tensor product algorithms for linear multivariate problems, J. Complexity **15**, 402–447.

Weba, M. (1991a), Quadrature of smooth stochastic processes, Probab. Theory Relat. Fields **87**, 333–347.

Weba, M. (1991b), Interpolation of random functions, Numer. Math. **59**, 739–746.

Weba, M. (1992), Simulation and approximation of stochastic processes by spline functions, SIAM J. Sci. Stat. Comput. **13**, 1085–1096.

Weba, M. (1995), Cubature of random fields by product-type integration rules, Comput. Math. Appl. **30**, 229–234.

Werschulz, A. G. (1991), The computational complexity of differential and integral equations, Oxford Univ. Press, Oxford.

Werschulz, A. G. (1996), The complexity of the Poisson problem for spaces of bounded mixed derivatives, in: The mathematics of numerical analysis, J. Renegar, M. Shub, S. Smale, eds., pp. 895–914, Lect. in Appl. Math. **32**, AMS, Providence.

Wittwer, Gisela (1976), Versuchsplanung im Sinne von Sacks-Ylvisaker für Vektorprozesse, Math. Operationsforsch. u. Statist. **7**, 95–105.

Wittwer, Gisela (1978), Über asymptotisch optimale Versuchsplanung im Sinne von Sacks-Ylvisaker, Math. Operationsforsch. u. Statist. **9**, 61–71.

Wloka, J. (1987), Partial Differential Equations, Cambridge Univ. Press, New York.

Woźniakowski, H. (1986), Information-based complexity, Ann. Rev. Comput. Sci. **1**, 319–380.

Woźniakowski, H. (1987), Average complexity for linear operators over bounded domains, J. Complexity **3**, 57–80.

Woźniakowski, H. (1991), Average case complexity of multivariate integration, Bull. Amer. Math. Soc. (N. S.) **24**, 185–194.

Woźniakowski, H. (1992), Average case complexity of linear multivariate problems, Part 1: Theory, Part 2: Applications, J. Complexity **8**, 337–372, 373–392.

Woźniakowski, H. (1994a), Tractability and strong tractability of linear multivariate problems, J. Complexity **10**, 96–128.

Woźniakowski, H. (1994b), Tractability and strong tractability of multivariate tensor product problems, J. Computing Inform. **4**, 1–19.

Yadrenko, M. I. (1983), Spectral theory of random fields, Optimization Software, New York.

Yanovich, L. A. (1988), On the best quadrature formulas in the spaces of random process trajectories (in Russian), Dokl. Akad. Nauk BSSR **32**, 9–12.

Ylvisaker, D. (1975), Designs on random fields, in: A survey of statistical design and linear models, J. Srivastava, ed., pp. 593–607, North-Holland, Amsterdam.

Zenger, Ch. (1991), Sparse grids, in: Parallel algorithms for partial differential equations, W. Hackbusch, ed., pp. 241–251. Vieweg, Braunschweig.

Zhensykbaev, A. A. (1983), Extremality of monosplines of minimal deficiency, Math. USSR Izvestiya **21**, 461–482.

Žilinskas, A. (1985), Axiomatic characterization of a global optimization algorithm and investigation of its search strategy, OR Letters **4**, 35–39.

Author Index

Subject Index

Index of Notation

4. Lecture Notes are printed by photo-offset from the master-copy delivered in camera-ready form by the authors. Springer-Verlag provides technical instructions for the preparation of manuscripts. Macro packages in T_EX, L^AT_EX2e, $L^AT_EX2.09$ are available from Springer's web-pages at

http://www.springer.de/math/authors/b-tex.html.

Careful preparation of the manuscripts will help keep production time short and ensure satisfactory appearance of the finished book.

The actual production of a Lecture Notes volume takes approximately 12 weeks.

5. Authors receive a total of 50 free copies of their volume, but no royalties. They are entitled to a discount of 33.3 % on the price of Springer books purchase for their personal use, if ordering directly from Springer-Verlag.

Commitment to publish is made by letter of intent rather than by signing a formal contract. Springer-Verlag secures the copyright for each volume. Authors are free to reuse material contained in their LNM volumes in later publications: A brief written (or e-mail) request for formal permission is sufficient.

Addresses:

Professor F. Takens, Mathematisch Instituut,
Rijksuniversiteit Groningen, Postbus 800,
9700 AV Groningen, The Netherlands
E-mail: F.Takens@math.rug.nl

Professor B. Teissier
Université Paris 7
UFR de Mathématiques
Equipe Géométrie et Dynamique
Case 7012
2 place Jussieu
75251 Paris Cedex 05
E-mail: Teissier@ens.fr

Springer-Verlag, Mathematics Editorial, Tiergartenstr. 17,
D-69121 Heidelberg, Germany,
Tel.: *49 (6221) 487-701
Fax: *49 (6221) 487-355
E-mail: lnm@Springer.de